FOUNDATIONS OF PROBABILITY THEORY,

STATISTICAL INFERENCE,

AND STATISTICAL THEORIES OF SCIENCE

VOLUME II

THE UNIVERSITY OF WESTERN ONTARIO
SERIES IN PHILOSOPHY OF SCIENCE

A SERIES OF BOOKS

ON PHILOSOPHY OF SCIENCE, METHODOLOGY,

AND EPISTEMOLOGY

PUBLISHED IN CONNECTION WITH

THE UNIVERSITY OF WESTERN ONTARIO

PHILOSOPHY OF SCIENCE PROGRAMME

Managing Editor

J. J. LEACH

Editorial Board

J. BUB, R. E. BUTTS, W. HARPER, J. HINTIKKA, D. J. HOCKNEY,

C. A. HOOKER, J. NICHOLAS, G. PEARCE

VOLUME 6

FOUNDATIONS OF PROBABILITY THEORY, STATISTICAL INFERENCE, AND STATISTICAL THEORIES OF SCIENCE

Proceedings of an International Research Colloquium held at the University of Western Ontario, London, Canada, 10–13 May 1973

VOLUME II

FOUNDATIONS AND PHILOSOPHY OF STATISTICAL INFERENCE

Edited by

W. L. HARPER and C. A. HOOKER

University of Western Ontario, Ontario, Canada

No Longer Property of
Phillips Memorial Library

D. REIDEL PUBLISHING COMPANY

DORDRECHT-HOLLAND / BOSTON-U.S.A.

PHILLIPS MEMORIAL
LIBRARY
PROVIDENCE COLLEGE

QA
276
A1
F67

Library of Congress Cataloging in Publication Data
Main entry under title:

Foundations and philosophy of statistical inference.

(Foundations of probability theory, statistical inference
and statistical theories of science ; v. 2) (University of Western
Ontario series in philosophy of science ; v. 6)
Second vol. of the proceedings of an international
research colloquium held at the University of Western Ontario,
London, Canada, May 10–13, 1973.
Includes bibliographies.
1. Mathematical statistics—Congresses. I. Harper,
William Leonard. II. Hooker, Clifford Alan. III. London,
Ont. University of Western Ontario. IV. Series. V. Series:
London, Ont. University of Western Ontario. Series in philosophy
of science ; v. 6.
QA276.A1F67   519.5   75–38667
ISBN 90–277–0618–2
ISBN 90–277–0619–0 pbk.

The set of three volumes (cloth) ISBN 90–277–0614–X
The set of three volumes (paper) ISBN 90–277–0615–8

Published by D. Reidel Publishing Company,
P.O. Box 17, Dordrecht, Holland

Sold and distributed in the U.S.A., Canada, and Mexico
by D. Reidel Publishing Company, Inc.
Lincoln Building, 160 Old Derby Street, Hingham,
Mass. 02043, U.S.A.

All Rights Reserved
Copyright © 1976 by D. Reidel Publishing Company, Dordrecht, Holland
No part of the material protected by this copyright notice may be reproduced or
utilized in any form or by any means, electronic or mechanical,
including photocopying, recording or by any informational storage and
retrieval system, without written permission from the copyright owner

Printed in The Netherlands by D. Reidel, Dordrecht

CONTENTS

CONTENTS OF VOLUMES I AND III

VOLUME I

Foundations and Philosophy of Epistemic Applications of Probability Theory

VOLUME III

*Foundations and Philosophy of Statistical Theories
in the Physical Sciences*

GENERAL INTRODUCTION

In May of 1973 we organized an international research colloquium on foundations of probability, statistics, and statistical theories of science at the University of Western Ontario.

During the past four decades there have been striking formal advances in our understanding of logic, semantics and algebraic structure in probabilistic and statistical theories. These advances, which include the development of the relations between semantics and metamathematics, between logics and algebras and the algebraic-geometrical foundations of statistical theories (especially in the sciences), have led to striking new insights into the formal and conceptual structure of probability and statistical theory and their scientific applications in the form of scientific theory.

The foundations of statistics are in a state of profound conflict. Fisher's objections to some aspects of Neyman-Pearson statistics have long been well known. More recently the emergence of Bayesian statistics as a radical alternative to standard views has made the conflict especially acute. In recent years the response of many practising statisticians to the conflict has been an eclectic approach to statistical inference. Many good statisticians have developed a kind of wisdom which enables them to know which problems are most appropriately handled by each of the methods available. The search for principles which would explain why each of the methods works where it does and fails where it does offers a fruitful approach to the controversy over foundations. The colloquium first aimed both at a conceptually exciting clarification and enrichment of our notion of a probability theory and at removing the cloud hanging over many of the central methods of statistical testing now in constant use within the social and natural sciences.

The second aim of the colloquium was that of exploiting the same formal developments in the structure of probability and statistical theories for an understanding of what it is to have a statistical theory of nature, or of a sentient population. A previous colloquium in this series has already examined thoroughly the recent development of the analysis of quantum mechanics in terms of its logico-algebraic structure and brought out many of the sharp and powerful insights into the basic physical significance of this theory which that formal approach provides. It was our aim in this colloquium to extend the scope of that inquiry yet further afield in an effort to understand, not just one particular idiosyncratic theory, but what it is in general we are doing when we lay down a formal statistical theory of a system (be it physical or social).

Our aim was to provide a workshop context in which the papers presented could benefit from the informed criticism of conference participants. Most of the papers that appear here have been considerably rewritten since their original presentation. We have also included comments by other participants and replies wherever possible. One of the main reasons we have taken so long to get these proceedings to press has been the time required to obtain final versions of comments and replies. We feel that the result has been worth the wait.

When the revised papers came in there was too much material to include in a single volume or even in two volumes. We have, therefore, broken the proceedings down into three volumes. Three basic problem areas emerged in the course of the conference and our three volumes correspond. Volume I deals with problems in the foundations of probability theory; Volume II is devoted to foundations of statistical inference, and Volume III is devoted to statistical theories in the physical sciences. There is considerable overlap in these areas so that in some cases a paper in one volume might just as well have gone into another.

INTRODUCTION TO VOLUME II

This second volume is devoted to problems in the foundations and philosophy of statistical inference. One of the main themes is the continuing controversy over Bayesian approaches to Statistics. The papers by Lindley and Jaynes present and defend Bayesian approaches, while those by Giere and Kempthorne are critical of Bayesian approaches and support approaches based on hypothesis testing. Giere contributes a flexible non-behaviorist analysis of accepting hypotheses as part of his case against the Bayesians.

A related theme that received considerable attention was a challenge to the coherence assumptions built into Savage's version of the Bayesian framework. The papers by Good, Kyburg, Shafer, and Suppes all advocate representations of rational belief that violate Savage's assumptions. Suppes criticizes the coherence assumptions on the basis of problems about measurement. Kyburg generates his representation on the basis of an objective epistemological account of randomness. Good has long advocated allowing for interval valued belief functions. His paper in this volume is the most extensive presentation of his current views available. Shafer's paper gives a general theory of support based on Dempster's rule of combination applied to non-additive belief functions.

Godambe and Thompson give attention to philosophical problems and philosophical lessons presented by survey sampling practice. Kalbfleisch and Sprott contribute some techniques on locating relevant information in data to the theory of hypothesis testing. Fraser and MacKay use Fraser's group-theoretic analysis of information procession to argue that, significance tests, likelihood analysis, and objective posterior distributions are all essentially equivalent methods where they overlap in application.

One of the most radical contributions is Finch's defence of a general approach to inference that does not use probability distributions at all.

P. D. FINCH

THE POVERTY OF STATISTICISM

A SUMMARY

The title of this paper was chosen with Karl Popper's 'The Poverty of Historicism' in mind. Popper's fundamental thesis was that belief in human destiny is sheer superstition and the corresponding thesis of this paper is not only that belief in chance is sheer superstition but that adherence to it is responsible for the continuing argument about the foundations of probability theory and statistical inference. But, put in so brief a form, it is a thesis which fails to catch the reservations and careful mathematical detail with which I would argue it.

To avoid acrimony in discussing these matters I have rejected the convention whereby common politeness and good scholarship demand the citation of sources and relevant authorship. In like manner I have invented a fictitious school of thought, to wit 'statisticism', to avoid argument about the correct attribution of ideas to frequentists, Bayesians and the like. I am free therefore to attribute to members of that school any views I wish to criticize; there being no question as to whether or not statisticists do hold those views, because they are fictional people created for just that purpose. But, of course, the similarity of outlook between statisticists and actual persons, whether alive or dead, is not purely coincidental.

However it would be misleading to convey the impression that I mean by statisticism only a collective palimpsest on which the names of others have been effaced to exhibit a contrived polemic. There is something I call the statisticist doctrine which is common to all the views I criticize, though the extent and nature of its presence varies with the season and circumstance of its occurrence.

The statisticist doctrine preaches that man is faced by uncertainty, that there are random processes at work in the world and that what is the case is, to a greater or lesser extent, the result of chance happenings. On occasion the preaching of this doctrine is quite explicit and then it is often

Harper and Hooker (eds.), Foundations of Probability Theory, Statistical Inference, and Statistical Theories of Science, Vol. II, 1–46. All Rights Reserved. Copyright © 1976 by D. Reidel Publishing Company, Dordrecht-Holland.

developed as a philosophical system in which the mysterious agency of
chance plays a central role. When this is so the statisticist doctrine may
well be presented as a consequence of the theory of probability: there are,
we learn, no causes; it is as if God were playing dice with the world. But,
of course, the argument goes on, there is no need to postulate a God, there
is only chance and what happens does so in accordance with the laws of
probability. At other times the statisticist doctrine is an implicit and
inadequately analysed assumption and then, according to the extent to
which this is so, the more difficult it is to recognise its presence and detect
its consequences. In such cases the statisticist doctrine may well be presented as part of a phenomenological basis of the theory of probability in
contrast to its alternative formulation as a consequence of that theory.

A notable feature of statisticism is its servile dependence on procedures
appropriate to pure mathematics. Although this dependence is not part of
the statisticist doctrine as such, it is so much a part of the systematic
development of that doctrine that it may fairly be considered an essential
part of the practice of statisticism. It is convenient to have a name for this
dependence, I call it 'mathematicism'. This term is introduced to emphasise
the similarity between the present context and that which led to the introduction of the word 'scientism' to describe methods of historical and
social enquiry which imitate what are taken to be the methods and language
of science. In the same way statisticism apes what it takes to be the methods and procedures of pure mathematics.

I am not, of course, suggesting that statistical practice has no need of
mathematics, but I am suggesting that statistics is science and not pure
mathematics. In pure mathematics one is concerned primarily with those
properties of formal systems which are incependent of any particular
interpretation of the terms being used. In science, on the contrary,
although one is again dealing with formal systems and, on occasion, even
those studied by the pure mathematician, one is now concerned with one
particular interpretation of its terms. But in scientific enquiry the relationship between the terms of the formal system and operationally defined
measurement procedures is incomplete and imprecise. In an ambiguous
sense, the formal system is just as much an interpretation of the empirical
system and it is in that ambiguous sense that we sometimes talk of an
incorrect interpretation of experimental facts. On the other hand, within
mathematics we give a technical meaning to the term 'interpretation';

there is no place for an incorrect interpretation of a formal system, what purports to be an interpretation either is one or it is not. One may reject a proposal to consider a particular formal system as an interpretation or model of an empirical system on the grounds that some of its theorems do not correspond to experimental facts. But one cannot reverse the argument to establish that a particular formal system provides a correct interpretation or description of reality because certain empirical statements are translated, under that interpretation, into theorems of the formal system. It is just such a reverse argument which characterises the role of mathematicism in the formal development of statisticism. For ease of reference we call this type of argument the mathematicist fallacy.

The controversy about the foundations of probability theory and the nature of statistical reasoning has its genesis in statisticist doctrine. What is being disputed is whether or not one particular procedure meets the criteria of what are, to a large, extent aims common to different kinds of statisticism. These aims are

(i) the formulation of a precise definition of probability,

(ii) the development from it of a formal probability calculus,

(iii) the specification of rules and procedures whereby the formal concepts are related to experimental facts, and

(iv) a justification of that specification.

What distinguishes the different kinds of statisticism is the way these aims are thought to be attained and not the question of whether the attainment of them is fruitful.

By way of contrast the preceding ideas are developed and brought together in a systematic account of the practice of statistics which denies the aims of statisticism in as much as it does not attempt to introduce a concept of probability at all. Our discussion proceeds from the observation that it is a recognition of the variability around us which gives substance to statements of uncertainty and emphasises that the primary role of practical statistics is the development of language codes to describe what we observe. However because of the current trend of some statistical practice more attention than would otherwise be merited is given to a discussion of Bayesian inference.

1. INTRODUCTION

This paper presents my own view of what I am doing when I practice statistics. But when I talk of practicing statistics I do not mean writing up papers for Statistical Journals nor the preparation of mathematically oriented papers arising out of some distant motivation of a probabilistic nature; neither do I mean simply giving statistical advice to scientists. I am referring to what I do when I collaborate with experimental and observational scientists in the subject matter of their disciplines. I mention collaboration and subject matter deliberately because I want to emphasise that when a statistician is working with a scientist he is doing work in that scientist's discipline, in physics, psychology, sociology or preventive medicine etc as the case may be; he is not doing 'statistics'.

I emphasise this because it seems to me that many statisticians do not do much more than pay lip service to such a statement. This is the first conference on foundations I have attended but I have, on occasion, studied the proceedings of other conferences and my overall impression is that some of the people who talk about the foundations of statistics do not practice statistics as an activity within an applied discipline. To be sure they advise applied scientists but to me that is something rather different.

I mention these things both to reveal my own bias and to emphasise that my own statistical experience has been limited by the type of problem on which I have been invited to collaborate. Thus I know of many practical problems only by hearsay and I have to admit the possibility of being faced one day with other practical situations which will force a change in my present view of what statistics is all about.

My own view is derived from the fact that in all the actual practical problems I have considered (as distinct from the hypothetical ones) the statistical analyses in question could have been reduced to simple arithmetic operations, counting and the calculation of observed frequencies. This suggested to me that one might be able to give an account of statistical practice which does not require a concept of probability but is expressed directly in terms of observed frequencies. Of course, such an account would not, in itself, explain the reasons for that practice. However the variability we do observe leads to the concept of a variability we might observe, and when this is taken into account one can discuss reasons for that practice without a concept of probability. Thus the approach adopted ignores one of the main aims of statisticism, namely the formulation of a precise definition of probability.

2. BASIC IDEAS: SITUATION SCHEMA,
CATALOGUES, REFERENCES AND CATALOGUING FUNCTIONS

Our immediate aim is the development of a formal language in which to discuss various aspects of experimental and observational investigations. However since this language is to be used to talk about matters in dispute we introduce a vocabulary which, though suggestive, is intended to be neutral in as much as it does not overlap with the conventional vocabulary of what is in dispute.

At the formal level the basic building block of our descriptive language is a triple $\mathscr{S} = (\Gamma, C, R)$ consisting of non-empty sets Γ and R and a function $C: \Gamma \to R$ with domain Γ and a codomain which is permitted to be a proper subset of R. Then $\mathscr{S}$ is called a situation schema, Γ is its catalogue, C is its cataloguing function and R is its reference set; the members of R are called references and the elements of Γ are called indices. Thus C assigns a reference to each index γ in the catalogue Γ. We do not suppose that each reference in the codomain of C is determined in this way by just one index in Γ (although this possibility is not excluded), but to avoid unnecessary mathematical complications in exposition it is to be understood that for each r in the codomain of C the set $C^{-1}(r)$ is finite.

This formal language is introduced to describe practical situations in which there is under consideration a set of alternative possibilities to each of which is associated an observation. The catalogue Γ lists these alternative possibilities, the reference set R names the observations in question and the cataloguing function assigns observations to possibilities. One can, of course, question the extent to which practical situations can be described in these terms but at this stage I am more interested in finding out if such a description is useful where it is applicable rather than in delineating the range of its applicability. However we will show shortly that it is applicable to any practical situation which can be discussed in terms of the standard model of probability theory.

Let $\mathscr{S} = (\Gamma, C, R)$ be a situation schema and for any finite set A let $N(A)$ be the number of its elements. The set

$$(R, C) \, \mathrm{Desc} = \{(r, N(C^{-1}(r))): r \in R\}$$

is called the (R, C)-description of $\mathscr{S}$. Note that $N(C^{-1}(r)) = 0$ precisely when the reference r does not belong to the codomain of the cataloguing function. The (R, C)-description of the situation schema $\mathscr{S}$ lists the observations in question together with the number of times each of them is associated with an index in the catalogue. It is a statement of the variability of the cataloguing function over the catalogue of the situation schema under consideration. It is convenient to have a special name for the (R, C)-description of a situation schema $\mathscr{S}$ with reference set R and cataloguing function C, we call it the standard description of $\mathscr{S}$.

3. MEASUREMENT PROCEDURES

Let $\mathscr{S} = (\Gamma, C, R)$ be a situation schema. Once a definite possibility γ in Γ has eventuated one measures certain things about the reference $C(\gamma)$ which it determines. We formalise the operation of measurement in the following way. A measurement procedure for $\mathscr{S}$ is an ordered pair (M, μ) consisting of a non-empty set M and a function $\mu: \Gamma \to M$ with domain Γ; we say that M is the measurement space and that μ is the measurement function. In the present framework the result of measurement is determined by the possibility which eventuates rather than by the associated reference, this allows for the possibility that the result of measuring what we observe may depend not only on what we do observe but on the way in which it came about.

A measurement procedure (M, μ) for $\mathscr{S}$ is said to have the finite property when, for each m in M, there are at most a finite number of γ in Γ with $\mu(\gamma) = m$. It is to be understood in what follows that we restrict the discussion to measurement procedures which do have the finite property. The set

$$(M, \mu) \text{ Desc.} = \{(m, N(\mu^{-1}(m))): m \in M\}$$

is called the (M, μ)-description of $\mathscr{S}$. Note that $N(\mu^{-1}(M)) = 0$ precisely when m does not belong to the codomain of the measurement function. The (M, μ)-description of the situation schema $\mathscr{S}$ lists all of the distinct results which could occur, together with the number of times each of them would occur, were the measurement procedure (M, μ) to be applied just once in the case of each possibility in Γ. It is a statement of the variability of the measurement function μ over the catalogue Γ.

The simplest example of a measurement procedure is the one which yields the possibility in question together with the reference it determines.

Its measurement function is C_0 defined by

$$C_0(\gamma) = (\gamma, C(\gamma)), \qquad \gamma \in \Gamma$$

and the associated description is essentially the situation schema itself. A related, but more interesting, measurement procedure is given by (R, C), the associated description of $\mathscr{S}$ is then the (R, C)-description defined earlier. For this reason we call (R, C) the counting measurement procedure.

A measurement procedure (M, μ) for $\mathscr{S}$ is (R, C)-bound when

$$\forall \gamma, \delta \in \Gamma : C(\gamma) = C(\delta) \Rightarrow \mu(\gamma) = \mu(\gamma).$$

Any (R, C)-bound measurement procedure is a function of the counting measurement procedure; this is the content of

PROPOSITION (3.1.). *Let (M, μ) be a measurement procedure for the situation schema $\mathscr{S}$ with catalogue Γ, reference set R and cataloguing function C. Then (M, μ) is (R, C)-bound if and only if there is a function $\phi : R \to M$ whose domain includes the codomain of the cataloguing function C and is such that*

(3.1) $\gamma \in \Gamma : \mu(\gamma) = \phi(C(\gamma)).$

It follows at once that for an (R, C)-bound measurement procedure (M, μ) one has

(3.2) $N(\mu^{-1}(m)) = \sum\limits_{r : \phi(r) = m} N(C^{-1}(r))$

where ϕ is the function in (3.1). Thus if one knows the (R, C)-description of $\mathscr{S}$ then one can obtain from it the (M, μ)-description for any given (R, C)-bound measurement procedure (M, μ).

In practice (R, C)-bound measurement procedures usually arise from a function $\phi : R \to M$ with domain R, the associated measurement function μ being defined by Equation (3.1). Such a function ϕ will be called a reference set measurement function.

More generally suppose that $C_j : \Gamma \to R, j = 1, 2, ..., n$ are a number of cataloguing functions with associated situation schemas

$$\mathscr{S}_j = (\Gamma, C_j, R), \qquad j = 1, 2, ..., n$$

each of which has the same catalogue and the same reference set. Let $(M, \mu_1), ..., (M, \mu_n)$ be measurement procedures for $\mathscr{S}_1, \mathscr{S}_2, ..., \mathscr{S}_n$, respectively, having a common measurement space M. We say that they are jointly (R, C)-bound when

$$\forall \gamma, \delta \in \Gamma \,\& \forall j, \quad k = 1, 2, ..., n :$$
$$C_j(\gamma) = C_k(\delta) \Rightarrow \mu_j(\gamma) = \mu_k(\delta).$$

Then we have

PROPOSITION (3.2). *Let* $(M, \mu_1), ..., (N, \mu_n)$ *be measurement procedures, for the situation schemas* $\mathscr{S}_1, \mathscr{S}_2, ..., \mathscr{S}_n$ *respectively. They are jointly* (R, C)-*bound if and only if there is a function* $\phi : R \to M$ *whose domain includes the codomains of each of the cataloguing functions* $C_1, C_2, ..., C_n$ *and is such that*

$$(3.3) \qquad \mu_j(\gamma) = \phi(C_j(\gamma))$$

for each $j = 1, 2, ..., n$ *and each* γ *in* Γ.

In practice jointly (R, C)-bound measurement procedures arise through a reference measurement function ϕ, the various μ_j being defined by Equation (3.3).

Instead of saying that a measurement procedure is (R, C)-bound we sometimes say that it is bound in $\mathscr{S}$, the situation schemc in question.

4. REPRESENTATIONS OF SITUATION SCHEMAS

In the summary I noted that part of my thesis is that belief in chance is sheer superstition and although I do not want to enter a plea on its behalf at this time, for our overall approach will be directed towards seeing how far one can go without concepts of chance and probability, I do want to indicate something of what I mean when I assert that thesis. To discuss them we need some preliminary results.

Let $\mathscr{S} = (\Gamma, C, R)$ and $\mathscr{S}' = (\Gamma, D, T)$ be two situation schemas with the same catalogue but possibly different reference sets and different cataloguing functions. Since D is a function on Γ the pair (T, D) is a measurement procedure for $\mathscr{S}$, moreover the (T, D)-description of $\mathscr{S}$ is just the standard description of $\mathscr{S}'$. Thus if our main interest is in the situation schema $\mathscr{S}'$ and its standard description these can be discussed in terms of the situation schema $\mathscr{S}$ and its (T, D)-description. In particular if it should be the case that (T, D) is bound for $\mathscr{S}$, that is

$$\gamma, \delta: C(\gamma) = C(\delta) \Rightarrow D(\gamma) = D(\delta),$$

then by Proposition (3.1) and Equation (3.2) we have $D = \phi \cdot C$, where $\phi: R \to T$ has a domain which contains the codomain of C, and

$$N(D^{-1}(t)) = \sum_{r: \phi(r)=t} N(C^{-1}(r)), \qquad t \in T.$$

Thus if ϕ is known we can calculate the (T, D)-description of $\mathscr{S}$, in other words the standard description of $\mathscr{S}'$, in terms of the standard description of $\mathscr{S}$. In such a case we say that $\mathscr{S}'$ is $\mathscr{S}$-representable.

This result is important because we are more familiar with some types of situation schema than with others. Thus if $\mathscr{S}'$ is a given situation schema then the discovery of another situation schema $\mathscr{S}$ which is both well-known and such that $\mathscr{S}'$ is $\mathscr{S}$-representable can be a useful way of reducing a new problem to an old problem. The more familiar types of situation schema seem to be those of the so-called games of chance and these therefore play an important representative role in the study of the variability encountered in practical situations. However there are a number of things about such representations which need emphasis.

In the first place we are not concerned here with the role of randomisation in experimental design but with the representation of variable quantities by 'random variables'. Secondly we note that the purpose of such a representation is simply to describe one type of variability in terms of another more familiar one. But the important thing about the more familiar situation is not that it comes from a 'game of chance' but that its variability is known – or at least thought to be known. However such games comprise just one of many types of physical situation which exhibit variability and, from the present viewpoint, they have no inherently special features except our familiarity with them. Finally, even if we do use a game of chance to describe some instance of variability in the real world we cannot conclude that 'chance' is a mysterious agency at work in the world, for 'chance' is only a term summarising the type of variability observed in certain special kinds of situation. To attribute 'chance mechanisms' to all processes exhibiting variability is to exhibit a misdirected reverence of the unknown and is, therefore, quite properly said to be a superstition.

5. THE STANDARD MODEL OF PROBABILITY THEORY

The standard model of probability theory is a triple $(M, \mathfrak{B}, P)$ where M is a non-empty set, $\mathfrak{B}$ is a σ-field of subsets of M and P is a countably additive non-negative normed measure on $\mathfrak{B}$. This general model comes from the simple case $M = \{m_1, m_2, ..., m_k\}$ is a finite set, $\mathfrak{B}$ is the set of all subsets of M and the measure P is derived from rational numbers $P_i = P(m_i)$, $i = 1, 2, ..., k$, which are non-negative and add to unity. We refer to this special case as the simple probability model on M, the general model arises only as a simplifying idealisation which leads to useful approximations.

A practical situation which can be described in terms of a situation schema with a finite catalogue can always be described in terms of the simple probability model. Thus if (M, μ) is a measurement procedure for the situation schema $\mathscr{S} = (\Gamma, C, R)$ where Γ is finite ard $M = \{m_1, m_2, ..., m_k\}$, then the (M, μ)-description of $\mathscr{S}$ is just the set

$$(M, \mu) \operatorname{Desc} \mathscr{S} = \{(m_i, N(\mu^{-1}(m_i))): i = 1, 2, ..., k\}.$$

One passes to the simple probability model on M by taking the 'probabilities' p_i to be given by the equations

$$p_i = N(\mu^{-1}(m_i))/\sum_{j=1}^{k} N(\mu^{-1}(m_j)),$$

for each $i = 1, 2, ..., k$. Such a reduction often leads to useful simplifications but it has to be kept in mind that there are aspects of the underlying situation which become blurred in the process of reduction. Thus many different situation schema lead to same simple probability model and that model, does not, in itself, distinguish between them.

Conversely any problem which is formulated in terms of the simple probability model on M can be discussed in terms of a suitable situation schema. Thus there exists a smallest positive integer N_0 for which there are non-negative integers $N_1, N_2, ..., N_k$ such that $N_0 p_i = N_i$, $i = 1, 2, ..., k$. Let b be a positive integer and let Γ be a set with bN_0 elements. Let $G_1, G_2, ..., G_k$ be disjoint subsets of Γ such that $N(G_i) = bN_i$, $i = 1, 2, ..., k$. Let $\mathscr{S}$ be a situation schema with catalogue Γ and define $\mu: \Gamma \to M$ by

$$i = 1, 2, ..., k; \qquad \gamma \in G_i \Rightarrow \mu(\gamma) = m_i.$$

Then (M, μ) is a measurement procedure for $\mathscr{S}$ and the procedure outlined above leads back to the simple probability model on M. Note that the situation schema $\mathscr{S}$ is not uniquely determined by the simple probability model, in particular the reference set R and the cataloguing function C can be chosen at will.

However, in practice, this additional freedom is to a large extent illusory because, when one examines the detail of a particular practical situation for which the simple probability model is being used, one finds that the nature of the situation schema in question is largely determined in an unambiguous way. It is in that sense that a situation schema is a description rather than a model.

6. THE CLASSIC CONCEPTION OF PROBABILITY

It is a recognition of the variability around us which gives substance to statements of uncertainty. This, that or one of several other things might be the case, sometimes it is one, sometimes it is another one of them, this much we observe; but at a particular instance we may not know which of the possibilities is in fact the case. The theory of probability purports to provide the means whereby one may account for this variability and describe its quantitative aspects. However, as noted above, any practical problem which can be formulated in terms of the theory of probability can be reformulated in terms of a situation schema and a measurement procedure. Moreover this reformulation is a natural one in the sense that it is not the setting up of an abstract formal system as a conceptual model of reality, it is rather the talking about reality in a particular formal language. But an important aspect of this formulation is the absence of familiar concepts of probability theory. To be sure an (M, μ)-description of a situation schema involves counting and the corresponding proportions behave; at the formal level, like probability measures; but although variability is taken into account there are no explicit concepts of uncertainty and chance.

It might be argued that the calculation of the (M, μ)-description of a situation schema assumes an equal weighting of alternatives, since only counting is involved, so that writing down a situation schema requires an implicit judgement as to the propriety of this equal weighting. It would seem, therefore, that the preceding sections comprise just a pedantic version of the so-called classic conception of probability theory whereby numerical probability was defined as the proportion of favourable cases among a number of equally likely alternatives. It could be argued, then, that one is subject to exactly the same criticisms which led to the abandonment of those ideas. We will argue, however, that this is not the case.

7. CRITICISMS OF THE CLASSIC CONCEPTION OF PROBABILITY

The classic conception of probability has been criticised on a number of grounds, we consider six of them.

(1) That probability is defined in terms of a relation of equiprobability and such a definition is circular unless equiprobability is independently defined.

(2) That to define equiprobability one has to appeal to the so-called 'principle of insufficient reason'. This principle is itself unsatisfactory because (a) it is only applicable by one who has had no experience relative to the situation at hand; (b) there is no criterion to determine whether or not it has been correctly applied; (c) it is difficult to apply in practice and even reputable mathematicians have misapplied it; moreover (d) there are situations in which there is lack of agreement on its correct application.

(3) That, according to the classic conception of probability, numerical probabilities must be rational numbers. But there are cases where irrational numbers occur as probabilities and these cannot be interpreted as ratios of alternatives. It has been noted, for instance, that the probability that two integers chosen at random are relatively prime is $6/\pi^2$. This cannot mean, it is said that there are π^2 equiprobable alternatives of which 6 are favourable to getting relative primes. (This is a criticism which attributes reality to the continuum and fails to recognise that, in relation to practical measurement, it is only a convenient method of approximation. We do not consider it further here).

(4) That there is no logical relation between the number of different ways in which something can happen and the actual relative frequency with which it does happen. Thus probability, as defined, cannot tell us about actual relative frequencies.

(5) That the classic conception of probability does not make it clear whether or not a numerical probability derived from judgements of equiprobability is a measure of what we actually do believe or of what we

ought to believe. If the former is meant then it is not clear that the axioms of probability are, in fact, satisfied; whereas if the latter is meant then the source of the imperative is not given.

(6) That, according to the classic conception, probability can be defined only when the underlying situation can be analysed into a finite set of equiprobable alternatives. But there are situations where we make numerical probability statements but such an analysis is without meaning. For instance one may make a statement such as 'The probability that a thirty-year-old man will live for another year is 0.945'. It has been said that it is absurd to suppose that this statement means there are a thousand possible courses to a man's career, and that exactly 945 of them are favourable to his survival for at least another year.

8. VARIETIES OF STATISTICISM

The varieties of statisticism arise from which of these criticisms is taken as the point of departure from which to rebuild the foundations of probability theory. Rather obviously, a start from (4) leads to the attempt to define probability directly in terms of relative frequency. A start from (1), (2) and (5) can proceed in two ways, according as probability judgements are thought of as relating to what we do believe or to what we ought to believe.

When probability judgements are thought of as relating to what we do believe the emphasis is placed on the act of making those judgements and, accordingly, probability is defined in terms of acts which, it is claimed, exhibit a person's qualitative probability judgements and enable them to be quantified. The axioms of probability are recovered by interpreting the imperative in terms of coherent behaviour. This approach leads to what is sometimes called the theory of personal probabilities.

When probability judgements are thought of as relating to what we ought to believe the emphasis is placed on the existence of probability judgements rather than on the act of making them. There exists, it is held, an objectively right judgement in relation to the available evidence, this is the source of the imperative and probability is thought of as a logical relation between propositions. We will not discuss such theories of so-called logical probability here because, although they are interesting to the extent to which they provide a metalanguage in which to converse about statistical practice, they do not provide a convenient framework within which to develop the practice of statistics.

The fourth criticism of the classic conception of probability raises the difficulty of relating probabilities to actual relative frequencies. The pro-frequency statisticist is, of course, motivated by the general aims of statisticism; in particular he wants to define probability. He does so by constructing a mathematical object within a formal system and, through a particular interpretation of its terms, talking about that object as an infinite sequence of independent events etc. Because certain theorems of the formal system appear to correspond, under that interpretation, to sentences true of the world, the pro-frequency statisticist predicates the existence in the world of processes corresponding to that mathematical object. Such a procedure involves the mathematicist fallacy noted earlier.

To a certain extent the formulation of this paper is related to the substance of the sixth criticism mentioned above, for it is based, in part, on the recognition of two things. Firstly, that although it may be absurd to suppose that there are a thousand courses to a man's career and that exactly 945 are favourable to his survival for at least another year, it is not absurd to state, as a matter of observation in respect of a particular life-table, that 94.5% of thirty year old men did survive for at least one year; and secondly, that statements such as this, rather than statements of probability, form the basis of the practice of statistics. The issue raised by the fourth criticism of the classic conception of probability also has a bearing on the formulation of this paper. The point will be taken up, inter alia, in the next section. For the present we emphasise the view that one finds out relative frequencies of happenings by observation and not by introspection, that the problem is not so much the derivation of actual relative frequencies from the number of ways something can happen but the fact that the number of ways something can happen takes its meaning from the number of ways in which it does happen.

9. SITUATION SCHEMAS AND EQUIPROBABILITY

We question, now, the extent to which a description of practical situations in terms of situation schemas depends, for its adequacy, on an implicit assumption of the classic conception of probability. There are two things to be made clear at the outset. Firstly the description in question is not adopted in pursuit of the aims of statisticism, in particular there is no attempt to define probability nor is there any appeal to a concept of chance. We are not, then, committed either to assertions or denials of equiprobability. However, and this is the second thing, although there are no statements about equiprobability, the criticisms of the classic conception may apply, mutatis mutandis, to the statements that are made. For, it might be said, counting the number of possibilities which correspond, under the cataloguing function, to a definite member of the reference set involves an implicit equal weighting of them; and so the determination of which situation schema is appropriate to an experimental situation involves a recognition of the propriety of this equal weighting. In other words, one has to make implicit judgements of equiprobability.

It is obvious, of course, that the act of setting up a situation schema for a practical situation does involve a judgement about its relevance; but what is not so clear is that the grounds for that judgement should be, in themselves, pertinent to a scientific enquiry which aims to say what is the case, rather than to elaborate on how one did, perhaps, come to know that it was so. The important question concerns the manner in which such judgements are substantiated, rather than whether or not they are judgements of equiprobability. The viewpoint here is that one learns more about the world by experimentation than introspection and casual observation. The extent to which a conjectured situation schema is appropriate to a particular experimental situation is answered by experiment and not by introspection. This outlook is dismissed by some statisticists who charge that it involves circularity in as much as, it is claimed, statistics is just that science which studies the extent to which one can declare that observed facts confirm conjectures. One cannot then, it is said, base statistics itself on judgements of the extent to which conjectures are verified.

However this dismissal is grounded in a number of confusions about the practice of science. Scientific enquiry does not proceed simply by verifying and discrediting conjectures in the light of the extent to which they fit the facts. Indeed a conjecture which fits the facts badly is useful when the only other ones fit them even more badly. Moreover one discards one conjecture, in favour of another, when the new one is more useful; at one extreme one may even adopt a procedure which fits the facts less accurately than another, because it is easier to apply and leads to a better understanding of the phenomena being studied. In other words,

the judgements of the scientist are, in substance, not absolute but relative;
they do not rest simply on the extent to which there is agreement with the
facts but rather on the extent to which one procedure is preferred to
another. When the direction of preference is not clear the question is
resolved through the accumulation of more facts and not, in general, by
introspection except possibly in the introduction of new paradigms.
Moreover grounds for preference vary with the circumstances of the
enquiry because they are bound up with the inter-relationships between
the various subdivisions of the discipline in question. There is seldom a
general appeal to universal principles of inference derived in some way,
as is sometimes claimed, from an analysis of the concept of experience.
Relevance and usefulness are assessed relative to the disciplines involved
and not in terms of abstract principles about the extent to which facts
confirm or discredit conjectures. It is for this reason that I started this
paper by emphasising that statistics is not a separate science and that the
statistician is a worker in the discipline in question.

However the belief that statistics is that science which studies the extent
to which observed facts confirm hypotheses has become an unquestioned
assumption in discussions of the foundations of statistics. It has led to a
basic premise of statisticist arguments in favour of personal probability
theory. For if statistics is what it is alleged to be then an experiment
should result in a numerical measure of the extent to which it does confirm various hypotheses, and this is, then, a measure of the strength of
our opinion about the relative credibility of those hypotheses in the light
of the experiment. But if the proper consequence of an experiment is a
measure of the strength of our opinions then we should approach an
experiment with opinions which are quantified to the extent that they have
been confirmed by earlier experimentation. It would seem, therefore, that
one could formulate the principles of statistical reasoning by specifying
the way in which we should change our opinions in the light of experimental results. This line of argument leads to the statisticist view that it is the
business of the scientific experimenter to revise his opinions in an orderly
way with due regard to internal consistency and the data.

We will return to some of these questions at a later stage for, judging by
the extent of the literature on the foundations of probability and statistics,
one could argue indefinitely about the merits and demerits of the various
kinds of statisticism without reaching a conclusion which would compel

universal assent. Instead we will develop the approach suggested earlier which is based on the view that, since one deals in practice with observed frequencies and not 'probabilities', one should be given an account of statistical practice which does not require a concept of probability but is expressed directly in terms of those observed frequencies. However I have chosen the topics to be discussed not for their intrinsic importance but simply in order to provide comparisons with some well-known statistical concepts.

10. ALTERNATIVE SITUATION SCHEMAS

In this section we embark on a programme whose aim is to indicate how some familiar statistical concepts can be introduced in the absence of a concept of probability. We start by considering a simple problem in hypothesis testing. However, in view of earlier remarks, it is only fitting that we explain the way in which this programme is meant to be understood. Thus, although we have questioned the propriety of views which emphasis the importance for science of the role played by the confirmation of hypotheses, we do not, of course, deny that hypothesis testing has some role to play; we question only the importance and precise nature of that role. It is to be kept in mind, however, that the present discussion is not meant to be an exhaustive analysis of the role of hypothesis testing in the practice of science. It is intended to exhibit a short route to some well-known concepts of statistical theory, and only that; we are not arguing that this route provides, in itself, a sufficient justification for using those concepts in a particular way, nor indeed, in any way at all.

We start by envisaging a practical situation in respect of which two possible situation schemas, $\mathscr{S}_1$ and $\mathscr{S}_2$, are being considered; the supposition is that the actual situation schema is either $\mathscr{S}_1$ or, if it is not, then $\mathscr{S}_2$. In more familiar language there are two hypotheses H_1 and H_2; H_1 being that $\mathscr{S}_1$ is the case and H_2 being that $\mathscr{S}_2$ is the case. To simplify the discussion we suppose that the situation schemas have the same finite catalogue Γ, the same finite reference set R but different cataloguing functions, C_1 and C_2, so that $\mathscr{S}_1 = (\Gamma, C_1, R)$ and $S_2 = (\Gamma, C_2, R)$.

The problem is to determine which of these hypotheses is the case in the light of measurement made when just one of the possibilities in Γ eventuates. To simplify matters we suppose that the measurement procedure in question arises through a reference set measurement function $\phi : R \to M$.

We define $\mu_1 : \Gamma \to M$ by

$$\mu_1(\gamma) = \phi(C_1(\gamma)), \qquad \gamma \in \Gamma$$

and $\mu_2 : \Gamma \to M$ by

$$\mu_2(\gamma) = \phi(C_2(\gamma)), \qquad \gamma \in \Gamma.$$

Then, in the terminology of Section 3, (M, μ_1) and (M, μ_2) are jointly bound measurement procedures.

We treat this problem in a conventional way, and introduce two subsets of M, say M_{12} and M_{21}, which are not necessarily disjoint but whose union is M; such a pair of sets will be called an acceptability basis for H_1 and H_2. Next we adopt the following 'decision procedure': if application of measurement to the reference we encounter yields m in M then we accept H_1 or H_2 or make no decision, according as m belongs to M_{21}-M_{12}, to M_{12}-M_{21} or to $M_{12} \cap M_{21}$. The problem is then reduced to find a 'suitable' acceptability basis.

However the usual language of hypothesis testing has some objectionable features. For instance, it is doubtful whether the scientist does, in general, accept or reject an hypothesis on the basis of just one experiment and, for this reason, some scientists have objected to the standard procedures of hypothesis testing. Moreover the terminology of hypothesis testing confuses an act of acceptance with possible grounds for that act. We will, therefore, adopt the following terminology: if the reference we encounter yields m in M_{12} we will say that H_1 is not more acceptable than H_2, whereas if the reference in question yields m in M_{21} we will say that it H_2 is not more acceptable than H_1. Derived terminology is used in an obvious way; if m is in $M_{12} \cap M_{21}$ then H_1 and H_2 are equally acceptable, similarly H_1 is less acceptable than H_2 when m is in $M_{12} - M_{21}$ and so on. All such statements of acceptability are to be understood as qualified by the phrase 'in the light of the measurement m'. This terminology leads to awkward sentence structure but it has the advantage that, whilst it does not preclude acceptance, it permits one to distinguish between an act of acceptance and possible grounds for that act.

The set $\mu_1^{-1} M_{12}$ consists of the possibilities in Γ for which we do not accord H_1 the greater acceptability, were one of them to be realised, even when H_1 was in fact the case. In like manner, the set $\mu_2^{-1} M_{21}$ consists of the

possibilities in Γ for which we would not accord H_2 the greater acceptability, were one of them to be realised, even when H_2 was, in fact, the case.
Thus the set

$$\mu_1^{-1}M_{12} \cup \mu_2^{-1}M_{21}$$

consists of the possibilities in Γ for which, were one of them to be realised,
the decision procedure based on the acceptability basis (M_{12}, M_{21}) would
not ensure that we reached the right decision. The complement in M of
the set (10.1) is the set

$$(10.2) \quad \mu_2^{-1}M_{12} \cap \mu_1^{-1}M_{21},$$

its elements are the possibilities in Γ for which the decision procedure
based on (M_{12}, M_{21}) does lead to the right decision, whenever one of
them is realised, irrespective of whether it is H_1 or H_2 which is, in fact, the
case.

For any m, m' in M write

$$\Gamma(m, m') = \mu_1^{-1}(m) \cap \mu_2^{-1}(m')$$

and

$$K_1(\mu_1, \mu_2) = \text{Min} \left[N\left(\mu_1^{-1}M_{12} \cup \mu_2^{-1}M_{21}\right) : (M_{12}, M_{21}) \in \mathcal{M} \right]$$

where $\mathcal{M}$ is the set of all acceptability bases on M.

An acceptability basis (M_{12}, M_{21}) will be said to be minimal when

$$N\left(\mu_1^{-1}M_{12} \cup \mu_2^{-1}M_{21}\right) = K_1(\mu_1, \mu_2).$$

Minimal acceptability bases have an obvious appeal. If we use one then
the associated decision procedure has the property that we cannot get a
smaller set of possibilities in Γ for which we are not sure of reaching the
right decision; equivalently such a decision procedure has the property
that we cannot get a larger set of possibilities in Γ for which we are sure of
reaching the right decision, whenever one of them is realised, whether it is
H_1 or H_2 which happens to be the case. (Of course there will be situations
where one would not want to use a minimal acceptability basis because,
for instance, one attached greater importance to a right decision when say
H_1 is the case; or again, for example, when greater importance is attached
to reaching a right decision for the possibilities in some specified subset
of Γ.)

By way of illustration consider a case in which Γ has 32 elements, M contains three elements m_1, m_2 and m_3, and the measurement functions μ_1, μ_2 are such that the 3×3 matrix $\|N(\Gamma(m_i, m_j))\|$, $i, j = 1, 2, 3$ is

$$\left\|\begin{matrix} 1 & 2 & 3 \\ 2 & 4 & 6 \\ 3 & 5 & 6 \end{matrix}\right\|.$$

A short calculation shows that there is a unique minimal acceptability basis, namely

$$M_{12} = \{m_3\}, \qquad M_{21} = \{m_1, m_2\}$$

corresponding to $K_1(\mu_1, \mu_2) = 23$. The set

$$\Gamma(m_1, m_3) \cup \Gamma(m_2, m_3)$$

contains the nine possibilites for which we are sure of reaching the right decision.

If the particular case just considered is modified slightly, so that Γ has 31 elements and the corresponding matrix is

$$\left\|\begin{matrix} 1 & 2 & 3 \\ 2 & 4 & 5 \\ 3 & 5 & 6 \end{matrix}\right\|$$

then there are two minimal acceptability bases, namely

$$M_{12} = \{m_3\}, \qquad M_{21} = \{m_1, m_2\}$$

and

$$M_{12} = \{m_1, m_2\}, \qquad M_{21} = \{m_3\}.$$

This modified case is interesting because the frequencies with which m_1, m_2, and m_3 occur, namely 6, 11 and 14 respectively, are the same whether it is H_1 or H_2 which happens to be the case. If this particular problem was formulated in terms of the simple probability model one would be 'distinguishing' between two identical probability distributions!

Usual discussions of hypothesis testing ignore the fact that the measurement functions μ_1, μ_2 are defined on the same set. They consider the hypothesis H_1 to be specified by the (M, μ_1)-description of $\mathscr{S}_1$ and, in like manner, the hypothesis H_2 is taken as specified by the (M, μ_2)-description of $\mathscr{S}_2$. There is, of course, a genuine practical problem to be faced when one does not know the relationship between the functions μ_1, μ_2 but only the (M, μ_1)-description of $\mathcal{S}_1$ and the (M, μ_2)-description of $\mathcal{S}_2$ and we discuss it in the next section. For the present we remark that if the relationship between the measurement functions μ_1 and μ_2 is known, at least to the extent that the numbers $N(\Gamma(m, m'))$ are known for each m, m' in Γ then the determination of a minimal acceptability basis is a computational problem. Of course, taking $M_{12} = M_{21} = M$ would give $N(\mu_1^{-1}M_{12} \cup \mu_2^{-1}M_{21}) = N(\Gamma)$ but the problem is to do better than that! The condition under which one can do better is given by the easily proved

PROPOSITION (10.1). *There exists an acceptability basis* (M_{12}, M_{21}) *such that*

$$N(\mu_1^{-1}M_{12} \cup \mu_2^{-1}M_{21}) < N(\Gamma)$$

if and only if $\mu_1 \neq \mu_2$, *that is there exists* γ *in* Γ *such that* $\mu_1(\gamma) \neq \mu_2(\gamma)$.

11. Universal acceptability bases

We consider now the decision problem formulated in the last section when, in addition to the result of the measurement, one does not know the exact relationship between the measurement functions μ_1 and μ_2 but only the (M, μ_1)-description of $\mathcal{S}_1$ and the (M, μ_2)-description of $\mathcal{S}_2$. Put another way, we do not know the individual entries in the matrix $\|N(\Gamma(m_i, m_j))\|$ but only the totals of each of the rows and the totals of each of the columns.

Let $M = \{m_1, m_2, ..., m_k\}$. We suppose that we are given two sets of non-negative integers $\{a_{11}, a_{12}, ..., a_{1k}\}$ and $\{a_{21}, a_{22}, ..., a_{2k}\}$ such that

$$\sum_{j=1}^{k} a_{1j} = N(\Gamma) = \sum_{j=1}^{k} a_{2j}.$$

and the additional supposition is that although the reference set measurement function ϕ remains fixed, the respective cataloguing functions C_1 and C_2 could be any ones which are such that

$$N(\mu_1^{-1}(m_j)) = a_{1j}, \quad j = 1, 2, ..., k,$$

and

$$N(\mu_2^{-1}(m_j)) = a_{2j}, \quad j = 1, 2, ..., k,$$

where, of course, $\mu_1 = \phi \circ C_1$ and $\mu_2 = \phi \circ C_2$. Such a pair of measurement functions will be said to be allowable and the set of allowable pairs will be denoted by $\mathscr{A}$.

[In more familiar language we are discriminating between two simple hypotheses, both of which are specified by a probability distribution, one being $\{a_{1j}/N(\Gamma), \ j=1, 2, ..., k\}$ and the other being $\{a_{2j}/N(\Gamma), \ j=1, 2, ..., k\}$. However in the present framework these 'simple' hypotheses are in fact inter-related composite hypotheses].

Each allowable pair (μ_1, μ_2) leads to a corresponding minimal acceptability basis which, of course, depends on the particular μ_1, μ_2 in question. Write

$$K_1(\mathscr{A}) = \text{Max}\left[K_1(\mu_1, \mu_2):(\mu_1, \mu_2) \in \mathscr{A}\right].$$

An acceptability basis (M_{12}, M_{21}) is said to be $\mathscr{A}$-universal when

$$\forall (\mu_1, \mu_2) \in \mathscr{A}: N(\mu_1^{-1} M_{12} \cup \mu_2^{-1} M_{21}) \leqslant K_1(\mathscr{A}).$$

Universal acceptability bases have an obvious appeal. If we do use one then for a particular allowable pair (μ_1, μ_2) we could, perhaps, obtain a larger set of possibilities in Γ for which we could be sure of reaching the right decision, by using the minimal acceptability basis for that pair, however there will always be at least one allowable pair for which could not get a larger set of such possibilities. In other words an $\mathscr{A}$-universal acceptability basis achieves the minimum value of the set of maxima of the sizes of those subsets of Γ for which we are sure of reaching the right decision. The question is therefore do $\mathscr{A}$-universal acceptability bases exist and, if they do, can they be determined. For our purposes the answer is provided by the following considerations.

Let

$$A_{12} = \{j: a_{1j} \leqslant a_{2j}\}, \qquad A_{21} = \{j: a_{2j} \leqslant a_{1j}\}.$$

We say that an acceptability basis (M_{12}^*, M_{21}^*) is $\mathscr{A}$-determined when

$$M_{12}^* = \{m_j: j \in A_{12}^*\}$$
$$M_{21}^* = \{m_j: j \in A_{21}^*\}$$

where $A_{12}^* \subseteq A_{12}$ and $A_{21}^* \subseteq A_{21}$ are disjoint sets with

$$A_{12}^* \cup A_{21}^* = A_{12} \cup A_{21}.$$

Then we have the following result

PROPOSITION (11.1). *If (M_{12}, M_{21}) is an $\mathscr{A}$-determined acceptability basis then it is $\mathscr{A}$-universal and, moreover,*

$$\forall(\mu_1, \mu_2) \in \mathscr{A}: N\left(\mu_1^{-1}M_{12} \cup \mu_2^{-1}M_{21}\right) = K_1(\mathscr{A}).$$

$\mathscr{A}$-determined bases are not necessarily minimal. For instance suppose that Γ contains 24 elements, that $M = \{m_1, m_2, m_3\}$ and that the 3×3 matrix $\|N\left(\Gamma(m_i, m_i)\right)\|$ is given by

$$(11.1) \qquad \begin{Vmatrix} 1 & 2 & 1 \\ 1 & 2 & 4 \\ 10 & 2 & 1 \end{Vmatrix}.$$

Then there is a unique minimal acceptability basis, namely

$$M_{12} = \{m_1, m_2\}, \qquad M_{21} = \{m_3\},$$

giving 12 elements of Γ for which we are sure of reaching the right decision. On the other hand if $\mathscr{A}$ is the set of all allowable pairs giving matrices with the same row and column sums as the matrix (11.1) there is a unique $\mathscr{A}$-determined acceptability basis, namely

$$M_{12} = \{m_1\}, \qquad M_{21} = \{m_2, m_3\},$$

giving 11 elements of Γ for which we are sure of reaching the right decision.

Decision procedures based on $\mathscr{A}$-determined acceptability bases do have an intuitive appeal, firstly because, as indicated above, they have the optimal 'max-min' property of $\mathscr{A}$-universality and secondly because they are easy to apply. Thus when one observes an m in M for which

$$N\left(\mu_1^{-1}(m)\right) > N\left(\mu_2^{-1}(m)\right)$$

one accords the greater acceptability to H_1. In like manner when one observes an m in M for which

$$N\left(\mu_2^{-1}(m)\right) > N\left(\mu_1^{-1}(m)\right),$$

the greater acceptability is accorded to H_2. The set of m in M for which

$$N\left(\mu_1^{-1}(m)\right) = N\left(\mu_2^{-1}(m)\right)$$

is partitioned into two subsets, observation of an element in one of them leads to H_1 whereas an observation in the other one leads to H_2; but, in so far as $\mathscr{A}$-universality is concerned, it is irrelevant which of the possible partitions is adopted.

Write

$$L_{12} = \{m\colon N(\mu_1^{-1}(m)) \leqslant N(\mu_2^{-1}(m))\}$$
$$L_{21} = \{m\colon N(\mu_2^{-1}(m)) \leqslant N(\mu_1^{-1}(m))\}.$$

Then an acceptability basis (M_{12}, M_{21}) is $\mathscr{A}$-determined if and only if M_{12}, M_{21} are disjoint subsets of M such that $M_{12} \subseteq L_{12}$, $M_{21} \subseteq L_{21}$ and $M_{12} \cup M_{21} = M$. It is convenient to discuss the decision procedures derived from the $\mathscr{A}$-determined acceptability bases in terms of the particular acceptability basis (L_{12}, L_{21}). We call this the symmetric acceptability basis and the associated decision procedure is referred to as the symmetric decision procedure.

12. Symmetric decision procedures and maximum likelihood

We suppose now that instead of just two hypotheses we have r of them, $H_1, H_2, \ldots, H_r$. More specifically we have an underlying practical situation in respect of which r possible situation schemas $\mathscr{S}_j = (\Gamma, C_j, R)$, $j = 1, 2, \ldots, r$, are under consideration. We suppose that each of these situation schemas has the same catalogue, the same reference set but differ in respect of their cataloguing functions. We are given a reference set measurement function $\phi\colon R \to M$ and associated measurement procedures (M, μ_j) defined by

$$\mu_j(\gamma) = \phi(C_j(\gamma)), \qquad \gamma \in \Gamma.$$

For each $i, j = 1, 2, \ldots, r$ we write

$$L_{ij} = \{m\colon N(\mu_i^{-1}(m)) \leqslant N(\mu_j^{-1}(m))\},$$

in particular $L_{jj} = M$ for each $j = 1, 2, \ldots, r$. Then for each $i \neq j$ the pair (L_{ij}, L_{ji}) leads to the symmetric decision procedure for H_i and H_j. There are $r(r-1)/2$ such decision procedures and they can be considered together by defining on the set $\{H_1, H_2, \ldots, H_r\}$ a binary relation $\bar{\alpha}_m$ for each m in M in the following way:

$$H_i \bar{\alpha}_m H_j \Leftrightarrow m \in L_{ij}, \qquad i, j = 1, 2, ..., r.$$

We read the left-hand side of this expression as 'the hypothesis H_i is not more acceptable than the hypothesis H_j in the light of the observation m'. The binary relation $\bar{\alpha}_m$ has the properties

(i) $H_i \bar{\alpha}_m H_i, \qquad i = 1, 2, ..., r$

(ii)   at least one of $H_i \bar{\alpha}_m H_j$ and $H_j \bar{\alpha}_m H_i$ holds for any m in M and any $i, j = 1, 2, ..., r$, and

(iii) $H_h \bar{\alpha}_m H_i \;\&\; H_i \bar{\alpha}_m H_j \Rightarrow H_h \bar{\alpha}_m H_j.$

It follows that, for each m in M, the binary relation $\bar{\alpha}_m$ is a linear ordering of the set $\{H_1, H_2, ..., H_r\}$. In other words, for each m in M there is a permuation $(1', 2', ..., r')$ of $(1, 2, ... r)$ such that

$$H_{1'} \bar{\alpha}_m H_{2'} \bar{\alpha}_m ... \bar{\alpha}_m H_{r'}.$$

We call this ordering of the hypotheses the symmetric acceptability order. ing in the light of the measurement m. This ordering is unique except for allowable interchanges between pairs of equally acceptable hypotheses- Thus if the reference measured yields m in M then the r hypotheses $H_1, H_2, ..., H_r$ can be ranked in non-decreasing order of acceptability in the light of that measurement.

This result can be expresses in the following way. If the measured reference yields m in M then the $r(r-1)/2$ pairwise comparisons of the r hypotheses, through the corresponding pairwise symmetric decision procedures, lead to an acceptability ranking of those hypotheses; this ranking is the natural ranking of the quantities $N(\mu_j^{-1}(m))$, $j = 1, 2, ..., r$. If one of these quantities is larger than all the others then the corresponding hypothesis is the most acceptable one in the light of the measurement m; it is that hypothesis for which the frequency with which m occurs is the greatest.

There is an obvious analogy to the likelihood ordering of hypotheses and the principle of maximum likelihood. However it has to be kept in mind that the symmetric acceptability ordering comes from a particular kind of decision procedure involving the use of a particular type of acceptability basis, its propriety stemming from that of the minimal acceptability basis. As noted earlier there may be reasons for not using minimal acceptability bases. Moreover in some specific situations one may be able to do 'better' than accept that hypothesis which is most acceptable under the symmetric decision procedure. For instance in the case of the matrix (11.1) the result m_2 occurs more frequently under the hypothesis H_1 than under the hypothesis H_2 but the minimal acceptability basis leads to a decision procedure which accepts H_2 when m_2 is observed.

The quantities $N(\mu_i^{-1}(m))$, $m \in M$, $i = 1, 2, ..., r$ can be regarded as a matrix with one row for each hypothesis and one column for each possible measurement result. An element of this matrix can be looked at in two ways, according as one does so from the viewpoint of its row or from the viewpoint of its column. Each row lists the frequencies of the possible measurement results in the situation schema corresponding to that row and each column exhibits the variation over the set of hypotheses of the frequencies in question. The symmetric acceptability ordering of the hypotheses in the light of m in M is just the natural ordering of the corresponding column; this is only an ordering, the entries in that column are not thereby quantitative measures of acceptability. Nevertheless it is convenient to define, for each m in M, an integer-valued function $A(\cdot \mid m)$ on the set $\{H_1, H_2, ..., H_r\}$ by the equations

$$A(H_j \mid m) = N(\mu_j^{-1}(m)), \qquad j = 1, 2, ..., r.$$

We read the left-hand side of this equation as the acceptability of the hypothesis H_j in the light of the measurement m. However, as noted above, the function $A(\cdot \mid m)$ does not yield a quantitative measure of acceptability, it is the ordering of the hypotheses according to the natural ordering of the quantities $A(H_j \mid m)$, $j = 1, 2, ..., r$ which has empirical meaning.

13. BAYES' THEOREM AND BAYES' POSTULATE

Consider two experimental situations in respect of each of which the same r hypotheses $H_1, H_2, ..., H_r$ are being considered. Suppose further that the two experimental situations are so related that whichever of these hypotheses is true of one of them is also true of the other. From the experimental situation $\mathscr{E}_1$ we obtain an acceptability ordering of the hypotheses in the way discussed in the last section. In like manner we obtain another acceptability ordering for the hypotheses from the experimental situation $\mathscr{E}_2$. In this section we study the way in which these two orderings are combined to provide a single acceptability ordering. In other words we examine how an acceptability ordering from one experiment is revised in the light of the result of another experiment. We will assume that the experiments in question are independent of each other, at the technical level this assumption is put in evidence by the way we combine the situation schemas of the individual experiments.

Let the r situation schemas for the experiment $\mathscr{E}_1$ be $\mathscr{S}_{1j}=(\Gamma_1, C_{1j}, R_1)$, $j=1, 2,..., r$, each of them having the same catalogue Γ_1 and the same reference set R_1. Similarly let the r situation schemas for the experiment $\mathscr{E}_2$ be $\mathscr{S}_{2j}=(\Gamma_2, C_{2j}, R_2)$, $j=1, 2,..., r$, each of them having the same catalogue Γ_2 and the same reference set R_2. The 'hypothesis' H_j is that $\mathscr{S}_{1j}$ is the situation schema for $\mathscr{E}_1$ and $\mathscr{S}_{2j}$ that for $\mathscr{E}_2$. We suppose that the measurements in question arise from a reference set measurement function $\phi_1: R_1 \rightarrow M_1$ in the case of $\mathscr{E}_1$ and from a reference set measurement function $\phi_2: R_2 \rightarrow M_2$ in the case of $\mathscr{E}_2$. For each $i=1, 2$ and each $j=1, 2,..., r$ we define the measurement procedure (M_i, μ_{ij}) for $\mathscr{S}_{ij}$ by the equations

$$\gamma_i \in \Gamma_i; \qquad \mu_{ij}(\gamma_i) = \phi_i(C_{ij}(\gamma_i)).$$

In the experimental situation $\mathscr{E}_1$ a result m_1 in M_1 leads, through the procedure of the last section, to an acceptability ordering of the hypotheses $H_1, H_2,..., H_r$, namely the natural ordering of the quantities

(13.1) $\quad A_1(H_j \mid m_1) = N(\mu_{1j}^{-1}(m_1)), \qquad j=1, 2,..., r.$

In like manner a result m_2 in M_2 leads, in the experimental situation $\mathscr{E}_2$, to another acceptability ordering of these hypotheses, namely the natural ordering of the quantities

(13.2) $\quad A_2(H_j \mid m_2) = N(\mu_{2j}^{-1}(m_2)), \qquad j=1, 2,..., r.$

The r situation schemas for the combined experiment $\mathscr{E}_{12}$ are taken to be $\mathscr{S}_{12,j}=(\Gamma_{12}, C_{12,j}, R_{12})$, $j=1, 2,..., r$, where $\Gamma_{12}=\Gamma_1 \times \Gamma_2$, $R_{12}= = R_1 \times R_2$ and $C_{12,j}$ is defined by

$$C_{12,j}(\gamma_1, \gamma_2) = (C_{1j}(\gamma_1), C_{2j}(\gamma_2)).$$

When H_j is the case $\mathscr{S}_{1j}$ is the situation schema for $\mathscr{E}_1$, $\mathscr{S}_{2j}$ is that for $\mathscr{E}_2$ and $\mathscr{S}_{12,j}$ is that for $\mathscr{E}_{12}$. The independence of the experiments $\mathscr{E}_1$ and $\mathscr{E}_2$ is taken to mean that the possibilities for the combined experiment are just the pairs formed by the individual possibilities in question. We introduce a system measurement function $\phi_{12} : R_{12} \rightarrow M_{12}$, where $M_{12} = M_1 \times M_2$, by the equations

$$\phi_{12}(r_1, r_2) = (\phi_1(r_1), \phi_2(r_2)).$$

Similarly for each $j = 1, 2, \ldots, r$ we introduce the measurement procedure $(M_{12}, \mu_{12,j})$ for $\mathscr{S}_{12,j}$ given by

$$(13.3) \quad \mu_{12,j}(\gamma_1, \gamma_2) = \phi_{12}(C_{1j}(\gamma_1), C_{2j}(\gamma_2)) = (\mu_{1j}(\gamma_1), \mu_{2j}(\gamma_2)).$$

Thus if we observe m_1 in the experiment $\mathscr{E}_1$ and m_2 in the experiment $\mathscr{E}_2$ we observe (m_1, m_2) in the combined experiment and the acceptability ordering of the hypotheses in the light of this combined result is the natural ordering of the quantities

$$(13.4) \quad A_{12}(H_j \mid (m_1, m_2)) = N(\mu_{12,j}^{-1}(m_1, m_2)).$$

But from (13.3)

$$N(\mu_{12,j}^{-1}(m_1, m_2)) = N(\mu_{1j}^{-1}(m_1)) N(\mu_{2j}^{-1}(m_2))$$

and so, from (13.1), we can write

$$(13.5) \quad A_{12}(H_j \mid m_1, m_2) = A_1(H_j \mid m_1) N(\mu_{2j}^{-1}(m_2))$$

for each (m_1, m_2) in $M_1 \times M_2$ and each $j = 1, 2, \ldots, r$.

Equation (13.5) shows how to calculate the acceptabilities of the combined experiment in terms of those from the first experiment $\mathscr{E}_1$ and the frequencies in the second experiment $\mathscr{E}_2$. There is an obvious analogy between (13.5) and the usual statement of Bayes' theorem in respect of hypotheses. The analogy can be made more explicit by defining relative acceptabilities

$$a_1(H_j \mid m_1) = A_1(H_j \mid m_1) / \sum_{i=1}^{r} A_1(H_i \mid m_1)$$

$$a_{12}(H_j \mid (m_1, m_2)) = A_{12}(H_j \mid (m_1, m_2)) / \sum_{i=1}^{r} \times$$
$$\times A_{12}(H_i \mid (m_1, m_2)),$$

and relative frequencies

$$v_2(m_2 \mid H_j) = N(\mu_{2j}^{-1}(m_2)) / \sum_{i=1}^{r} N(\mu_{2i}^{-1}(m_2)).$$

Equation (13.5) can then be put in the form

(13.6) $a_{12}(H_j \mid (m_1, m_2)) = Ka_1(H_j \mid m_1) \, v_2(m_2 \mid H_j),$

where

$$K^{-1} = \sum_{i=1}^{r} a_1(H_j \mid m_1) \, v_2(m_2 \mid H_j).$$

If the relative acceptabilities from the first experiment are thought of as prior probabilities of hypotheses in respect of the second experiment, and if the relative frequencies for the second experiment are thought of as probabilities conditional on hypotheses, then reading (13.6) as Bayes' theorem it asserts that the relative acceptabilities on the left-hand side are the posterior probabilities of the hypotheses after the second experiment.

From (13.6) one can deduce a form of Bayes' postulate. Thus (13.6) is an algorithm which shows how to calculate the relative acceptabilities in a combination of independent experiments. But the second experiment by itself gives the relative acceptabilities

(13.7) $a_2(H_j \mid m_2) = v_2(m_2 \mid H_j).$

This in algorithm for calculating relative acceptabilities from a single experiment. Thus (13.6) and (13.7) are algorithms applicable in different types of situation. Nevertheless, one can ask whether or not it is possible to extend the field of applicability of (13.6) to include (13.7) as a special case, in other words what 'prior' acceptabilities in (13.6) correspond to the absence of the first experiment so that (13.6) then reduces to (13.7)? To find out we equate (13.6) and (13.7) to obtain

(13.8) $Ka_1(H_j \mid m_1) \, a_2(H_j \mid m_2) = a_2(H_j \mid m_2).$

But if we wish to extend the applicability of the algorithm (13.6) independently of the result of the second experiment and the associated relative acceptabilities we can assume that $a_2(H_j \mid m_2) > 0$ for each $j = 1, 2, ..., r$ and (13.8) gives

$$Ka_1(H_j \mid m_1) = 1.$$

So $a_1(H_j \mid m_1)$ does not depend on j and since these quantities add to unity we deduce that

$$a_1(H_j \mid m_1) = 1/r, \qquad j = 1, 2, ..., r.$$

Thus in the absence of the first experiment one can still use the algorithm (13.6) provided one adopts the convention that the fictitious acceptabilities for the first experiment are all equal. One could, perhaps, regard this result as establishing a form of Bayes' postulate but it would be more apt, in the present framework, to say that it is a clarification of the question Bayes' postulate attempted to answer.

It is to be emphasized however, that in Equation (13.5) the acceptabilities from the experiment $\mathscr{E}_1$, namely the quantities $A_1(H_j \mid m_1)$, are quantities which result from experimentation; they are not a quantitative statement of one's personal belief about which of the hypotheses is the case. However, to calculate the right-hand side of (13.5), after the second experiment $\mathscr{E}_2$ has been performed, we do not need to know the frequencies in respect of the first experiment nor, indeed, the actual outcome of that experiment. All that we need to know are the acceptabilities which result from the first experiment $\mathscr{E}_1$. Thus the acceptabilities for the combined experiment $\mathscr{E}_{12}$ could still be calculated if, for instance, the details of the first experiment had been lost and all that was available was a report of the acceptabilities which resulted from it. Thus a scientist performing the second experiment might feel justified in substituting, for the acceptabilities from the first experiment, numerical values which were not obtained by actual observation of the results of that experiment, but which were derived in some other way; provided, of course, that he is willing to take these substitute quantities as a good enough guess at the results of previous experimentation. In such a case he might, for instance, declare that he is calling on his previous experience. However even when an experimenter does behave in this way it is still the case that he is trying to approximate to the results of experimentation, that is to happenings in the real world; he is not simply reporting what he personally believes to be the case about the world.

14. THE BAYESIAN APPROACH TO STATISTICAL INFERENCE

The argument of the preceding section is not presented as a vindication of the Bayesian approach to statistical inference but as a brief analysis, in non-Bayesian language, of one aspect of that approach. However that aspect is sufficiently important to motivate the development of the analysis, in greater detail, with a view to a better understanding of what is involved in the use of Bayesian methods in statistics.

The Bayesian approach to statistical inference is based on an argument of the following form:

(i) It is the business of the scientific experimenter to revise his opinions in an orderly way with due regard to internal consistency and the data, and so

(ii) one has to develop techniques for the orderly expression of opinion with due regard to internal consistency and the data, but

(iii) the only orderly expression of opinion with due regard to internal consistency is the Bayesian one, and

(iv) the only orderly revision of those opinions with due regard for the data is through Bayes' theorem, therefore

(v) the Bayesian approach to statistical inference is the only 'correct one'.

It is characteristic of detailed expositions of the Bayesian approach that great attention is paid to the fact that requiring a person to quantify his strength of opinion, in an orderly and consistent way, leads to the conclusion that the desired quantitative measure of strength of opinion is a normed additive measure. For the purposes of this discussion we concede that much, without argument, because it is a technical but relatively unimportant aspect of the Bayesian approach. But the fact that the Bayesian's quantitative measure of strength of opinion is a normed additive measure does not, in itself, entail that it is practically useful, nor is it reason, in itself, for calling that measure 'probability'. In one sense it is, of course, little more than a linguistic quibble to cavil at the name given to this quantitative measure of strength of opinion, but the Bayesian's readiness to pre-empt the use of the word 'probability' in this context has tended to obscure the point at issue.

Thus it has been said that what characterises the modern Bayesian is the principle that it is legitimate to use the axioms of probability even when this involves the use of probabilities of hypotheses. But what the experimental scientist wants to know is not whether it is legitimate to manipulate probabilities of hypotheses but whether a quantitative measure of strength of opinion is of any practical use to him in the design and analysis of his experiments. If it is of use to him he will care little about the name given it by the theoretical statistician; if it is not of use to him he will care even less about the relevant statistical nomenclature. Let us, then, use a word other than 'probability' for the Bayesian's quantitative measure of strength of opinion. From the point of view of this discussion any other word would do as well, even a nonsense one, but to avoid frivolity we will use the word 'credibility' because it has some, at least, of the connotations of the Bayesian's probability of an hypothesis.

When the quantitative measure of strength of opinion is called probability the need to question its practical usefulness does not seem urgent. There are so many instances of the practical usefulness of probability conceived in terms of relative frequency that one tends to assume that if this quantitative measure of strength of opinion is indeed probability, then, by that fact alone, it must be useful. Using a different word for the measure prevents one from assuming usefulness by association without further consideration. Thus a non-Bayesian statistician might brand as nonsense a statement about the probability of an hypothesis, but this does not mean he would dismiss the corresponding statement about its credibility as being nonsense; provided that by credibility one meant no more than a quantitative measure of strength of opinion derived from principles of orderliness and internal consistency. Yet such a statistician could still attempt to formulate inference about hypotheses without reference to their credibilities, simply because he fails to see either their usefulness or their relevance to the problem of making such inferences. It is the question of the relevance and practical usefulness of credibilities which separates Bayesian and non-Bayesian statisticians and not whether it is legitimate to talk of probabilities of hypotheses and manipulate them according to the axioms of probability.

The arguments which Bayesians put forward to illustrate the practical usefulness of credibilities are little more than recommendations about how one ought to change one's opinion in the light of an experimental result. But if opinions are not relevant in the first place it is difficult to see why changes of opinion should have any more relevance. The only sort of post hoc justification put forward by Bayesians is the fact that if opinions are revised in the prescribed way, namely through Bayes' theorem, then, in general, a large enough experiment will force dissimilar initial opinions into similar final opinions. But this seems to be an argument which points as much to the irrelevance as to the relevance of initial opinion. That a quantitative measure of strength of opinion is relevant to the making of inferences about hypotheses is a characteristic attitude of Bayesians, moreover it is an attitude in respect of which they seldom recognise the need for justification.

There is, of course, a colloquial, but ambiguous, sense in which an experimenter ought to revise his opinions in an orderly way, with due regard to internal consistency and the data. But what is in doubt is whether or not the precise formulation of the Bayesians is an adequate copy of the imprecise sense in which this is so. Doubt arises because the Bayesian approach requires that a high credibility for an hypothesis should indicate a correspondingly high strength of opinion in favour of that hypothesis; for brevity we refer to this aspect of the Bayesian approach as the 'high credibility requirement'. We will see, in the next section, however, that this requirement is less straightforward than appears at first sight.

15. ACCEPTABILITY AND THE HIGH CREDIBILITY REQUIREMENT

It could be argued that the acceptabilities introduced in Section 12 correspond to Bayesian probabilities of hypotheses or, as we are calling them, credibilities. However the function $A(\cdot \mid m)$ does not yield a quantitative measure of acceptability but only an ordering of the hypotheses considered, according to the natural ordering of the acceptabilities in question. None the less it is hard to resist the temptation to use the function $A(\cdot \mid m)$ as a quantitative measure of acceptability and to go on to equate relative acceptability with Bayesian credibilities. Whether or not it is generally meaningful to do so there is one case in which such a procedure seems unobjectionable and we will therefore restrict ourselves to that case. The case in question is the one in which the Bayesian would attach equal prior credibilities to the hypotheses being considered. In such a case experimentation would lead, through Bayes' theorem, to credibilities which take on exactly the same values as the relative acceptabilities defined above, thus whether or not these relative acceptabilities are conceptually the same as the Bayesian posterior credibilities, they are certainly the same numerically.

Suppose then that r hypotheses $H_1, H_2, ..., H_r$ are being considered in terms of the symmetric decision procedure according to the formulation of Section 12. Experimentation gives a result m and acceptabilities $A(H_j \mid m)$, $j = 1, 2, ..., r$ whose natural ordering yields a corresponding acceptability ordering of the hypotheses, say

$$H_1 \cdot \bar{\alpha}_m H_2 \cdot \bar{\alpha}_m \ldots \bar{\alpha}_m H_{r'}.$$

The hypothesis $H_{r'}$ is, then, a most acceptable hypothesis, in the sense that none of the hypotheses is more acceptable than $H_{r'}$ in the light of the observation m. The basic question to be discussed concerns the extent to which this acceptability ordering provides grounds for believing $H_{r'}$ to be the hypothesis which actually is the case. In particular, are the grounds the more cogent the higher the value of $a(H_{r'} \mid m)$, the relative acceptability of that hypothesis?

To discuss this question we return to the simple case involving only two hypotheses H_1 and H_2; this simplification entails no real loss of generality because the symmetric decision procedure for many hypotheses results from a number of such pair-wise comparisons. We suppose then that $M = \{m_1, m_2, ..., m_k\}$, $k \geqslant 2$, and write

$$G_{ij} = N\left(\Gamma(m_i, m_j)\right), \qquad i, j = 1, 2, ..., k,$$

where,

$$\Gamma(m_i, m_j) = \mu_1^{-1}(m_i) \cap \mu_2^{-1}(m_j).$$

We put

$$G_{i.} = \sum_{j=1}^{k} G_{ij},$$

$$G_{.j} = \sum_{i=1}^{k} G_{ij},$$

and

$$G = \sum_{i=1}^{k} \sum_{j=1}^{k} G_{ij}.$$

Then $G = N(\Gamma)$ and

$$A(H_1 \mid m_i) = G_{i.}, \qquad i = 1, 2, ..., k,$$
$$A(H_2 \mid m_j) = G_{.j}, \qquad j = 1, 2, ..., k.$$

The corresponding relative acceptabilities are

$$a(H_1 \mid m_i) = G_{i.}/(G_{i.} + G_{.i}),$$

and

$$a(H_2 \mid m_j) = G_{.j}/(G_{j.} + G_{.j}).$$

We will suppose, for simplicity of exposition, that no two of the relative acceptabilities $a(H_1 \mid m_i)$, $i = 1, 2, ..., k$ are the same, and, in like manner, that no two of the relative acceptabilities $a(H_2 \mid m_j)$, $j = 1, 2, ..., k$ are the same, so that for example, one obtains a relative acceptability for H_2 with value $a = a(H_2 \mid m_j)$ if and only if m_j is the result of the application of measurement to the situation being considered. It is convenient to focus attention on one of the hypotheses and one of the measurement results, let these be H_2 and m_j respectively. With this in mind we write $\beta = G_{.j}/G$, $\alpha = G_{j.}/G$ and $a = a(H_2 \mid m_j)$. Then

$$(15.1) \quad a = \frac{\beta}{\alpha + \beta}.$$

Note that α is the proportion of elements in Γ, the underlying situation catalogue, which lead to the result m_j when H_1 is the case; in like manner β is the proportion of elements in Γ which lead to the result m_j when H_2 is the case.

The hypothesis H_2 will be more acceptable than the hypothesis H_1 in the light of the observation m_j if and only if $a > \frac{1}{2}$. Suppose then that a does exceed one-half. We ask: in what sense, if any, is this grounds for for an opinion which is more strongly in favour of H_2 than H_1? In particular, are the grounds the more cogent the closer a is to unity? The discussion of earlier sections tells us that, in such circumstances the policy of taking H_2 to be the case has certain properties in respect of the number of right decisions. For instance, if $k = 2$ the symmetric decision procedure leads to a minimal acceptability basis and the policy of taking the hypothesis with the greater acceptability to be the one which is the case ensures that the number of elements in the situation catalogue for which such a decision must be right is as large as possible. But this fact is not, in itself, related to opinion about which of the hypotheses is the case; nor does it, by itself, provide an imperative to guide the formation of opinion. Moreover it is by no means clear what bearing, if any, an opinion, so formed, would have on the problem being considered.

None the less there is a superficial sense in which the high credibility requirement seems to be satisfied. Thus from Equation (15.1) one finds that

$$a > 1 - \varepsilon \Rightarrow \alpha < \varepsilon(1 - \varepsilon)^{-1},$$

and

$$a > 1 - \varepsilon \Rightarrow \beta > \alpha(1 - \varepsilon)\,\varepsilon^{-1}.$$

Thus if a is very close to unity, so that ε is very small, then α will be very small and β will be very much larger than α. In other words, a very high relative acceptability for H_2 in the light of m_j implies both that m_j has only a low relative frequency of occurrence when H_1 is the case and that its relative frequency of occurrence is very much greater when H_2 is the case. Thus a very high relative acceptability for H_2 could be taken as discrimination in favour of H_2 in the classical sense. If H_2 is not the case a very rare type of possibility has eventuated, whereas if H_2 is the case then a very much more common type of possibility has eventuated. To that extent, at least, a very high relative acceptability for H_2 would seem to add cogency to the belief that H_2 is the case. But this cogency comes from what that high value implies about the situation being considered and not simply from the actual numerical value of the relative acceptability in question.

Indeed the situation is quite different when one considers values for a which, though greater than one-half, are not very close to unity. This may be seen quite easily through numerical examples. For instance, consider a case in which the relative acceptability for H_2 in the light of m_j is quite high, say $a = 0.8$, and where, moreover, the possibilities which lead to m_j when H_2 is the case are rather common, say $\beta = 0.9$. From (3.10.1) we find that $\alpha = 0.225$. In other words, getting an acceptability of 0.8 for H_2 would not be uncommon even when H_1 is the case. All that one can say now, is that if H_2 is not the case then a not uncommon type of possibility has eventuated; whereas if H_2 is the case a rather common type of possibility has eventuated. It is clear that such a statement does not discriminate between H_1 and H_2 in the classical sense. Part of the reason there is no such discrimination is that the relative frequencies of observing m_j under the two hypotheses are of comparable magnitude; it is not only their absolute magnitudes which are relevant. Thus Equation (3.10.1) determines the ratio β/α in terms of a, so that if, for example, we retain $a = 0.8$ but take $\beta = 0.009$ we find that $\alpha = 0.00225$. For this example, then, all that one can say is that if H_2 is not the case then a very uncommon type of possibility has eventuated, where as if H_2 is the case then a type of possibility which is only four times as common has eventuated. One can construct many examples of type, we mention just one of them. Suppose that $\beta=0.9$ and $a=0.6$, then from (15.1) $\alpha=0.6$ so that even under H_1 60% of the possibilities lead to a relative acceptability of 0.6 for H_2.

The preceding comments may or may not have bearing on the nature of the opinion we ought to hold in the situations being considered. However they do indicate that the use of acceptabilities as quantitative measures of degrees of belief is not, by itself, very helpful to the experimental scientist. For instance, take the example considered above, in which $\alpha=0.6$, $\beta=0.9$ and $a=0.6$. A scientist who maintained, on the basis of such an experiment, that one ought to assign '60% credibility' to the truth of the hypothesis H_2, in contrast to that of its rival H_1, would surely have difficulty convincing his colleagues of the usefulness and relevance of the concept of '60% credibility' in the face of the fact that, if indeed, the rival hypothesis was the case, then more than half of the possibilities being considered would lead to that degree of credibility in a false hypothesis. In such a case we wouldn't waste time arguing about the legitimacy of the credibility concept, nor would we discuss whether the credibilities satisfied this or that set of axioms; but we might point out that the experiment did not seem a useful one to perform and that a statement of 60% credibility in H_2 added nothing to our comprehension of the import of the experimental result.

The fact is, of course, that even a minimal acceptability basis may not prove useful in deciding between two hypotheses. For, in some situations, the best that one can do may be just not good enough to be practically useful; indeed, it is part of the art of experimentation to devise situations in which the best that one can do is practically useful. Practical usefulness is achieved when the experiment is such that one is assured of reaching the right decision for most of the possibilities being considered. In other words one needs to choose not only an acceptability basis (M_{12}, M_{21}) with optimum properties but, in practice, one has *first* to choose the situation catalogue Γ and the measurement functions μ_1 and μ_2 so that the number of elements in the set

$$\mu_1^{-1}M_{21} \cap \mu_2^{-1}M_{12}$$

is large relative to the total number of possibilities in the situation catalogue. Such a situation is achieved in practice, by performing a large
enough experiment and it is this fact, of course, which underlies the
Bayesian post hoc justification of the relevance of opinion, whereby the
use of Bayes' theorem forces dissimilar initial opinions into similar final
opinions. At the mathematical level this fact asserts the consistency of the
method of maximum likelihood and it is a well-known result. A simple
account in terms of the present framework, is given in the next section.

16. ACCEPTABILITIES IN LARGE EXPERIMENTS

Let $\mathscr{E}_1, \mathscr{E}_2, ..., \mathscr{E}_n, ...$ be a sequence of experimental situations in respect
of each of which the same two hypotheses are being considered. Suppose
further that these experimental situations are independent but such that
whichever of the hypotheses holds for any one of them must also hold for
all the others. For each $t = 1, 2, ...$ let the two situation schema for $\mathscr{E}_t$ be

$$\mathscr{S}_{t, i} = (\Gamma_t, C_{t, i}, R_t), \qquad i = 1, 2,$$

and let $\phi_t: R_t \to M_t$ be a reference set measurement function. Define the
measurement procedures $(M_t, \mu_{t, i})$ for $\mathscr{S}_{t, i}$ by

$$\gamma_t \in \Gamma_t: \mu_{t, i}(\gamma_t) = \phi_t(C_{t, i}(\gamma_t)).$$

In the experimental situation $\mathscr{E}_t$ a result m_t in M_t leads to an acceptability
ordering of the hypotheses H_1 and H_2, namely the natural ordering of the
quantities

$$A_t(H_i \mid m_t) = N(\mu_{t, i}^{-1}(m_t)), \qquad i = 1, 2.$$

This equation gives the acceptabilities of the two hypotheses in respect
of the individual experiments. But if the experiments $\mathscr{E}_1, \mathscr{E}_2, ..., \mathscr{E}_n$ are
considered together as one experiment $M_{12 ... n}$ and independence is taken
to mean that any possibility in any of the catalogues $\Gamma_1, \Gamma_2, ..., \Gamma_n$ can
eventuate without restriction with any of the possibilities in the other
catalogues, then the catalogue for the combined experiment is $\Gamma_{12 ... n} =$
$= \Gamma_1 \times \Gamma_2 \times \cdots \times \Gamma_n$ and the corresponding situation schemas are

$$\mathscr{S}_{12 ... n, i} = (\Gamma_{12 ... n}, C_{12 ... n, i}, R_{12 ... n}), \qquad i = 1, 2$$

where $R_{12\ldots n} = R_1 \times R_2 \times \cdots \times R_n$ and

$$C_{12\ldots n,\,i}(\gamma_1, \gamma_2, \ldots, \gamma_n) = (C_{1,\,i}(\gamma_1), C_{2,\,i}(\gamma_1), \ldots, C_{n,\,i}(\gamma_1)).$$

Let $M_{12\ldots n}$ be $M_1 \times M_2 \times \cdots \times M_n$ and let

$$\mu_{12\ldots n,\,i} : \Gamma_{12\ldots n} \to M_{12\ldots n}$$

be defined by

$$\mu_{12\ldots n,\,i}(\gamma_1, \gamma_2, \ldots, \gamma_n) = (\mu_{1,\,i}(\gamma_1), \mu_{2,\,i}(\gamma_2), \ldots, {}_{n,\,i}(\gamma_n)),$$

so that for $(m_1, m_2, \ldots, m_n)$ in $M_{12\ldots n}$

$$(16.1) \quad N(\mu_{12\ldots n,\,i}^{-1}(m_1, m_2, \ldots, m_n)) = \prod_{t=1}^{n} N(\mu_{t,\,i}^{-1}(m_t)).$$

If we observe $m_1, m_2, \ldots, m_n$ in the experiments $\mathscr{E}_1, \mathscr{E}_2, \ldots, \mathscr{E}_n$ respectively then the acceptabilities for the hypotheses H_1 and H_2 from the combined experiment are given by

$$A_{12\ldots n}(H_i \mid (m_1, m_2, \ldots, m_n)) = N(\mu_{12\ldots n,\,i}^{-1}(m_1, m_2, \ldots, m_n))$$

and so, from (16.1), these are just the products of the corresponding acceptabilities from the individual component experiments.

In what follows we will assume for simplicity that for each $(\gamma_1, \gamma_2, \ldots, \gamma_n)$ in $\Gamma_{12\ldots n}$ neither

$$N(\mu_{12\ldots n,\,1}^{-1}\, \mu_{12\ldots n,\,1}(\gamma_1, \gamma_2, \ldots, \gamma_n))$$

nor

$$N(\mu_{12\ldots n,\,2}^{-1}\, \mu_{12\ldots n,\,1}(\gamma_1, \gamma_2, \ldots, \gamma_n))$$

is zero.

For each $c > 0$ define $\Gamma_{12\ldots n}(c)$ to be the subset of $\Gamma_{12\ldots n}$ consisting of the elements $(\gamma_1, \gamma_2, \ldots, \gamma_n)$ for which

$$\frac{N(\mu_{12\ldots n,\,1}^{-1}\mu_{12\ldots n,\,1}(\gamma_1, \gamma_2, \ldots, \gamma_n))}{N(\mu_{12\ldots n,\,2}^{-1}\mu_{12\ldots n,\,1}(\gamma_1, \gamma_2, \ldots, \gamma_n))} \leqslant c.$$

Then, if one uses the symmetric decision procedure, $N(\Gamma_{12\ldots n}(1))$ is just the number of possibilities in the situation catalogue $\Gamma_{12\ldots n}$ which, when H_1 is the case, lead to a result for the experiment $\mathscr{E}_{12\ldots n}$ in the light of which the true hypothesis H_1 is not accorded the greater acceptability.

Write $B^n(c)$ for $N(\Gamma_{12...n}(c))$ and

$$\rho^t(m) = \frac{N(\mu_{t,2}^{-1}(m))}{N(\mu_{t,1}^{-1}(m))}, \qquad m \in M_t, \quad t = 1, 2, \dots.$$

Then from (16.1) we deduce that

$$B^n(c) = \sum_{m \in M_r} B^{n-1}(c\rho^n(m)) N(\mu_{n,1}^{-1}(m)).$$

Putting

$$b^n(c) = B^n(c)/N(\Gamma_{12...n})$$

we obtain

$$(16.2) \quad b^n(c) = \sum_{m \in M_n} b^{n-1}(c\rho^n(m)) \left[\frac{N(\mu_{n,1}^{-1}(m))}{N(\Gamma_n)}\right].$$

Note that $b^n(1)$ gives the relative frequency of possibilities in the situation catalogue of the combined experiment $\mathscr{E}_{12...n}$ which, when H_1 is the case, lead to a result in the light of which the true hypothesis is not accorded the greater acceptability.

Suppose now that the experimental situations $\mathscr{E}_1, \mathscr{E}_2, \dots, \mathscr{E}_n, \dots$ are all replications of a single experimental situation $\mathscr{E}$, by this we mean firstly that there is an experimental situation $\mathscr{E}$ with associated situation schema (Γ, C_i, R), $i = 1, 2$ and reference set measurement function $\phi: R \to M$ and secondly that in the formulation above we have, for each $t = 1, 2, \dots$, $\Gamma_t \simeq \Gamma$, $R_t \simeq R$, $C_{t,i} \simeq C_i$. When this is so $\mu_{t,i}$ is effectively $\mu_i = \phi \circ C_i$ and Equation (16.2) takes on a much simpler form. Thus taking $M = \{m_1, m_2, \dots, m_k\}$ and writing

$$\rho_j = \frac{N(\mu_2^{-1}(m_j))}{N(\mu_1^{-1}(m_j))}, \qquad j = 1, 2, \dots, k$$

Equation (16.2) becomes

$$(16.3) \quad b^n(c) = \sum_{h=1}^{k} b^{n-1}(c\rho_h) g_h, \qquad n > 1$$

where $g_h = G_{h\cdot}/G$ and $G_{h\cdot}$, G are as defined in the last section. Note that $\rho_h G_{h\cdot} = G_{\cdot h}$ so that

$$(16.4) \quad \sum_{h=1}^{k} \rho_h g_h = 1.$$

Then one can establish

PROPOSITION (16.1). *Under the conditions which lead to Equation (16.3): If $\rho_h \neq 1$ for some $h = 1, 2, ..., k$ then for any $c > 0$,*

(16.5) $\lim\limits_{n \to \infty} b^n(c) = 0$.

Remark. It is clear that (16.5) cannot hold without some restriction. Indeed suppose that $\rho_h = \rho$ for each $h = 1, 2, ..., k$, then, from (16.4), $\rho = 1$ and (16.3) gives

$$b^n(c) = b^{n-1}(c) = \cdots = b^1(c).$$

Proposition (16.1) has a number of consequences. We note firstly

PROPOSITION (16.2). *Let $\mathscr{E}_{12...n}$ be an n-fold replicated experiment with situation catalogue $\Gamma_{12...n}$. If H_1 is true then the relative frequency in $\Gamma_{12...n}$ of those possibilities for which H_1 is accorded the greater acceptability tends to unity as n tends to infinity. In like manner if H_2 is true then the relative frequency in $\Gamma_{12...n}$ of those possibilities for which H_2 is accorded the greater acceptability tends to unity as n tends to infinity. In other words, if (M_{12}^n, M_{21}^n) is the symmetric acceptability basis for $\mathscr{E}_{12...n}$ then*

$$\lim\limits_{n \to \infty} \left[N(\mu_{12...n, 1}^{-1} M_{21}^n)/N(\Gamma_{12...n}) \right] = 1$$

and

$$\lim\limits_{n \to \infty} \left[N(\mu_{12...n, 2}^{-1} M_{12}^n)/N(\Gamma_{12...n}) \right] = 1.$$

COROLLARY

$$\lim\limits_{n \to \infty} \left[N(\mu_{12...n, 1}^{-1} M_{21}^n \cap \mu_{12...n, 2}^{-1} M_{12}^n)/N(\Gamma_{12...n}) \right] = 1.$$

Note that the corollary asserts that the symmetric decision procedure for the two hypotheses in respect of the combined experiment $\mathscr{E}_{12...n}$ has the property that the relative frequency in $\Gamma_{12...n}$ of the possibilities for which we are sure of reaching the right decision tends to unity as the number of replications increases without bound.

Secondly we note

PROPOSITION (16.3). *If H_i is true and $0 < a < 1$ then the relative fre-*

quency in $\Gamma_{12...n}$ of the possibilities which lead to a relative acceptability $a(H_i)$ for H_i such that $a(H_i) > a$ tends to unity as n tends to infinity. In other words, the relative acceptability of the true hypothesis will be close to unity for most of the possibilities being considered provided that the number of replication is very large.

This proposition is proved by noting that, when H_1 is true, $b^n(c)$ is the relative frequency in $\Gamma_{12...n}$ of those possibilities which lead to a relative acceptability $a(H_1)$ for H_1 such that

$$a(H_1) \leqslant c/1 + c.$$

The mathematical content of Proposition (16.1) is not new but it is perhaps worthwhile to put on record the following elementary proof which proceeds through a number of contributory lemmas.

LEMMA (16.1). *If $\rho_h \neq 1$ for at least one $h = 1, 2, ..., k$ then*

$$0 < \rho_1^{g_1} \rho_2^{g_2} ... \rho_h^{g_n} < 1.$$

This result follows at once from the well-known inequality between the geometric and arithmetic means.

LEMMA (16.2). *Let $f_1, f_2, ..., f_k$ be non-negative real numbers and let $\delta \geqslant \mathrm{Max}\{|f_j - g_j|: 1 \leqslant j \leqslant k\}$. Then*

$$\rho_1^{f_1 - g_1} \rho_2^{f_2 - g_2} ... \rho_k^{f_k - g_k} \leqslant \rho^{k\delta}$$

where $\rho \geqslant 1$ is the maximum of the quantities

$$\rho_1, \rho_2, ..., k, \rho_1^{-1}, \rho_2^{-1}, ..., \rho_k^{-1}.$$

This result is an immediate consequence of the fact that

$$\rho_j^{f_j - g_j} \leqslant \rho^{|f_j - g_j|}$$

for each $j = 1, 2, ..., k$.

LEMMA (16.3). *If $\rho_h \neq 1$ for at least one $h = 1, 2, ..., k$ then, for any ε such that*

(16.6) $0 < \varepsilon < 1 - \rho_1^{g_1} \rho_2^{g_2} ... \rho_k^{g_h}$,

there exists a $\delta > 0$ such that for any non-negative numbers $f_1, f_2, ..., f_k$ with

$|f_j - g_j| \leqslant \delta$, one has, for each $j = 1, 2, \ldots, k$

$$\rho_1^{f_1} \rho_2^{f_2} \ldots \rho_k^{f_k} \leqslant 1 - \varepsilon.$$

Proof. Let $\rho > 1$ be as defined in the previous Lemma and choose δ so that

(16.7) $\quad \rho^{k\delta}(1 - \varepsilon) \rho_1^{-g_1} \rho_2^{-g_2} \ldots \rho_k^{-g_k}.$

By Lemma (16.2)

$$\rho_1^{f_1} \rho_2^{f_2} \ldots \rho_k^{f_k} \leqslant \rho^{k\delta} \rho_1^{g_1} \rho_2^{g_2} \ldots \rho_k^{g_k} \leqslant 1 - \varepsilon$$

whenever $|f_j - g_j| \leqslant \delta$ for each $j = 1, 2, \ldots, k$.

LEMMA (16. 4). *For all* $c > 0$, $0 \leqslant b^1(c) \leqslant \operatorname{Min}(1, c)$.
   Proof. Clearly $0 \leqslant b^1(c) \leqslant 1$. But, in addition,

$$b^1(c) \leqslant \sum_{h:\, G_{h.} < cG_{.h}} g_h$$
$$\leqslant c \sum_{h=1}^{k} (G_{.h/G}) \leqslant c.$$

LEMMA (16.5).

(16.8) $\quad b^{n+1}(c) = \sum \frac{n!}{r_1!\, r_2! \ldots r_k!} \, b^1\left(c\rho_1^{r_1} \rho_2^{r_2} \ldots \rho_k^{r_k}\right) g_1^{r_1} g_2^{r_2} \ldots g_k^{r_k}$

where the summation is over all non-negative integers $r_1, r_2, \ldots, r_k$ *with* $r_1 + r_2 + \cdots + r_k = n$.
   This result is an immediate consequence of (16.3). Finally we note without proof,

LEMMA (16.6). *Write*

$$Q(r_1, r_2, \ldots, r_k) = \frac{n!}{r_1!\, r_2! \ldots r_k!} \, g_1^{r_1} g_2^{r_2} \ldots g_k^{r_k}$$

and let S_δ *be the set of k-tuples of non-negative integers* $(r_1, r_2, \ldots, r_k)$ *with* $r_1 + r_2 + \cdots + r_k = n$ *and*

$$\left| \frac{r_j}{n} - g_j \right| > \delta$$

for at least one $j = 1, 2, ..., k$. Then

$$\sum_{S_\delta} Q(r_1, r_2, ..., r_k) \leqslant \frac{1}{n\delta^2} \sum_{j=1}^{k} g_j(1 - g_j).$$

To prove Proposition (16.1) we choose ε so that (16.6) holds and then choose δ so that (16.7) holds. Let S_δ be as defined in Lemma (16.6) and let T_δ be the set of k-tuples which are not in S_δ. From Lemma (16.3)

$$(r_1, r_2, ..., r_k) \varepsilon T_\delta \Rightarrow \rho_1^{r_1} \rho_2^{r_2} ... \rho_k^{r_k} \leqslant (1 - \varepsilon)^n$$

and hence, from Lemma (16.4),

$$b^1 (c\rho_1^{r_1} \rho_2^{r_2} ... \rho_k^{r_k}) \leqslant \mathrm{Min} \left[c(1 - \varepsilon)^n, 1 \right]$$

whenever $(r_1, r_2, ..., r_k)$ belongs to T_δ. Thus from (16.8)

$$0 \leqslant b^{n+1}(c) \leqslant \sum_{S_\delta} Q(r_1, r_2, ..., r_k) + \mathrm{Min} \left[c(1 - \varepsilon)^n, 1 \right].$$

But from Lemma (16.6) the first term tends to zero as n tends to infinity and, quite obviously the same is true of the second term. This is the desired result.

Monash University, Melbourne, Australia

DISCUSSION

Commentator: Kempthorne: Is this theory used for ordering models or ordering hypotheses?

Finch: For ordering hypotheses. It seems to me that someone who does not wish to adopt the Bayesian viewpoint and dislikes talking about probabilities of hypotheses can nevertheless use Bayesian methods within statistics provided that he is willing to interpret the mathematical manipulations in another way. The point at which I would disagree with the Bayesian approach as such is that first of all I am not particularly interested in the waffling talk about the probability of the sun rising tomorrow and so on, I am only interested in concrete problems faced by statisticians, and in that situation it seems to me you can define quantities which play the role of the probability of hypotheses.

Giere: If it is a case that your sampling of a statistical hypothesis is only looking at the ratios there will be no data points, as it were, for which you can be sure that you have made the right decision. That is, there will be no analogue of the class of situations for which you can be sure that your decision is the right one in these cases.

Finch: Yes, in general what you can do is simply to enumerate all the possible acceptability bases and see which one gives you the best results. But in practice you will seldom be in this position.

Harper: As I see it, you have set up an apparatus from a God's eyeview as it were within which you can represent the approaches of various views concerning statistical inference, e.g. the Bayesian approach, and you have argued that these approaches have deffects, under certain conditions they do not perform very well. But why should not a Bayesian, e.g., reply, of course, if he had access to all of that information he would proceed differently, but in general he does not have that information – and similarly for other approaches?

Finch: I am not saying anything more than we normally say in statistics. A common type of practical problem is that you have a population of individuals and associated with each individual w you have some property

$f(w)$, it might be his response to a certain test, it might be a whole vector of things, even a complex of responses and what you want to know is the distribution of this quantity $f(w)$ over the population w. This is all I am saying, one hypothesis is that you have one kind of distribution, another hypothesis is that you have a different distribution. What you want to do is to compare the hypotheses.

Godambe: What is the data?

Finch: The data is the observation that you make. It is in the light of that data that you construct the ranking of the hypothesis. There is in fact another way of getting the same result. Suppose you say to your self that I must only consider those situations in which my algorithm for carrying out the test is invarient under name changes between the two hypotheses. That is, you don't want the decision to depend on which hypothesis you happen to call by which name. Then you can show in fact that the corresponding regions are just those to be expected. For only two hypotheses use first of all maximum likelihood and then from that basis you can get pair-wise comparisons for sets of them and you get integers, acceptabilities in that way.

I might also add that these susceptabilities arise naturally within in-formation theory – one can see that if you take an information-theoretic approach to the experiments which I have described.

D. A. S. FRASER AND JOCK MACKAY

ON THE EQUIVALENCE OF
STANDARD INFERENCE PROCEDURES

1. SUMMARY AND INTRODUCTION

Multiparameter statistical models lead to some difficult problems in statistical inference. The usual approach is to factor the inference procedure into a sequence of procedures each involving a small subset of the parameters. The classic example is the analysis of variance separation of the components of the linear model with normal error.

This paper examines the three inference procedures, tests of significance, likelihood analyses, and objective posterior distributions, as applied both to the full parameter and to factored component parameters of statistical models. The range of applications of tests of significance is very wide and requires only a model under the hypothesis being tested. The range for likelihood analyses is somewhat less and requires a model with density under all alternative hypothesis being considered. The range for objective posterior distributions is considerably less and requires a structural or probability space model. This paper examines the standard inference procedures for the common range of applications – the structural models. It goes further and examines the procedures as applicable to component parameters; this requires the component parameters to be based on a semidirect factorization. It is shown that the three inference procedures are essentially equivalent on the common range of applications, both for the full parameter and for component parameters. Thus where the standard procedures overlap they are in essential agreement.

The structural model is described briefly in Section 2 and the analysis of data with the model in Section 3.

The factorization of the parameter of the structural model and the resulting analyses are examined in Section 4. The corresponding factorization of the probability density is surveyed in Section 5.

In Section 6 the three standard procedures are examined for component parameters of the structural model. This is illustrated in Section 7 using a simple location-scale model.

Harper and Hooker (eds.), Foundations of Probability Theory, Statistical Inference, and Statistical Theories of Science, Vol. II, 47–62. *All Rights Reserved. Copyright © 1976 by D. Reidel Publishing Company Dordrecht-Holland.*

2. The structural model

The structural model (Fraser, 1968) in its simplest form has two basic components: (i) an N dimensional variable Z with known probability distribution P; (ii) a class G of transformations acting as an exact group on the space S of the variable Z. In an application the variable Z describes the underlying variation affecting the response of the system, and the observable response Y of the system is obtained as $Y = \theta Z$ from some one θ of the transformations in G as applied to the underlying variation Z.

In the structural model all the variation derives from a single distribution – it derives from the *probability space* (S, B^N, P). Accordingly the structural model is also called a *probability space model*. This emphasizes the distinction from an ordinary model which has a class of distributions, a statistical space.

In some applied contexts a complete specification of the distribution for Z may not be possible. In these cases the model can be extended by introducing a class of probability spaces $\{(S, B^N, P_\lambda): \lambda \in \Lambda\}$ where λ is a parameter inducing the possible distributions for variation. Then for any given value of λ we obtain a probability space model for the system being investigated.

A limitation of probability space models is that they do not cover all applications. For an application, the probability space must be identifiable in the physical system being investigated; some aspects of this were discussed in Fraser (1971). The presence in itself of a classical statistical model with some invariance characteristics is not in itself grounds to conclude that a probability space model is appropriate – although it may very well be appropriate in terms of the experimental evidence.

In most instances it is relatively easy to accept the applicability of a classical statistical model. It is, however, very difficult to justify the artificial 'reduction' criteria (sufficiency, invariance, ancillarity, etc.) needed in the classical analysis. On the other hand, the analysis of the structural model is entirely necessary analysis – based directly on the definition of the model and the rules of probability – no further assumptions or reduction principles are required.

As an illustration consider the linear model. The response is an unknown linear function of input variables with a scaling of the underlying error.

For n responses we have

$$\mathbf{Y} = X\boldsymbol{\beta} + \sigma\mathbf{z}$$

where $\mathbf{z}$ is a sample from a distribution with density f and X is a full rank (say) $n \times r$ design matrix. The group action can be expressed as matrix multiplication using the following notation:

$$\begin{pmatrix} X' \\ \mathbf{y}' \end{pmatrix} = \begin{pmatrix} I & \mathbf{0} \\ \boldsymbol{\beta}' & \sigma \end{pmatrix} \begin{pmatrix} X' \\ \mathbf{z}' \end{pmatrix} = \theta \begin{pmatrix} X' \\ \mathbf{z}' \end{pmatrix}.$$

The probability space is $(\mathbb{R}^n - L(X),\ B^n,\ P^n)$ and the group is $G = \{\theta\colon \boldsymbol{\beta} \in \mathbb{R}^r, \sigma \in \mathbb{R}^+\}$. The distribution given by f is typically symmetric about 0, say normal, or Cauchy, or t with specified degrees of freedoms. The model can be extended by introducing a parameter λ for the density f, for example the degrees of freedom of a t distribution.

3. THE ANALYSIS

The analysis of the probability space model follows directly from the model and realized response and uses the definition of conditional probability. It is important to note that the analysis is performed on the single probability space of the model.

Let Y be the observed response. According to the model, the observed Y has been produced by a transformation θ applied to a realized value Z on the probability space. The information concerning Z is that it is an element of the set

$$\{g^{-1}Y\colon g \in G\} = \{hY\colon h \in G\} = GY$$

on the space S. No further information is available to discriminate among the possible values within this set. Group theory shows that the sets GW with W in S form a partition of the space. Thus the conditions are fulfilled for the applicability of conditional probability.

At this point the question of why the set of transformations G must be a group is raised. Classically, invariance is used to reduce the set of solutions of a statistical problem. The role of the group on a probability space is more fundamental. The set of transformations acts as an information processor (Fraser, 1972) transferring information from the observed response to the probability space and vice versa. To preserve the identity of the information it is necessary and sufficient that the set of transformations be effectively a group. If further information is available that restricts the class of transformations to a set which is not a group then the conditional probability analysis may no longer be justifiable.

To proceed with the analysis note that the assumption of exactness ensures that there is a natural one-one transformation between the elements of G and the elements of GY for given Y. Let $[\cdot]$ be a continuously differentiable function from S to G such that $[gZ]=g[Z]$ for all g in G and Z in S. Let $D(Z)=[Z]^{-1}Z$ and $Q=\{D(Z): Z \in S\}$. Then it follows that $Z \leftrightarrow ([Z], D(Z))$ is a one-one correspondence from S to $(G \times Q)$. With this notation the observed set GY for Z can alternatively be designated by $D(Z)=D(Y)$; thus we have the information that the function $D(Z)$ on the probability space has taken the value $D(Y)$. It follows from conditional probability that the description of Z is given by the conditional distribution of $[Z]$ (the unobserved component of the realization) given $D(Z)=D(Y)$. This conditional distribution can be derived easily for any density function f; see Fraser (1968) for details.

Sometimes it is convenient to simplify the notation used with the conditional distribution given a *particular* observed response Y. For this we choose a surface Q passing through this observed response. Then

$$Y = iY = \theta [Z] Y$$

and thus $\theta [Z]=i$, the identity element of the group. This gives a simple relationship between the possible values for θ and the possible value for $[Z]$, namely $\theta = [Z]^{-1}$.

4. THE FACTORED PARAMETER

When the parameter θ is multidimensional the problems of statistical inference become cumbersome and difficult. The typical solution is to partition θ into components and to analyze the components separately and usually sequentially. With a probability space model a multidimensional parameter means a many dimensional group G.

As a simple first case suppose that θ can be represented as a unique product

$$\theta = \theta_2 \theta_1$$

where $\theta_1 \in H_1$, $\theta_2 \in H_2$, and H_1 and H_2 are subgroups of the group G; then $G = H_2 H_1$ is called a semidirect product of the component groups H_1 and H_2. We now factor $[Z]$ in the reverse order

$$[Z] = h_1 h_2$$

where $h_1 \in H_1$, $h_2 \in H_2$, and the reverse factorization $G = H_1 H_2$ is justified on the basis of the mapping $g \leftrightarrow g^{-1}$. Now consider the simplified notation for describing conditional distributions as mentioned at the end of Section 3:

$$\theta_2 \theta_1 = \theta = [Z]^{-1} = (h_1 h_2)^{-1} = h_2^{-1} h_1^{-1}.$$

We thus obtain the relationships

$$\theta_2 = h_2^{-1}, \qquad \theta_1 = h_1^{-1};$$

these simple relationships are one of the advantages of the simplified notation.

The order of the inference procedure is first the analysis of θ_2 based on the distribution of h_2 given the observed $D(Z) = D(Y)$ and then the analysis of θ_1 based on the distribution of h_1 given h_2 and $D(Z)$. In the second stage h_2 or correspondingly θ_2^{-1} is given assumed values based in whole or part on inferences from the first stage of the analysis.

The order of this sequential procedure deserves some comment. At the second stage we assign a value $\theta_2 = \bar{\theta}_2$ on reasonable grounds. In terms of the original transformation on the probability space we have

$$Y = \bar{\theta}_2 \theta_1 Z \quad \text{or} \quad \bar{\theta}_2^{-1} Y = \theta_1 Z.$$

In this $\bar{\theta}_2$ is fixed and θ_1 is in a group H_1. We thus have the group properties needed for the information processor referred to in Section 3. This in part emphasizes the need for the semi-direct factorization required in this section. Complications arise if we attempt a sequential analysis of θ_1 then θ_2; the conditions for information processing are not fulfilled. For further argument and discussion see Fraser (1968, §2.6–2.8).

The particular factorization of the parameter is assumed to arise from the application. Of course a factorization may occur which does not satisfy the requirements presented above; the necessary analysis discussed in this paper is then not available and recourse typically would be to methods suggested by the various branches of inference theory for the standard statistical model.

In the typical application a nuisance parameter will be on the right in the factorization $\theta = \theta_2 \theta_1$ and may not be analyzed at all: the inferences would be restricted to the essential remaining parameter θ_2.

More generally we will consider more than two factors. For this we need a sequence of subgroups of G:

$$G = G_r \supset G_{r-1} \supset \cdots \supset G_0 = \{i\}$$

where i is the identity element. We suppose that $H_r, \ldots, H_1$ is a generating set of subgroups such that

$$G_s = H_s G_{s-1}$$

is a semidirect product. We then have

$$G = H_r H_{r-1} \ldots H_1$$

and θ can be factored uniquely as

$$\theta = \theta_r \theta_{r-1} \ldots \theta_1$$

with θ_s in H_s.

As an example consider the ordinary linear model with the factorization:

$$\theta = \begin{pmatrix} I & 0 \\ \beta & \sigma \end{pmatrix} = \begin{pmatrix} I & 0 \\ 0 \cdots 0 \ \beta_r & 1 \end{pmatrix} \cdots \begin{pmatrix} I & 0 \\ \beta_1 \ 0 \cdots 0 & 1 \end{pmatrix} \begin{pmatrix} I & 0 \\ 0 \cdots 0 & \sigma \end{pmatrix}$$
$$= \theta_r \cdots \theta_1 \theta_0 .$$

The order of the inference sequence is $\beta_r, \ldots, \beta_1, \sigma$. At each stage we have a one-dimensional problem conditional on values of preceding parameters. Typically β_r would represent a high order interaction or high power of some input variable, $\ldots, \beta_1$, the general level relative to the $\mathbf{1}$ vector, and σ the error scaling factor.

Now consider in general the inference sequence for the factorization

$$\theta = \theta_r \theta_{r-1} \ldots \theta_1 .$$

In terms of the simplified notation we first factor $[Z]$ in the reverse order

$$[Z] = h_1 h_2 \ldots h_r$$

with h_s in H_s. We then obtain

$$\theta_r \ldots \theta_1 = \theta = [Z]^{-1} = h_r^{-1} \ldots h_1^{-1}$$

and thus have

$$\theta_r = h_r^{-1}, \qquad \theta_1 = h_1^{-1}.$$

The inference procedure is to first analyze θ_r given the observed $D(Z) = D(Y)$, then θ_{r-1} given in addition an assumed value for $\theta_r = h_r^{-1}$, and so on to the analysis of θ_1 given in addition an assumed value for $\theta_1 = h_1^{-1}$.

5. THE FACTORIZATION OF THE DENSITY

The factorization of the density on the probability space has been discussed briefly in Fraser (1968, §2.7) and in major detail in Fraser and MacKay (1975).

Let f be the density for Z taken with respect to Euclidean volume in S. Let μ be the left invariant measure on the group G and m be the quotient measure on the space Q. Then we have

$$dz = dgD = d\mu(g)\, dm(D)$$

and correspondingly we obtain

$$f(Z)\, dZ = f(g:D)\, d\mu(g) \cdot k(D)\, dm(D)$$

where $k(D)$ is determined as the norming constant of the conditional distribution of g given D.

Now consider the factorization $g = h_1 \dots h_r$ on the group G. Let

$$\mu_{1\dots s}(h_1 \dots h_s) = \mu_{(s)}(h_{(s)})$$

be the left invariant measure on the group $H_1 \dots H_s$ and

$$\mu_{s+1,\dots,r}(h_{s+1} \dots h_r) = \mu_{[s+1]}(h_{[s+1]})$$

be the quotient measure obtained from μ over $\mu_{(s)}$. Then we have

$$dZ = d\mu_{(s)}(h_{(s)})\, d\mu_{[s+1]}(h_{[s+1]})\, dm(D)$$

and correspondingly we obtain

$$f(Z)\, dZ = f^{(s)}(h_{(s)}: h_{[s+1]}, D)\, d\mu_{(s)}(h_{(s)}) \cdot$$
$$\cdot f^{[s+1]}(h_{[s+1]}: D)\, d\mu_{[s+1]}(h_{[s+1]}) \cdot k(D)\, dm(D).$$

The complete factorization is then made possible by introducing the quotient measure $\mu_{[s]}$ over $\mu_{[s+1]}$ in terms of the left invariant measure μ_s on the group H_s:

$$d\mu_{[s]}(h_{[s]}) = d\mu_s(h_s) \cdot \frac{\varDelta_{(s)}(h_s)}{\varDelta_s(h_s)} d\mu_{[s+1]}(h_{[s+1]})$$

where $\varDelta_{(s)}$ and $\varDelta_s$ are the modular functions respectively in $G_s = H_{(s)}$ and H_s. We thus obtain a complete factorization of the density f:

$$
\begin{aligned}
f(Z)\, dZ = f^1(h_1 : h_{[2]}, D)\, d\mu_1(h_1) \cdot f\ (h_2 : h_{[3]}, D) \times \\
\times \frac{\varDelta_{(2)}(h_2)}{\varDelta_2(h_2)} d\mu_2(h_2) \cdots f^2(h_r : D) \times \\
\times \frac{\varDelta_{(r)}(h_r)}{\varDelta_r(h_r)} d\mu_r(h_r) \cdot k(D)\, dm(D).
\end{aligned}
$$

For the details see Fraser and MacKay (1975).

6. THE STANDARD INFERENCE PROCEDURES

Now consider the three standard inference procedures as applied sequential to the various components of the factored parameters.

(i) *Tests of significance.* A test of significance is used to assess the weight of evidence against a particular hypothesized value for a parameter. Suppose a value $\theta_s = \bar{\theta}_s$ is to be tested conditional on assumed values for $\theta_r = \bar{\theta}_r, \ldots, \theta_{s+1} = \bar{\theta}_{s+1}$. The hypothesized value gives access – by a proper information processor – to the value $h_s = \bar{\theta}_s^{-1}$ for the corresponding factor of $[Z]$. This can be assessed in the standard way by calculating the tail area probability based on the conditional distribution of h_s:

$$\alpha(\bar{\theta}_s) = \int\limits_{h_s = \bar{\theta}_s^{-1}}^{\infty} f^s(h_s : \bar{\theta}_{s+1}^{-1}, \ldots, \bar{\theta}_r^{-1}, D) \frac{\varDelta_{(s)}(h_s)}{\varDelta_s(h_s)} d\mu_s(h_s).$$

(ii) *Likelihood analysis.* A likelihood function is used to assess the relative plausibility of any two possible parameter values. The determination of likelihood with probability space models is complicated by the fact that an appropriate support measure must be determined for the marginal density of what has been observed; the most highly structured of the likelihood determinations is the transit likelihood of Fraser (1972b). In the present context the transit likelihood of θ_s given $\bar{\theta}_{s+1}, ..., \bar{\theta}_r$ is proportional to $f^s(\theta_s^{-1} : \bar{\theta}_{s+1}^{-1}, ..., \bar{\theta}_r^{-1}, D)$ and the relative plausibility by the ratio of the likelihood values. This can be assessed for a hypothesized value $\bar{\theta}_s$ by calculating an integrated tail area likelihood. This gives

$$\alpha(\bar{\theta}_s) = \int\limits_{\theta_s^{-1} = \bar{\theta}_s^{-1}}^{\infty} f^s(\theta_s^{-1} : \bar{\theta}_{s+1}^{-1}, ..., \bar{\theta}_r^{-1}, D) \times$$

$$\times \frac{\Delta_{(s)}(\theta_s^{-1})}{\Delta_s(\theta_s^{-1})} d\mu_s(\theta_s^{-1}).$$

(iii) *Objective posterior.* A probability distribution describing possible values for a parameter is used to assess the plausibility of, typically, an interval of values. With a probability space model the possible values for θ_s are identified with the possible values of h_s^{-1} – all conditional on values for $\theta_{s+1}, ..., \theta_r$. We then obtain the posterior distribution for θ_s directly from the conditional distribution of h_s without added assumptions. The probability distribution for θ_s is given by the probability differential

$$f^s(\theta_s^{-1} : \bar{\theta}_{s+1}^{-1}, ..., \bar{\theta}_r^{-1}; D) \frac{\Delta_{(s)}(\theta_s^{-1})}{\Delta_s(\theta_s^{-1})} d\mu_s(\theta_s^{-1});$$

integration of this over a tail area formed from a contemplated value $\bar{\theta}_s$ gives the probability $\alpha(\bar{\theta}_s)$.

Thus the three standard methods of reference all lead to the same numerical assessment and in terms of the $1-1-1$ correspondences are thus essentially equivalent.

7. AN EXAMPLE

For most distributions of variation, evaluation of the conditional distributions will involve numerical integration. For illustrative purposes we consider a very simple example in which the distributions are easy to derive.

Suppose we have a location-scale model with exponential errors

$$y_i = \mu + \sigma z_i$$

where

$$f(z_i) = \exp\{-z_i\}, \qquad z_i > 0, \qquad i = 1, ..., n.$$

The group can be represented by matrix multiplication using the notation

$$\begin{pmatrix} 1' \\ y' \end{pmatrix} = \begin{pmatrix} 1 & 0 \\ \mu & \sigma \end{pmatrix} \begin{pmatrix} 1' \\ z' \end{pmatrix}.$$

We consider the following two factorizations:

$$\theta = \begin{pmatrix} 1 & 0 \\ \mu & \sigma \end{pmatrix} = \theta_2 \theta_1 = \begin{pmatrix} 1 & 0 \\ \mu & 1 \end{pmatrix} \begin{pmatrix} 1 & 0 \\ 0 & \sigma \end{pmatrix},$$

$$\theta = \begin{pmatrix} 1 & 0 \\ \mu & \sigma \end{pmatrix} = \alpha_2 \alpha_1 = \begin{pmatrix} 1 & 0 \\ 0 & \sigma \end{pmatrix} \begin{pmatrix} 1 & 0 \\ \mu/\sigma & 1 \end{pmatrix}.$$

The first factorization would ordinarily be the more natural factorization, it leads to inference first for μ then for σ.

The transformation function $[\cdot]$ can be chosen in a straightforward manner and then adjusted to give the simplified notation discussed in Section 3. Let $z_{(1)}$ be the first order statistic and $s(z) = \sum (z_i - z_{(1)})$; then

$$[z] = \begin{pmatrix} 1 & 0 \\ z_{(1)} & s \end{pmatrix}$$

which satisfies the general relation $[gZ] = g[Z]$. The conditional distribution is easily derived and can be expressed by

$$(z_{(1)}, s) = (\chi_2^2/2n, \chi_{2n-2}^2/2)$$

in terms of independent chi-squares with degrees of freedom as subscribed.

Now consider the analysis as based on the first factorization of the parameter. The simplified notation with transformation taken relative to the point Y has the following factorization

$$[z][y]^{-1} = \begin{pmatrix} 1 & 0 \\ z_{(1)} & s \end{pmatrix} \begin{pmatrix} 1 & 0 \\ y_{(1)} & s(y) \end{pmatrix}^{-1}$$

$$= \begin{pmatrix} 1 & 0 \\ z_{(1)} - sy_{(1)}/s(y) & s/s(y) \end{pmatrix} =$$

$$= h_1 h_2 = \begin{pmatrix} 1 & 0 \\ 0 & s/s(y) \end{pmatrix} \begin{pmatrix} 1 & 0 \\ z_{(1)}s(y)/s - y_{(1)} & 1 \end{pmatrix}.$$

Thus for testing μ we have

$$\mu = -\left[\frac{z_{(1)}}{s}s(\mathbf{y}) - y_{(1)}\right],$$

$$\frac{y_{(1)} - \mu}{s(\mathbf{y})} = t = \frac{z_{(1)}}{s} = \frac{\chi_2^2/2n}{\chi_{2n-2}^2/2} = \frac{F_{2\ 2n-2}}{n^2 - n}$$

where $F_{2\ 2n-2}$ indicates an F variables with degrees of freedom as sub-scribed. And then for testing σ given μ we have

$$\sigma = [s/s(\mathbf{y})]^{-1},$$
$$\frac{s(\mathbf{y})}{\sigma} = s$$

where the conditional distribution of s given t is expressed by

(1) $\qquad s(1 + nt) = \chi_{2n}^2/2$.

Now consider the analysis as based on the second factorization of the parameter. We have

$$[\mathbf{z}]\,[\mathbf{y}]^{-1} = \begin{pmatrix} 1 & 0 \\ z_{(1)} - sy_{(1)}/s(\mathbf{y}) & s/s(\mathbf{y}) \end{pmatrix} = u_1 u_2 =$$
$$= \begin{pmatrix} 1 & 0 \\ z_{(1)} - sy_{(1)}/s(\mathbf{y}) & 1 \end{pmatrix}\begin{pmatrix} 1 & 0 \\ 0 & s/s(\mathbf{y}) \end{pmatrix}.$$

Thus for testing σ we have

$$\sigma = s/s(\mathbf{y}),$$
$$\frac{s(\mathbf{y})}{\sigma} = s$$

where s is $\chi_{2n-2}^2/2$. And then for testing μ/σ given σ we have

$$\mu/\sigma = z_{(1)} - sy_{(1)}/s(\mathbf{y}),$$
$$\frac{y_{(1)} - \mu}{\sigma} = z_{(1)}$$

where the conditional distribution of $z_{(1)}$ given s is just $\chi_2^2/2n$.

For each factorization the second stage of the inference sequence depends on the assumption of a plausible value for the first stage parameter component. For the first factorization the dependency is found in the scaling factor $1 + nt$ in (1), which involves a value for μ; and for the second factorization the dependency is found in the explicit use of σ in (2).

University of Toronto
and
University of Manitoba

REFERENCES

Fraser, D. A. S.: 1968, *The Structure of Inference*, Wiley, New York.
Fraser, D. A. S.: 1971, 'Events, Information Processing, and the Structural Model', in V. P. Godambe and D. A. Sprott, *Foundations of Statistical Inference*, pp. 32–40, Holt, Rinehart and Winston, Toronto and Montreal.
Fraser, D. A. S.: 1972a, 'Bayes, Likelihood, or Structural', *Ann. Math. Statist.* **43**, 777–790.
Fraser, D. A. S.: 1972b, 'The Determination of Likelihood and the Transformed Regression Model', *Ann. Math. Statist.* **43**, 898–916.
Fraser, D. A. S. and MacKay, Jock: 1975, 'Parameter Factorization and Inference based on Significance, Likelihood and Objective Posterior', *Ann. Statist.* **3**, 559–572.

DISCUSSION

Commentator Lindley: I have not understood today's paper. Professor Fraser has built a theory which he has developed to the stage where few can follow him. In my commentary today I will merely draw your attention to a feature of the theory which disturbs me greatly, and leads me to reject it. I think it will do the same for you. The same feature at first disturbs a Bayesian, but he has a way out which turns out to strengthen that theory. I hope it will do the same for you.

It seems to me easier to describe the objection in general terms. In a particular example one can easily get confused with details. Anyone wanting to see an example should refer to the paper by A. P. Dawid, M. Stone and J. V. Zidek to appear in a future issue of the *Journal of the Royal Statistical Society*, Series B, from which almost all the ideas in this commentary are taken. (θ, below, could be the correlation coefficient in a normal bivariate sample: x is r.)

Consider a model in which data, that, for a reason that will appear in a moment, we write as an ordered pair (x, y), depends on a pair of parameters (θ, ϕ) in a way that satisfies Fraser's requirements for a model. It can happen that when the Fraser analysis is carried out and the marginalization of θ completed, the inference about θ depends only on x, and that y is irrelevant. This being so it might occur to someone to ask what happens when only x, and not y, is available. Integrating out y it can happen that the resulting distribution of x is found to depend only on θ, and, moreover, that the resulting model again satisfies Fraser's requirements. This being so, we can apply his analysis (simpler now, involving only x and θ) and make inferences about θ. The disturbing feature is that the two inferences – one based on (x, y) the other on x – can differ, even though the first depends only on x. Put it the other way round: starting from x, the addition of an irrelevant y provides a different answer.

The Bayesian version is only slightly different. Bayesian one (B_1) with x and y at hand chooses a prior for θ and ϕ that makes his inference match with Fraser's and obtains a posterior for θ that depends only on x. A lazy

Bayesian (B_2) starts with x, having a distribution that depends only on θ and uses the prior for θ obtained by formal integration of ϕ from B_1's prior. The disturbing feature is that B_1 and B_2 disagree in their final inference about θ.

For me to see what is happening here it is easiest to consider the Bayesian aspect. Firstly, it is easy to show that if the prior for θ and ϕ is *proper*; that is, its integral over the whole parameter space is one, then the conflict between B_1 and B_2 cannot arise. It arises in the Bayesian theory only when some impropriety is present. This sheds light on Fraser's approach, for this is often related (in the sense that it produces the same answer) to a Bayesian one with an *improper* prior – usually being an invariant measure. This understanding shows that the Bayesian has an important escape route, namely to abandon imporpriety. No such device is available to Fraser because his group structure imposes the method of analysis, which analysis leads to the disturbing situation just described.

I conclude by making two remarks about Bayesian statistics that are suggested by the above commentary. Firstly, the use of improper distributions is surely unsatisfactory on practical grounds, quite apart from paradoxes of the type just discussed. If, as was suggested in my own paper, we think of θ as a physical quantity, and not just as a Greek letter, our views of such a quantity are never satisfactorily described by an improper distribution. The reality of θ almost imposes propriety. For example, suppose θ is an integer produced at random by a specific computer, say the IBM 360/65 at my own College. It might be argued that all values of θ are equally likely – an improper distribution. But we know that the machine is of finite capacity, works for a finite (and usually small) time in producing θ, and, for example, there are some numbers so large that it would take the machine a whole day to state. For me (and for you?) these are less likely than a 20-digit integer. Often the mere statement of the units in which θ is measured will tell us a lot about it: for example, if the unit is parsecs, then θ is rather large. Improper distributions just do not agree with practice.

My second point concerns the criticism that has been made that Bayesians are always changing their position: in this case over improper distributions. (Previously, for example in my 1965 book, I had advocated their use.) This change seems to me to be a strength rather than a weakness. Bayesian theory, being well-defined and reasonably complete, is easily criticized. (Contrast with fiducial theory, which is ill-defined and hard to criticize.) Each criticism that it survives strengthens the theory. (This is almost pure Popper!) I admit the strength of these criticisms, but the theory withstands them and is thereby strengthened.

Fraser and MacKay: We welcome Professor Lindley's discussion in response to our paper.

Lindley says he does not understand our paper. Consider the topics in the paper: traditional tests of significance, likelihood, objective posterior. All three are topics proscribed by the Bayesian frame of reference – yes, even likelihood, for it is *lost* in the calculation of the posteriors. And Lindley *is* a committed Bayesian.

The question then is the degree to which a Bayesian can examine and comment on the objective aspects of the analysis in our paper. Lindley chooses not to.

Rather Lindley raises questions concerning (structural) theory or, – more correctly, concerning the analysis of structural models. A first question of course centres on the mathematical validity of the model; a second on the appropriateness or relevance of the model for applications. For discussion on these points see Fraser (1968) and Fraser (1971).

The analysis of the model leads to estimation, tests of significance, and confidence intervals; for some recent discussion see Fraser (1975). One extension of the analysis presents a case for objective posteriors. This is what Lindley comments on. And it is of course the one element of the analysis that gives results that compete or seem to intersect with Bayesian posteriors.

Consider Lindley's comments on this one specific extension of the analysis of structural models. He cites the paper by Dawid, Stone, and Zidek (1973) and mentions a problem discussed in that paper. Interestingly the main mathematical properties underlying the Dawid, Stone, and Zidek paper had been presented publicly at the Third International Symposium on Multivariate Analysis in Dayton, Ohio, June 1972 and reproduced as Fraser (1973a). The specific points raised by Dawid, Stone, and Zidek, including the particular problem mentioned by Lindley, were answered (Fraser, 1973b) in the comments on the Dawid, Stone, Zidek paper. Indeed the particular problem is closely related to the subject of our paper.

Lindley also takes the opportunity to comment on the current Bayesian position. Recent criticisms of the Bayesian position have embraced the difficulties with improper priors. Lindley feels the solution is to avoid that 'impropriety' by restricting attention to proper priors. This may solve the particular difficulties but it leaves the Bayesian in an excessively tight position and most of the criticisms of the Bayesian position as in say Fraser (1972, 1974) remain effective. In a sense the Bayesian has jumped from the pan into the fire.

As an illustration of the Bayesian position without flat priors consider a primary sample $(y_1 \ldots y_n)$ from $N(\mu, \sigma^2)$ against a background of measurements $x_1, \ldots, x_r$ from $N(\mu_1, \sigma^2), \ldots, N(\mu_r, \sigma^2)$. No choice of proper prior can eliminate the irrelevant past measurements $x_1, \ldots, x_r$ from a current Bayesian inference concerning μ. Clearly an unacceptable position from a scientific point of view!

We summarize: Professor Lindley has chosen not to comment on our paper but to discuss a problem examined in a recent paper. The problem, however, as it bears on structural analysis, had been disposed of elsewhere. Lindley also argues that the current Bayesian withdrawal from improper priors is a change of position that is a sign of strength. A realistic assessment of the limitations with proper priors however shows that the Bayesian is in desperate retrenchment.

REFERENCES

Dawid, A. P., Stone, M., and Zidek, J. V.: 1973,' Marginalization Paradoxes in Bayesian and Structural Inference', *J. Roy. Statist. Soc.* B35, 189–213.

Fraser, D. A. S.: 1968, *The Structure of Inference*, John Wiley and Sons, New York.

Fraser, D. A. S.: 1971, 'Events, Information Processing and the Structural Model', in V. P. Godambe and D. A. Sprott, *Foundations of Statistical Inference*, Holt, Rinehart, and Winston of Canada, Limited, Toronto.

Fraser, D. A. S.: 1972, 'Bayes, Likelihood, or Structural?', *Ann. Math. Statist.* 43, 777–790.

Fraser, D. A. S.: 1973a, 'Inference and Redundant Parameters', in P. R. Krishnaiah, (ed.), *Multivariate Analysis III*, Academic Press, New York.

Fraser, D. A. S.: 1973b, 'Comments on the Dawid-Stone-Zidek Paper', *J. Roy. Statist. Soc.* B35, 225–228.

Fraser, D. A. S.: 1974, 'Comparisons of Inference Philosophies', in G. Menges, (ed.), *Information, Inference, and Decision*, D. Reidel Publishing Company, Dordrecht-Holland.

Fraser, D. A. S.: 1975, 'Necessary Analysis and Adaptive Inference', to appear in *J. Am. Statist. Assoc.*

Fraser, D. A. S. and MacKay, Jock: 1975, 'Parameter Factorization and Inference based on Significance, Likelihood, and Objective Posterior', *Ann. Statist.* 3, 559–572.

RONALD N. GIERE

EMPIRICAL PROBABILITY, OBJECTIVE STATISTICAL METHODS, AND SCIENTIFIC INQUIRY*

1. INTRODUCTION: WHAT MIGHT A THEORY OF STATISTICS BE?

Current controversies over the foundations of statistics are both extensive and deep, and are even beginning to affect the everyday operations of working statisticians and scientists.[1] In this context it seems appropriate for a philosopher, i.e., one who's business is analyzing the conceptual discomforts of others, to begin by considering just what a theory of statistics might be.

One of the reasons the Bayesian viewpoint has attracted so much attention is that it satisfies a very simple but comprehensive picture of what a theory of statistics should be. On this account, statistics is just a special application of the general Bayesian theory of rational belief and action. Moreover, there is little ambiguity as to what the theory claims, for it can be developed as a set of axioms governing the behavior of an ideally rational agent – just like geometry or mechanics. The theoretical simplicity and comprehensiveness of a Bayesian approach are so great that some advocates now seem to view it as self-evidently correct and are inclined to regard criticism as evidence of incoherence in the critic.[2] But the view of the ideal scientist as a Bayesian decision maker is far from self-evident and is indeed contrary to commonly held conceptions of the scientific enterprise. Explaining why I think this is so will provide an introduction to the body of this paper which is an attempted theoretical justification of an 'objectivist' approach to statistical inference, and to scientific inference in general.

I take it as obvious that a Bayesian theory of rational belief and action cannot be *merely* a mathematical system, a pure geometry, any more than mechanics can be pure mathematics. The conclusions of both theories have more than mathematical content. But unlike mechanics, Bayesian decision theory cannot be a purely empirical theory either. The differences between ideal point masses and real masses are not like the differences between ideally rational agents and real agents. Deviations in the behav-

Harper and Hooker (eds.), Foundations of Probability Theory, Statistical Inference, and Statistical Theories of Science, Vol. II, 63–101. All Rights Reserved. Copyright © 1976 by D. Reidel Publishing Company, Dordrecht-Holland.

ior of real masses are dealt with by modifications in the theory. But deviations of real agents are dealt with by calling the agent irrational. It is the agent that should conform, not the theory. The Bayesian 'theory', then, can only be regarded as a systematization of certain normative principles of rationality. Since these principles cannot be sufficiently justified either by mathematics or empirical procedures alone, and since few empiricists would be willing to take them as given a priori, they may be questioned. Here it is their application in the context of scientific inquiry that is primarily at issue.

Any Bayesian theory begins with an agent contemplating a set of possible events and a set of possible actions. The conditions of rationality are then shown to entail that the agent have a coherent belief structure over possible events and a utility measure over possible outcomes, i.e., event-action pairs. The rule of rational action then reduces to one: Maximize Expected Utility.

The 'actions' of Bayesian theory are generally thought of as bits of behavior performable by human agents. The 'events' or 'states of nature' are taken to be local, identifiable physical happenings. Bayesians indeed take pride in the fact that the objects of their theory are so concrete and empirical. (If only there were a good, independent 'operational' definition of utility as well.) An extreme form of such empiricism would reduce all actions, events, etc., to the immediate sensory impressions of some particular agent. Indeed, in this respect I am inclined to regard de Finetti as the philosophical descendent of Hume, Mill, Mach, Russell and many early twentieth-century positivists.[3] The difference is that now Hume's 'impressions' are given a coherent probabilistic structure. But even without going to such philosophical extremes, it is clear that the view of science that emerges is instrumentalistic. Theories themselves are not objects of belief, but only formal devices for organizing events and their associated degrees of belief.[4] There is no need to regard the supposed objects of theories as 'real', and to do so invites the charge of being unempirical or 'metaphysical'.

The most popular alternative view of science *begins* with the 'metaphysical' assertion that the world consists of complex physical systems, and that the aim of scientific inquiry is to produce theories that give true causal descriptions of these systems. Moreover, we realize that our interactions with physical systems are confined to a sample of macroscopic outputs. It is these that constitute the data for our theoretical conclusions. This, of course, is where statistical inference enters the picture.

Now it might be replied that there is no necessary opposition between a Bayesian approach to inference and a realistic metaphysics. The fact that a theory is true may be a state of nature and accepting or rejecting a theory as being true may be actions. The bearing of evidence on theories is then measured by coherent degrees of belief which evolve according to Bayes' theorem. This reply, however, is almost entirely programmatic. Except for some work by Abner Shimony (1970), a philosopher, there is no Bayesian account that applies to theories considered as possible descriptions of reality.[5] Moreover, it is not possible simply to extend current treatments. The state of a theory being true is not the kind of thing one could make the object of a bet. It would not be possible even to settle such a bet, and a bet without a possible payoff is no bet at all. Similarly, the action of accepting a theory as true is not something to which we know how to assign utilities in various possible states. Combining a Bayesian theory of inference with scientific realism will not be an easy task.

The modern basis for a theory of inference that may go comfortably together with scientific realism was given by Neyman and Pearson. Of course they were building on the work of others, notably Fisher. But what they did that others had not was to make explicit the idea that statistical tests result in the rejection or acceptance of a statistical hypothesis (Fisher considers only rejection[6]) and that the 'goodness' of a test is a function of the probabilities of mistaken rejection and mistaken acceptance. In this framework it is possible to take accepted statistical hypotheses (and eventually all scientific hypotheses) as truly describing a general feature of a real physical system – so long as one remembers that the method that leads to acceptance has a certain probability for leading us astray. The important thing is that the objectivist approach to inference satisfies the realist idea that the role of data is to help us to accept true hypotheses and to reject false ones.

As should be expected, the objectivist conception of what a theory of statistics is differs substantially from the Bayesian account. For an objectivist there is no 'theory' of statistics that could be set out axiomatically and developed deductively. Rather, there is a conception of a statistical test (or an interval estimate) based on the concept of a sample space divided into two regions – acceptance and rejection. A 'good' test is one with 'good' error characteristics. From here on the work of the theoretical statistician consists in proving more or less general theorems concerning the error characteristics for various classes of tests, for various distributions, etc. It is just such work that fills the journals of mathematical statistics. The most famous general result of this type is of course the Neyman-Pearson lemma.

Anyone familiar with the actual history of statistical theory since 1920 will realize the amount of hindsight and conceptual reconstruction that has gone into making the neat double dichotomy, Bayesian-instrumentalist *vs.* objectivist-realist. While insisting that this classification is not merely polemical and has a sound conceptual basis, I must admit that the historical record fails to be anywhere nearly so clear. This can be explained, however, by the fact that the objectivist approach achieved its modern maturity during a period in which the dominant view of science among both scientists and philosophers was in fact highly instrumentalistic. Thus the instrumentalism common among current Bayesians was already present in the thinking of their objectivist predecessors. Indeed, this instrumentalism may be seen clearly in Neyman's distinction between inductive *inference* and inductive *behavior*.[7] According to Neyman, there is no such thing as inferring the truth of a hypothesis from inconclusive data; there is only a choice of one action over others. By thus reducing 'accepting a hypothesis' to 'choosing an action', Neyman set the stage for Wald's formulation of statistical inference as decision making under uncertainty. And with the apparent non-existence of a satisfactory general criterion for decision making under complete uncertainty, Savage's introduction of subjective probabilities becomes a decisive, though radical, way of simply eliminating the problem. The way to rescue objectivist methods from this progression, I think, is to go back and provide a less behavioristic interpretation of what it is to 'accept' a hypothesis. Moreover, I think this can be done without appeal to the subjective beliefs of individual investigators. That, at least, is the main objective of this paper.

There is another doctrine which nearly all objectivist statisticians share with instrumentalist and positivist philosophers, namely, the interpretation of empirical probability in terms of relative frequencies. Though well-established for nearly a century, this interpretation makes it difficult if not impossible to understand the application of probability in inferences concerning particular systems at specified times. A great many contemporary objections to objectivist methods are just specific versions of this general point. Such objections are becoming commonplace and cannot be ignored by any defender of an objectivist approach to statistical inference. In the following section I will argue that most of these objections can be met by assuming the existence of real causal tendencies ('propensities') in individual physical systems. Once these questions about the nature of empirical probability have been resolved, it is possible to move on to more direct questions about the nature of objectivist methods.

Finally, since unlike Bayesian theories, my understanding of statistical inference applies in 'scientific' contexts only, a separate account of practical decision making is required. The final section, then, is devoted to a sketch of objective decision making and of relations between scientific inquiry and practical decision making.

2. FREQUENCIES, PROPENSITIES. AND CONDITIONAL TESTS

The identification of empirical probability with relative frequencies is so much a part of objectivist statistical theory that non-Bayesians are commonly referred to as 'frequentists' – just as Bayesians are commonly called 'subjectivists' or, as Savage insisted, 'personalists'. Now there is a fundamental difficulty with the frequency interpretation which has been recognized by frequency theorists from Venn (1866) down to Reichenbach (1949) and Salmon (1967). If one wishes to associate a probability with a particular event (e.g., a test result), one must imbed the event in a set or sequence of similar events. But there may be more than one possible reference set, each with *different* relative frequencies for the characteristic in question. Which reference set is the 'right' one? A great many criticisms of objectivist methods are specific versions of this general problem. I will begin by giving two common examples from the statistical literature which illustrate the general problem in the context of hypothesis testing. I will then argue that these examples do not necessarily count against the orthodox understanding of hypothesis testing. Rather they merely exhibit the inadequacies of the frequency interpretation of empirical probability. By adopting a sufficiently radical 'propensity' account of empirical probability, one can accept the point of the examples without giving up the essentials of the objectivist approach, i.e., the importance of error probabilities in experimental design. In effect one ends up providing a non-Bayesian justification for conditionalizing on the values of certain variables in statistical tests.

The first example is adopted from Cox (1958). Imagine two distinct physical processes, both producing outcomes normally distributed with mean zero, i.e., $\mu_1 = \mu_2 = 0$. The variance of the first process, however, is much greater than the second, i.e. $\sigma_1^2 \gg \sigma_2^2$. Now suppose we wish to test the hypothesis $\mu = 0$ vs. $\mu = \mu' \approx \sigma_1$ given one observation which may come from either process depending on the outcome of a random device with equal probabilities. Once we get the result, however, we know which of the two it came from. Intuitively the best 5% test would seem to have the following critical region:

(A) $x > 1.64\,\sigma_1$ if x is from process one
           $x > 1.64\,\sigma_2$ if x is from process two .

The expected relative frequency of Type I errors with this test is 5%. However, the expected frequency of Type II errors for the alternative $\mu' = \sigma_1$ will be greater for the first process than the second. This suggests that the overall expected frequency of Type II errors could be lowered without raising the overall expected frequency of Type I errors. Consider the critical region:

(B) $x > 1.28\,\sigma_1$ if x is from process one
           $x > 5\,\sigma_2$ if x is from process two.

Here the probability of mistaken rejection is 10% for process one but negligible for process two so the overall relative frequency of mistaken rejection remains at 5%. But the overall frequency of mistaken acceptances will be lower since it is appreciably less for outputs from the first process and only negligibly greater for outputs from process two.

The point of the example is familiar and obvious. Given, for example, that we know our observation came from the first process, it seems un-tuitively silly to apply Test B which has a 10% probability for Type I error. Reference to the long run or average error characteristics of Test B seems irrelevant. It is the particular inference we care about, and we know we are faced with process one. Why should we ignore this seemingly very relevant fact? Cox concludes that "the sample space must not be deter-mined solely by consideration of power, of by what would happen if the experiment were repeated indefinitely", and he recommends that we conditionalize with respect to all facts "which do not give a basis for discrimination between the possible values of the unknown parameter of interest". Before examining this example further let us look at a second example which has been discussed by Basu (1964) and Birnbaum (1969) and no doubt many others.

Consider a chance process with a single unknown parameter, and suppose we wish to test a simple hypothesis against either a simple or composite alternative. Our sample consists of either 10 or 100 trials depending on the outcome of a binomial random process. The average error characteristics of the mixed test are clearly superior to those of the 10-trial test and inferior to those of the 100-trial test. Yet the characteristics of the mixed test seem scant justification for a conclusion when in fact only 10 trials are made. Similarly, why should we 'throw away' data by using the mixed text characteristics if in fact we have made 100 trials? Again it seems that we should condition our test on the actual sample size and thus eliminate the mixed sequence from consideration.

I take it as undeniable that the intended point of these examples is sound. We should not use the mixed test. The problem is to find a systematic, rational basis for this judgment. Unfortunately frequentist philosophers who have written at length on 'the problem of the reference class', e.g., Reichenbach (1949) and Salmon (1967), have not considered examples like the above. If they did, what they would want to show is that there is a basis for choosing a particular reference sequence in terms of which the operating characteristics of a test are to be determined. I doubt that this can be done satisfactorily, but committed frequentists will want to try. Statisticians faced with the above examples tend either to dismiss them (because, after all, the frequency interpretation is the only 'objective' interpretation) or take them as indications of flaws in the received view of testing and estimation (as in the quote from Cox above). In particular, increasing numbers of theoretical statisticians are beginning to take such examples as supporting a general 'likelihood principle': The total evidential weight of data is contained in the *likelihood* (or likelihood ratio) of the data actually obtained. This principle, which also follows from a Bayesian analysis, has even been taken, e.g., by A. W. F. Edwards (1972), as providing a sufficient basis for all statistical inference. It is clearly incompatible with the use of error probabilities which are defined relative to some sample space of possible outcomes, only one of which is actually obtained. The question is whether, in examples like the above, it is possible to justify conditionalizing on the result of the randomizer without going so far as to conditionalize on the actual data, thus abandoning error probabilities altogether.[8] I think this is possible if one abandons relative frequencies in favor of causal (though not invariable) propensities.

Except for those statisticians who have given up objective probabilities altogether, the identification of probability with relative frequency seems obviously correct. This is a natural consequence of the long association of statistics with sampling from finite populations.[9] Taking the binomial case, if a proportion p of the population exhibits property F, then obviously the probability of a randomly selected member being F is p. But the natural view may not be the best, nor even conceptually coherent. Indeed, difficulties begin as soon as one asks what constitutes 'random selection'. Can random selection be defined without circularity in terms of relative frequencies? Is it sufficient that we can give a mathematical definition of a random *sequence* in terms of computability, e.g., Church's (1940)? Or is it that mathematical definitions of random sequences (infinite sequences) have little to do with the statistician's concept of random selection?[10] The lack of clear answers to such questions is sufficient evidence of conceptual problems with the simple identification of probability with relative frequency. But rather than attempt to resolve such difficulties within the given framework, I suggest we abandon the framework – in spite of its historical significance for statistics as a whole.

The basic trouble with the frequency framework is that it does not give probability a sufficiently fundamental place in our scientific ontology. The objects in any population are regarded as having completely definite characteristics. If only we could examine every one we would have no need of probability. But we can't, so we introduce probability as a way of dealing with our ignorance.

Suppose we forget about populations for a moment and concentrate on physical systems, e.g., atomic systems, planetary systems, genetic systems, nervous systems, economic systems, even social systems. One way of classifying systems is as being deterministic or indeterministic. Roughly speaking, a *deterministic* system is one for which given values of all the state variables uniquely determine the total state of the system at all later times. An *indeterministic* system has more than one physically possible later total state. A *probabilistic* system is indeterministic but in addition has causal 'tendencies' with a probability measure distributed over possible later states. If one is willing to be sufficiently metaphysical, he may say that in a deterministic system there is a relation of causal necessity between initial and final states. The corresponding relation for probabilistic systems is one of causal propensity. (The word 'propensity' is usually attributed to Karl Popper (1959)).

Now the traditional empiricist (meaning positivist) view has been that strictly speaking there is no such thing as a relation of causal necessity between two temporal states of a single system. It is just the case that the one state as a matter of fact always follows the other. For probabilistic systems the corresponding doctrine is that the one state follows the other in a certain proportion of cases (perhaps in an infinite sequence of cases). The realist view of causal necessity, on the other hand, regards them as operative in individual processes. Similarly, causal propensities are to be understood as operating during individual trials of probabilistic systems. Thus when a probability value is assigned to an outcome, that value attaches directly to the particular trial – no reference to any set of trials is necessary. Of course any direct *test* of a propensity hypothesis will require frequency data on a number of trials with identical initial conditions. But no amount of frequency data, even a countably infinite sequence, can exhaust the *meaning* of a probability hypothesis construed as a statement about the strength of a propensity.

The statistical paradigm of sampling from a finite population may now be seen as a special case of operating a probabilistic system, i.e., a system consisting of a finite population plus a sampling mechanism. A random sampling mechanism is simply one which has equal propensity to select any member on each trial.[11] The ideal would be a random number generator using quantum noise effects for input. In practice we use random number tables and other devices that give a good approximation to the ideal.

It is unfortunate that more statisticians are not familiar with quantum theory, for micro-phenomena provide the most convincing examples of causal propensities. Consider a radioactive nucleus sitting in a counter. It is very plausible to interpret current quantum theory as saying that the individual nucleus is a probabilistic system and has, for example, a definite propensity to decay in the next minute. Moreover, if the theory is correct, there is no variable that strictly determines whether or not the decay will take place. The indeterminancy is not just a matter of our ignorance. But of course this example cannot be definitive as the foundations of quantum theory are about as controversial as the foundations of statistics.

The above is not enough to convince anyone who has not had a similar idea himself that propensities exist, or even that the idea makes sense.[12] Perhaps a better argument is just seeing how thinking in terms of propensities provides a basis for our intuitive judgments about the Cox and Birnbaum examples. Taking Cox's example first, let us follow in detail the course of the single trial yielding the data point x. First is a turn on the randomizer which has equal propensities for each of two possible outcomes. One outcome is linked with the activation of process one; the other with process two. Whichever process is activated then produces a single value, x, according to a normal propensity distribution (μ, σ). Assume we desire the best 5% test of the hypothesis $\mu = 0$ against $\mu = \mu' \approx \sigma_1$. Given that we *know* which process has been activated, we know whether the test hypothesis refers to the propensity distribution (μ_1, σ_1) or (μ_2, σ_2). Thus we know which process was causally operative in producing the data, and, consequently, the decision to accept or reject H. Equally we know which process was *not* operative. This provides the needed basis for declaring that features of the non-operative system must be irrelevant to our conclusion concerning the mean value of the chosen process on this particular trial. That other system was not a causal part of this trial. It is still true, of course, that the test we perform will be more powerful when process two is activated since that distribution has the smaller variance. We, however, have no choice between these two tests – that is up to the randomizer.

Now consider an example like Cox's except that the outcome of the randomizer remains unknown. That is, we have a black box containing the randomizer and the two processes with distributions (μ_1, σ_1), (μ_2, σ_2). Here the randomizer and the two processes together constitute a complex propensity producing a value, x. However, the output of the black box is not x, but one of the two statements, 'Reject H' or 'Accept H', where H is $\mu = 0$ and μ is the exclusive disjunction μ_1 or μ_2. Our problem is to program the black box so that its output represents the best 5% test of H against $\mu = \mu' \approx \sigma_1$. The program may make internal use of which process gets activated. In this case we should build in the *mixed* text (Test B) over the 'conditional' test (Test A) because then the overall propensity for mistaken 'Accept' outputs is lower. The crucial difference between this case and the previous one is that here all we know about the propensity producing the output is its overall structure – we don't know the output of the randomizer.

Before attempting any further generalization, let us look briefly at the Basu-Birnbaum example. Here there is only one physical process which we will assume to have a constant propensity distribution over all trials. However, for any specified form of the sample space, a set of 10 independent trials generates a different propensity distribution over that space than a set of 100 trials. So considering a set of trials as a complex process, there are really two different processes under consideration, the one providing far better operating characteristics. In order to decide which of the two tests to apply one need only determine which physical process produced the data. This is determined as soon as the output of the randomizer is known. Once we know it was the 10 trial process, possible results of a 100 trial process become irrelevant and vice versa.

Again it is illuminating to consider the similar case in which the outcome of the randomizer is *not* known, i.e., everything takes place in a black box that is limited to the outputs 'Accepts H' or 'Reject H'. Here we have no choice but to include the randomizer as part of the physical process that produces the end result. Thus the operating characteristic to be built into the box will have to be that for a 50–50 mixture of the 10 and 100 trial experiments. This will give the most favorable propensities for mistaken acceptance and rejection by the black box.

I am afraid that the above treatment of the Cox and Birnbaum examples is bound to disappoint a mathematical statistician. What he sees in Cox's case, for example, is the distribution

(M) $f(x) = \frac{1}{2}(\mu_1, \sigma_1) + \frac{1}{2}(\mu_2, \sigma_2).$

It is in terms of this whole distribution that he seeks an optimal test of $\mu = 0$ vs. $\mu = \mu' \approx \sigma_1$. If he is to be told that the only distribution he need consider is either (μ_1, σ_1) or (μ_2, σ_2), depending on which is chosen, he feels that should be stated and justified as a *formal* principle relating the three distributions. What I have given, however, is an informal physical (even metaphysical!) principle which may be stated roughly (for hypothesis testing) as follows:

> The operating characteristics of a test depended only on the propensity distributions of processes actually operative in producing the decision to accept or reject the test hypothesis.

It must be understood, of course, that which process we take to be the 'actually operative' process depends on our knowledge of the system. We cannot design a test which assumes knowledge we do not have.[13] The justification for the above principle is simply that our immediate goal is always to accept true and reject false hypotheses concerning individual physical processes that actually occur. Of course ultimately we want to generalize beyond individual experiments. In no case, however, are we concerned only with our success rate for a series of conclusions that need never exist.[14]

But is it not possible to give a more formal statement of the needed principle? I think not, and for very general reasons. At some point in the application of mathematics to empirical systems it must be indicated just what system is being studied. This cannot be done by formal conditions alone – no matter how extensive. For example, the formal axioms of finite probability spaces do not distinguish relative frequencies of different colored balls in an urn from relative masses of planets in a planetary system. Similarly, formal conditions on probability spaces cannot distinguish causal tendencies of an individual probabilistic system from relative frequencies in a set of deterministic systems. This must be an extra-mathematical distinction. Once it is realized that some such determination is necessary one can begin thinking constructively about ways to avoid the unwelcome conclusions of examples like Cox's. Introducing the idea of causal propensities is one way.

Finally, it should be noted explicitly that while I have attempted to justify restricting attention to *experiments* that actually occur, nothing in the above supports the full likelihood principle that would restrict attention to the *sample point* that actually occurs. Indeed, causal tendencies like causal necessities refer as much to possible outcomes as to actual ones, i.e., to outcomes that could have occurred as well as those that did occur. Thus there is every reason to think that a statistical inference should be sensitive to a whole sample space, not just to the outcome that happened to be realized this time.

3. STATISTICAL INFERENCE IN EXPLORATORY (THEORY-SEARCHING) INQUIRY

The second major problem for a realist reconstruction of 'objectivist' methods is to give an account of the 'acceptance' (and rejection) of statistical hypotheses which is consistent with the goal of discovering true hypotheses about real physical systems. As indicated earlier, Neyman implicitly gave up a realist approach when he reduced the idea of inferring the truth of a hypothesis to choosing an action. But it must be realized that one cannot set matters right simply by insisting that 'accept H', means 'accept H as being true'. Some further analysis, though not necessarily Neyman's, is necessary. The reason for this has been forcefully set out by a number of philosophers, e.g., Rudolf Carnap (1963), Richard Jeffrey (1956), and Isaac Levi (1967). If 'accept H' is taken to mean 'accept H as being true' without any further restrictions, then presumably this would commit one to *acting* on the assumption that H is true in any context whatsoever. But H must have been accepted on the basis of a test with some definite operating characteristic. And no matter how stringent this test might have been, one could always imagine a practical situation, e.g., where many human lives are involved, in which even a very small chance that H was accepted even though false makes acting on the truth of H unacceptably risky. This is a very simple observation, but to my mind it constitutes decisive proof that the acceptance of a hypothesis as true must be restricted in some way.

Neyman's analysis restricts acceptance to a definite decision making context so that the parameters of the test are determined by the uilities of that context. The problem of unacceptable risks cannot arise. Thus Neyman's treatment is *sufficient* to avoid the difficulty. But it goes too far. Any general concern with the truth of the hypothesis disappears completely in favor of concern over the choice of an action in a particular, well-defined decision context. The question is whether one can find something between restriction to a particular decision context and unrestricted acceptance as true.

The suggestion to be explored in this and the following section is that acceptance as true be restricted to a *context of scientific inquiry*.[15] This means that anyone working in that context could assume the truth of H in the course of that inquiry – unless later evidence shows H to be mistaken. This program will require some analysis of the nature of scientific inquiry and the role of statistical inference in scientific investigations. It also requires that there be an independent account of practical decision making and of relations between scientific inquiry and practical decision making. These are again not the kinds of analyses that the mathematical statistician expects in the foundations of statistics. I can only reply that it seems to me that the current controversies are not resolvable by formal analysis and that something more like the present investigation is needed – at least at the present time.

Scientific inquiry is obviously not a completely homogeneous enterprise. For present purposes I will distinguish only two kinds, or stages, of inquiry: *exploratory* (theory searching) and *confirmatory* (theory testing). Even exploratory research does not begin in an empirical vacuum. One always has some ideas about the kind of system under investigation. One may even have some vaguely formulated theories which can suggest more precise hypotheses to be investigated. What is lacking, however, is an explicitly formulated, reasonably developed, and reasonably complete theory of the domain in question. Indeed, formulating such a theory is the primary goal of exploratory research. The first step, of course, is just thinking of quantities that *might* be important variables in an eventual theory. Next is designing experiments to determine which of these variables may be causally related. Finally one may want to estimate degrees of association among various variables. To keep the discussion to manageable proportions, I will concentrate on null hypothesis testing which, for better or worse, is still the primary tool of exploratory research.[16]

Exploratory inquiry revolves about simple causal hypotheses of the form: changes in variable X produce changes in variable Y. The classic experimental test (I will not discuss non-experimental research here) consists of experimental and control groups, with interest focusing on whether or not there is any difference in some relevant function of the variable Y. So we have a test of the point null hypothesis, no difference, against the composite negation, i.e., some difference. Now significance testing has been widely criticizied on the ground that almost any two variables will admit some small causal connection. Thus whether the null hypothesis is rejected or not depends mainly on the size of the sample. But this criticism ignores the crucial role of power, or precision, in designing a test in an exploratory context. If one is looking for connections that would have to be explained by a relatively crude, 'first order' theory, then clearly it is inefficient to invest great effort in detecting small 'second order' effects. Of course it requires some professional judgment on the part of experienced investigators to determine both which variables are likely to figure prominently in an eventual theory, and also how great a difference in these variables would be worth even considering at this stage of the inquiry. Given such judgments, however, we will want low power against unimportant alternatives 'near' the null hypothesis and high power against 'interesting' alternatives 'far' from the null hypothesis. As will become clearer below, more exact specifications of 'high' and 'low' will vary from field to field. But in general low power would be something less than 25% and high power something around 90%. These numbers reflect the fact that at the exploratory stage it is not worth much effort detecting small, unimportant differences, but highly desirable to detect large, important ones because these will have to be accounted for by any proposed theory, even a crude one. Not to detect an important difference would mean losing an opportunity for theoretical advance.

Thus within rough limits the aims of exploratory inquiry put constraints on the power function of a statistical test. This already contradicts the impression given by many textbooks of mathematical statistics that the sample size and significance level are given, and all that remains is to optimize the power function. But there is no reason to think that any power function obtained in this fashion will have the appropriate power either for important or unimportant possible differences. This is a good illustration of the general point that to understand the foundations of statistics requires some consideration of experimental design and the general aims of inquiry. Theorems proving the existence of 'uniquely best' operating characteristics are necessary, but not sufficient.

If looking at the aims of exploratory inquiry tells us something about desirable power functions, what does it tell us about significance levels? The effect of rejecting a null hypothesis when there is little or no real difference is to lead the exploratory process down a blind alley – at least temporarily. This is always undesirable, so when data are cheap and quickly gathered, one should always make the probability of Type I error negligibly small, e.g., 0.001. But it is rare that data come so cheaply, either in money, or time, or energy. This is especially true in the social sciences and psychology, though perhaps less often the case in biology,

chemistry and physics. In general one must balance off obtaining excep-
tionally favorable error characteristics against the need to conserve
resources for other experiments. A successful program of exploratory
research may require years of experimentation before a recognizable
theoretical structure begins to emerge. How then is one to strike a balance
between resources (sample size), precision (power) and reliability (significance level)?

There are a few obvious rules of thumb. At the beginning of a program of inquiry, when very little is yet known, one can afford to sacrifice
some precision and reliability simply to have some results to think about.
A partly misleading basis for theoretical speculation is better than no
basis at all. Later, as the bits and pieces of a genuine theory begin to fall
into place, one can be more selective, getting more mileage out of few
experiments performed with greater reliability and precision. In the end,
however, the judgment concerning just what balance is appropriate at any
given stage of a particular inquiry will have to be made by people working
in the area. But of course they may judge wrongly and the inquiry drag on
inconclusively until it is simply abandoned. That is a risk one cannot
avoid.

The above analysis has focused on hypothesis testing. Yet a number of
methodologists have recommended abandoning null hypothesis testing
altogether, replacing it by interval estimation (if not by something even
farther removed from 'classical' statistics).[17] Now once a test has revealed
a potentially important difference, it may then be theoretically useful to
estimate the amount of the difference. But nothing is gained simply by
replacing tests with interval estimates. The same balance among resources
(sample size), precision (size of interval) and reliability (confidence level)
must be resolved. And it is still true that it is a waste of resources to get a
very precise estimate of a small, unimportant difference. Our efforts
should go into estimating differences with potential theoretical importance.

Sometimes when methodologists recommend estimation over testing
they really have in mind that one should report a whole family of confidence curves cum confidence levels.[18] This is similar to the suggestion to
report the *obtained* p-levels of tests – i.e., the minimal level at which the
observed difference would have been judged 'statistically significant'. But
on this sort of approach one never does 'accept' a conclusion that goes
beyond the data. One is simply reporting the data in a summary form.

According to my view, on the other hand, one really does accept statistical hypotheses (and the underlying causal hypotheses) as being true for the purposes of the inquiry. The 'operational meaning' of this acceptance as true is, first, that the hypothesis is regarded as something that must be explained by any proposed general theory of the relevant kind of system, and, secondly, that an accepted hypothesis is assumed to be true when one designs new experiments to test further hypotheses. This means that there are two more or less distinct ways in which a mistaken acceptance or rejection can mislead and divert the course of an exploratory inquiry. Of course no such mistake need be irreparable, and the whole process of inquiry is (or should be) designed so that serious mistakes will eventually be found out.

At several key junctures I have said that certain things must be left to the professional judgment of experienced investigators, e.g., determining which differences have potential theoretical importance and striking a balance between the demands of precision, reliability and limited resources. It is unfortunate that even the principle advocates of objectivist methods, e.g., Neyman, have passed off such judgments as merely 'subjective'.[19] This way of talking has provided much encouragement to Bayesians who then argue for the virtues of making one's subjectivism an explicit part of statistical theory.[20] But the judgments in question are not 'subjective' in a way that provides any reason to think explicit introduction of subjective degrees of belief would be at all desirable. In the first place, these judgments are not necesarily linked to any particular investigator in the way that a degree of belief must be. They are the kinds of judgments on which most experienced investigators should agree. More precisely, professional judgments should be approximately invariant under interchange of investigators. This is not true of degrees of belief. Secondly, these judgments are not in the form of degrees of belief in events or hypotheses. They are categorical assertions, however tentative, concerning the design of particular experiments. They may even prove to have been mistaken.

In general, charges of 'subjectivism' and 'objectivism' have tended to be battles between straw men. Science, after all, is a human activity – there is no science without scientists. So there is no question *whether or not* human judgment enters into the scientific process – it obviously does. The question is *how* it enters and why. The Bayesian personalist puts the individual investigator right at the center of things; the end result of all investigation is a modification of his distribution of beliefs. The objective realist, on the other hand, recognizes the essential role of the investigator in the design of experiments, the development of research programs, and the creation of theories. But once the experiment has been set in the context of some inquiry, it is nature that decides the outcome. And the conclusion is an assertion about a real physical system – not a report of a belief distribution.

It remains true nevertheless that accepting or rejecting a statistical hypothesis is in a broad sense a decision. Moreover, experimental design in exploratory inquiry is guided by considerations of utility, albeit 'scientific' rather than 'practical'.[21] Thus mistaken rejection of a null hypothesis may lead both to confusing results in later experiments and to theoretical frustration trying to explain a non-existent effect. Similarly, failure to detect an important effect is a clear opportunity loss. Should not these 'utilities' be fit into a formal decision theoretic framework of some kind? I have already given my reasons for being skeptical of a Bayesian decision analysis, but what about a classical decision structure? This suggestion immediately raises two questions: (i) How can the relevant utilities be determined? (ii) What decision strategy should be used? So long as we insist on a strategy that seeks to maximize or optimize some function of precisely determined utilities, it will, I think, be impossible to see hypothesis testing in explatory inquiry as a rational decision process. But if we are willing to relax our understanding of what constitutes a rational strategy, there is some hope. One way of doing this is to adopt a strategy of 'satisficing'. To satisfice one need not look for an optional decision; it is sufficient to find one that is ' good enough' for the purposes at hand. The immediate purpose of exploratory inquiry is to discover potentially important causal connections using statistical tests. Now in any particular case the suspected important differences either exists or not. On the above analysis, a test is designed so that the probabilities of deciding there is a difference when there is and not when there is not are 'good enough' for the ultimate purpose of the inquiry, namely, the construction of a comprehensive theory. Moreover, if the strategy is one of satisficing rather than optimizing, there is less need for precision in determining the relevant utilities – something that is anyway unattainable. We don't seek the best test, only one that is 'good enough'. Thus informed professional judgment may be sufficient.[22]

Finally, the emphasis on experimental design in the above reconstruction of objectivist statistical methods has an important general consequence for the concept of evidence sometimes employed by both statisticians and philosophers. It is natural to think that evidence comes in all degrees, with large samples generally providing better evidence. This idea certainly holds for a Bayesian analysis since one can conditionalize on any amount of data whatsoever. But if the goal of an experiment is always the acceptance or rejection of a hypothesis, then unless there is enough data to insure the appropriate precision and reliability for the inquiry at hand, the result is neither acceptance nor rejection. No relevant test has been performed. Thus insufficient evidence is not much better than no evidence at all. The idea that there must always be an evidential relation between any hypotheses and any amount of data must be rejected as a rationalist myth.

4. STATISTICAL INFERENCE IN CONFIRMATORY (THEORY-TESTING) INQUIRY

In the actual history of any scientific investigation there may or may not be a definite time when everyone recognizes that now a coherent, comprehensive theory has been formulated and that the inquiry has shifted from theory searching to theory testing. For a long period the processes of creating a theory and testing it may seem to be simultaneous. Yet if one is to have a realistic ontology and an objective epistemology, such a distinction must be made. The central question for the scientific realist is: Does the theory give a true (or approximately true) description of essential aspects of the systems in question? And the central question for the objective epistemologist is: Does the method by which the theory is to be tested against data have sufficiently low probabilities for rejecting a theory close to the truth or accepting one that is quite far off. The trouble with the exploratory process from the standpoint of theory testing is that we have almost no idea what these error probabilities are. How likely is it that a theory designed by someone to explain a number of exploratory results should be far from the truth? Perhaps if we knew enough about the exploratory process, including the capacities of human investigators, we could answer such questions. But we don't, so we can't.[23] We have no choice but to examine the theory-testing context in its own right.

The systems we seek to capture with our theories are complex and are connected only indirectly with things we can observe, count or measure. This is true not only of micro systems, but of astronomical systems, genetic systems, learning systems, etc. We interact with these systems only through intermediate systems. Schematically it is convenient to think of any test of a theory as involving three systems: the primary system (the one we are investigating), an intermediate system (one we assume to be well-understood), and a measuring system (also assumed well-understood). Now everyone agrees that a central step in testing a theory is deducing from the theory, together with what might be a vast network of intermediate theory, something to be observed, i.e., measured, or counted. What is not usually appreciated, however, is that in most real-life scientific cases what is deduced is a *statistical hypothesis*. This is obvious when the theory is itself probabilistic, like quantum mechanics or genetics, but it also holds for paradigms of deterministic theories like classical celestial mechanics or general relativity. The reason for this is simple. At their limits of accuracy all instruments produce a distribution of readings under conditions as near alike as we can make them. At this point we can lay aside the ontological issue whether our total system is really probabilistic or deterministic but with unknown and therefore uncontrolled variables. Methodologically our problem is the same. We face a distribution of readings and must decide whether or not to accept the predicted statistical hypothesis. And in addition we must decide what implications accepting or rejecting the predicted statistical hypothesis has for our acceptance or rejection of the *theory*. It turns out that these two questions are intimately connected, and the connection provides the basis for 'objective' determinations of the parameters of classical statistical tests. Here the emphasis on hypothesis testing, as opposed to estimation, is deliberate. A theory, together with theories of intermediate and measuring systems, predicts a statistical hypothesis, which in the best cases is a simple hypothesis, but which may also be composite. The question to be answered is: Is the prediction correct? This forms a testing context. For convenience we will assume the test is of a simple hypothesis versus a composite alternative, e.g., $\mu = 0$ vs. $\mu \neq 0$, with σ known.

Half a century ago the American pragmatist C. S. Peirce (1934–58) suggested that testing a theory is a matter of sampling its consequences to determine what proportion are true. A true theory, of course, has only true consequences. This is the kind of approach I wish to develop, though not in a sampling framework but in the framework of probabilistic systems. Consider the process of taking a clearly formulated theory and deducing from it (with appropriate auxiliary theories) a statistical hypothesis. Methodologically this process may be viewed as the operation of a binomial system which produces either true or false statistical hypotheses. We may, in fact, regard this process as having an unknown binomial propensity for producing true statistical hypotheses. If the theory (and the auxiliaries) are true, the propensity is a necessity, i.e., has value 1.

The trouble is, of course, that we cannot directly observe whether or not a predicted statistical hypothesis is true. What we have is a finite set of data which we must use first to judge the statistical hypothesis and then the theory. Let us begin by considering a test of a theory that consists of *one* predicted statistical consequence. The only reasonable decision rule is that the theory is to be rejected (or modified) if the consequence is rejected. If the predicted statistical consequence is accepted, the theory is provisionally accepted and subjected to further tests. The error probability for mistaken *rejection* of the *theory* is the same as that for the statistical hypothesis, namely, the significance level of the statistical test – assuming no gross mistakes in auxiliary assumptions. Determining error probabilities for mistaken *acceptance* is more difficult.

In exploratory inquiry there is no need to seek arbitrarily high precision in a test because there is no reason to bother with small unimportant differences. In confirmatory inquiry, not only is there no need for arbitrarily high precision, one doesn't want it. This is because the intermediate and measuring systems will introduce some uncontrolled fluctuations and systematic biases into all measurements. Surely we want to avoid rejecting the prediction, and thus the theory, simply because of noise or bias we have introduced in the testing process. Understanding the intermediate and measuring processes means that we should know how much they are likely to distort the 'true' values. More precisely, we should know a region of variation around the predicted simple hypothesis such that the probability is, say, 10% or less, that a greater variation could be noise or bias from the non-primary systems.[24] In this region we want to keep the power of our statistical test low, i.e., close to the significance level. Conversely, we want our statistical test to have high power against any divergences from the predicted value that have a low probability of coming solely from the secondary systems. A power of roughly 90% for such clearly divergent alternatives would seem a good rule of thumb.

Now the probability of rejecting a false theory depends not only on the power of the tests of its statistical predictions, but also so to speak, on how false the theory is. The best way I know of dealing with this difficult, and philosophically suspect, notion is to think in terms of the propensity of the testing process to yield a clearly false statistical prediction, i.e., one that produces a difference with a low probability (10%?) of having been introduced by the non-primary systems. Suppose, for example, that the theory is such that the chance of our deducing a clearly false prediction is 50%. Then if the power of the corresponding statistical test against clearly divergent alternatives is at least 90%, our chance of detecting the falsity of the theory could be as low as 45%, which seems hardly satisfactory. But we are considering only *one* prediction. What happens if we consider several, say five? Here one begins to see clearly the range of constraints imposed upon statistical tests in the theory testing context.

Since there is always some probability of mistakenly rejecting a true statistical prediction (Type I error), it would be possible not to reject the theory even though a prediction is rejected.[25] But the most powerful tests of the theory are those for which the decision rule is to reject the theory if *any* prediction is rejected. With five independent statistical predictions it follows that the significance level of the individual statistical tests must be no greater than 0.01. In this case the probability of mistakenly rejecting a true theory is only about 5%. This seems an acceptable risk. If the significance level of the individual statistical tests were as great as 0.05, the chance of at least one mistaken rejection would jump to almost 23%. Clearly the amount of effort, resources, etc., that goes into constructing a plausible theory is too great to allow nearly one chance in four of mistakenly rejecting the theory. As for mistaken acceptance of the theory, if the power of our statistical tests against clearly divergent alternatives is about 90% and if the propensity of the whole testing process to generate clearly false predictions is 50%, then our chances of rejecting a false theory would be at least 95%. This seems sufficiently rigorous. This probability drops, of course, if the propensity for generating false predications is lower.

This is a good place to make clear that the probability of producing a false consequence from a false theory depends not only on differences between reality and what the theory says about reality. It depends also on the procedures we employ in selecting consequences to be tested. To take an extreme, but instructive, case, suppose one deduced for testing only statistical hypotheses that had already been accepted as true in prior exploratory inquiry. Most scientists, many statisticians, and some philosophers would reject this procedure. But the reasons given for this rejection are varied and often obscure. Here the reason is clear. Theories are *designed* to account for hypotheses accepted in exploratory inquiry. Thus even if the theory is far from being generally true, the probability that this will be detected by testing such consequences is very low. Indeed this will happen only if there was a serious Type II error in the exploratory inquiry. To obtain a sufficiently powerful test of the theory we should test new consequences, ones not previously known to be true, or at least ones not considered in designing of the theory. Indeed, a good methodological rule is to test consequences that experienced investigators would judge likely to be false. Using such judgments, which need not be interpreted as expressing subjective degrees of belief, may increase the power of the test of the theory, though by how much will be unknown.

We have seen that significance levels of 0.05, 0.10, and even 0.20 might be appropriate in exploratory inquiry. In confirmatory inquiry, on the other hand, more stringent levels are needed. And this is not a matter of mere convention or subjective beliefs, but follows from the nature and aims of theory testing. Nevertheless there is still a need for further professional judgments. Resources for theory testing are never unlimited, and a balance may have to be found between the number of predictions tested and the reliability and precision of individual tests. The guiding principles of such judgments can only be the general aim of confirmatory inquiry – to reject clearly false theories and to accept ones that are true (or approximately true). To 'accept (or reject) a statistical hypothesis' in a confirmatory context means to take it to be true (or false) for the purpose of judging the truth or falsity of the theory under consideration. 'Accepting a theory' means taking it to be true for the purposes of explaining diverse phenomena in the domain and for guiding further inquiry. And in so far as the acceptance of a theory is viewed as a decision, the overall decision strategy might again be regarded as one of something like 'satisficing'. The theory is either far from true or approximately true. If it is far from true, we have a reasonably high (90% or so) probability of detecting this. If it is approximately true there is again a high probability that we will accept it as such. This is good enough for science.

In considering the example of testing a theory with five independent predictions, I do not mean to imply that scientists in any sense ever pick a fixed sample of predictions. Rather one looks hard to find any good, clear consequences for which the relevant auxiliary theories are well established. But the operating characteristics of the five-prediction binomial test described above are reasonably good. And as a matter of fact it seems to be the case that it only takes a few good successful predictions to establish a theory as the working paradigm for a scientific field. After that one is not so much testing the theory as using it to probe new phenomena in the same domain. What are then being tested are elaborations and new auxiliary theories. By and large the factors determining the parameters of statistical tests in this phase of inquiry, which corresponds roughly to Kuhn's (1962) conception of 'normal science', would seem to be very similar to those in confirmatory inquiry. But this question needs more study. Indeed investigation of the parallels between the above account of scientific inquiry and recent accounts by philosopher-methodologist-historians like Kuhn, Feyerabend, and Lakatos should prove quite fruitful.[26] This would be especially true if the investigation could include some actual case histories of exploratory inquiries that became confirmatory and then 'normal'.[27] That would provide some empirical flesh for the above logical skeleton.

5. Scientific inquiry and practical decision making

As noted much earlier, one of the main attractions of The Bayesian Way is that it provides a unified, global approach to inference and decision making. The objectivist approach outlined above rules out any such facile unification. On this account the same hypothesis might be acceptable as true in an exploratory context but not acceptable as true for the purpose of theory testing. In neither case would there be any implication that the hypothesis might or might not be acceptable as true in some specified practical decision context. The objective factors that determine the appropriate test characteristics in scientific contexts have no direct connection with the utilities relevant to any particular practical decision.

How then does one make practical decisions? And what is the relation between scientific knowledge and practical action? I cannot say much about these questions here, but I feel something must be said both because the Bayesian view is so neat and because the issue is so important.

Let us begin with the classical decision making framework consisting of a set of actions, a set of states, and a utility function defined over all possible outcomes (action-state pairs). A deeper treatment would include an analysis of how the matrix came to be generated, i.e., why these particular actions and states are being considered. Now the classical view treats three cases generated by our possible knowledge situations: certainty, uncertainty, and risk. Here I agree with the Bayesians that the important case is decision making with risk. Not, however, because there always exists a probability distribution, but because the other two cases are vacuous or unresolvable. No future state is ever known with certainty, though the risk might in some cases be so small as to make this schema a useful approximation. Similarly, we are rarely if ever in a state of complete uncertainty, and even if one considers this possibility, there seems to be no general solution forthcoming.[28] The only sure thing is the sure thing principle. When this fails to apply there seem to be no convincing arguments that one strategy is necessarily more rational than others, e.g., that maximin is better or worse than arbitrarily assigning equal probabilities and using Bayes' Rule. Let us therefore concentrate on decision making with risk.

If one genuinely believes that the world is a probabilistic system, then the probabilities attached to future states of the world will be physical propensities. But of course decisions are made not on the basis of what the probabilities are but on what we take them to be. Suppose for each probability we had a 99% confidence interval. This would enable us to calculate an upper and lower physically expected utility for each state – ignoring the 1% probability of error for each interval. Now if the lower expected utility of one action were clearly greater than the upper expected utility of any other, then the decision would be relatively obvious. But if the confidence levels were low, say only 75%, or the intervals overlapped, or both, then it is not at all clear what the optimal strategy would be. Perhaps this would be a good point to forget about optimizing in favor of 'satisficing'. Is there at least one action whose lower expected utility is 'good enough' for our purposes? If so, take it. If more than one, take any of these. But even this approach seems dubious if the confidence level that gives a just satisfactory lower expected utility is itself very low, say only 75%. What then?

At this stage I suggest that we should be prepared to allow that no 'rational' solution is possible without more evidence – in this case needed to raise the confidence level of the interval estimate. This is a radical suggestion which goes contrary to classical as well as Bayesian decision theory. Decision theorists of both schools have assumed that there should always be a uniquely best action no matter what our state of knowledge. This is the decision theory analog of the assumption – denied above – that some inference is possible on any evidence whatsoever.

Another conclusion suggested by the above sketch of objectivist decision making is that the connection between scientific inquiry and practical decision making – or, more generally, between science and technology – is quite tenuous. We have already seen that there is no basis for taking as true in a practical decision context a hypothesis accepted as true in a scientific context. The most one could do is reanalyze the scientific data in a form applicable to the particular decision problem at hand. But there is no reason why just that data needed for the practical decision ever should have been gathered in a scientific investigation, or in the right amount, with the right controls, etc. In short, practical decision makers need their own research base. The simple idea that technology is merely applied science must be discarded.[29]

The above conclusions are reinforced if one also considers a cost-benefit, or expected cost-benefit, model of practical decision making. Consider a simple toxicity test on human volunteers, where toxicity is measured by the probability, p, of a specific reaction on a given dose. If p is zero, widespread use of the drug may a measurable expected benefit, say in sick days avoided. Similarly, for each value of p greater than zero there would be an expected loss if the drug were marketed. Here one can construct a uniformly most powerful test of the hypotheses, $H: p=0$, vs. K: $p>0$. Using this test, the expected loss function is reduced by the probability of rejecting H if false. Given enough volunteers, the net expected loss may be kept well below the expected gain for all possible values of p. Here all the probabilities involved may be regarded as physical propensities and the test involved is purely objective. It is also immediately obvious that the practical decision process requires an experiment designed expressly for the particular decision at hand. That just the data needed for this test should have been gathered in any scientific inquiry would have been a happy coincidence.

In sum, an objectivist approach to practical decision making is possible. It turns out to be rather untidy, but that should have been expected. Moreover the connections between the fruits of scientific inquiry and practical decision making are tenuous and indirect. This remains true in spite of much overlap in concepts and mathematical techniques.

Dept. of History and Philosophy of Science
Indiana University, Bloomington, Indiana, U.S.A.

NOTES

* This work has been supported in part by a grant from the National Science Foundation.
[1] This is clear from the recent publication of several collections of papers highly critical of statistical methods and practice in sociology and psychology. See, for example, Morrison and Henkel (1970) and Leiberman (1971). For a review of Morrison and Henkel aimed at a philosophical audience see Giere (1972).
[2] Widespread contemporary interest in Bayesian ideas by the Anglo-American statistical community began in 1954 with the publication of Savage's *Foundations of Statistics* (1954). Since Savage's death the mantle of leadership has fallen to Denis Lindley who, significantly, now holds R. A. Fisher's chair at University College, London. Lindley has recently published the best overall review of Bayesian Methods (1971a) and an introductory text on decision making (1971b). I. J. Good (1965) maintains his position as a chief expositor of a pragmatic (non-dogmatic?) Bayesianism.
[3] This, at least, is the impression I get from his papers, especially (1936), and from Donald Gilles' review (1972) of the Italian version of his recent monograph (1970).
[4] Gilles (1972) comes to the same conclusion regarding De Finetti. Even Savage ((1967), p. 604) has written: "Since I see no grounds for any specific belief beyond immediate experience, I see none for believing a universal other than one that is tautological, given what has been observed..."
[5] In a recent article ((1973), p. 126) Brian Ellis, a philosopher with strong Bayesian sympathies, writes: "We have as yet no adequate logic of subjective probability, and with the failure of the Dutch book argument even to apply to most probability claims concerning propositions of real scientific or historical interest, no such system is in sight".
[6] Thus Fisher: "Every experiment may be said to exist only in order to give the facts a chance of disproving the null hypothesis." ((1935), fourth ed., p. 16)
[7] Neyman first used the term 'inductive behavior' in several papers, e.g., (1938), published around 1938. A later, extended discussion of the concept occurs in (1957).
[8] Allan Birnbaum (1962), (1969) has given the best theoretical discussions of relations between error probabilities, likelihood and related concepts. One need not share his pessimism regarding the prospects of developing a unified conception of statistical inference in order to appreciate the beauty of his analysis.

[9] Godambe (1969) apparently holds that the conceptual basis of all fundamental concepts of probability and statistical inference is to be found in sampling from *finite* populations. He identifies the view that physical probabilities are characteristics of chance mechanisms with the concept of sampling from infinite hypothetical populations, a concept much used by Fisher and others.

[10] This is Hacking's (1965) conclusion.

[11] In practice such a strong conception of random sampling is unnecessary. It is usually sufficient that the sampling mechanism exhibit no bias for or against individuals with characteristics relevant to the hypothesis in question.

[12] I have given a more extensive account of objective single-case propensities in Giere (1973a).

[13] This principle is not intended to rule out tests using a decision rule which randomizes for one point of a finite sample space. In Birnbaum's (1969) terminology it restricts the 'confidence concept' by adding a 'conditionality concept' but does not incorporate a 'sufficiency concept'. This latter omission is essential since sufficiency together with conditionality implies a full likelihood principle – which is incompatible with the confidence concept (error probabilities).

[14] I should note that my analysis implies a rejection of Herbert Robbins's (1963) 'compounding approach' according to which the operating characteristics of tests may be improved by relating tests in *different fields*! Tests of hypotheses concerning unrelated physical systems must be unrelated. Robbins's idea, by the way, has both been hailed as a great 'breakthrough' in statistical theory (by Neyman (1962)) and rejected as a *reductio ad absurdum* of the decision theoretic approach to statistical inference (by Birnbaum (1969)).

[15] Among philosophers, Isaac Levi (1967) comes closest to developing a similar view. Levi, however, assumes a very restricted framework in which one has both a probability and content measure for all relevant hypotheses. The connection between such a framework and real scientific inquiry is remote.

[16] Unfortunately, use of the terms 'exploratory' and 'confirmatory' is not standardized. My usage is determined by the fact that it is *theories*, not merely single statistical or causal hypotheses, for which we explore and which we hope to confirm. When someone like Tukey (1961), for example, speaks of 'exploratory data analysis', he is much more concerned with descriptive statistics, with displaying data in a form that may reveal interesting relationships. For me this is certainly part of exploratory inquiry, but so is hypothesis testing, which I think Tukey and others would classify as 'confirmatory'.

[17] See the essays by Bakan, Rozeboom and others in Morrison and Henkel (1970), for examples of this recommendation.

[18] This would be Birnbaum's recommendation.

[19] Thus concerning the relative importance of the two kinds of error in a test of two simple hypotheses, Neyman writes: "This subjective element lies outside the theory of statistics". ((1950), p. 263).

[20] This is a recurring theme in Savage's less formal writings, e.g., (1962).

[21] Several philosophers, e.g., Carl Hempel (1962) and Isaac Levi (1967), have explicitly discussed scientific inference in terms of 'epistemic utilities'.

[22] The concept of 'satisficing' was originally introduced by Herbert Simon. See pp. 196–206 and Chapters 14 and 15 of (1957). Unfortunately the idea does not seem to have had much impact outside of the theory of organizations. I know of one philosophical critic (Michalos (1973)) and no references in the 'standard' statistical literature. I

hope in the future to explore more fully the possibilities of using this notion in understanding the rational basis of inference in scientific inquiry.

[23] The common philosophical idea that the needed probabilities can be found by examining the formal structure of the theory and the data could only make sense if one embraced a concept of logical probabilities that would have no connection with real (i.e., physical) errors probabilities. No scientist or statistician would even seriously consider such an idea, and I think rightly so.

[24] Here it becomes very clear that to test one theory one needs already to have accepted some *other* theories, i.e., the theories of the auxiliary and measuring systems. This opens up the possibility of an epistemological regress and thus raises the whole philosophical problem of the 'ultimate basis' of knowledge. I have dealt at length with this problem, within the framework of hypothesis testing, in Giere (1975).

[25] Traditional philosophical discussions of the 'hypothetico-deductive method' allow for this possibility only by allowing that some auxiliary assumption may be false. But when the predictions from theory are recognized as being statistical hypotheses, it is possible that a prediction be rejected as false even though the theory and all auxiliary hypotheses are correct.

[26] During the past fifteen years a whole school of historically oriented philosophy of science has developed. Recent papers by Feyerabend (1970) and Lakatos (1970) will provide some entry into this literature. For a partial critique of the historical school, especially on the issue of choosing among theories, see my review essay, Giere (1973b).

[27] Among statisticians, Allan Birnbaum (1971) has been the most forceful in urging cooperation among statisticians, scientists, philosophers and historians of science. He sees case histories, of which his (1974) is an example, as providing a vehicle for communication among these all too isolated groups.

[28] Chapter 13 of Luce and Raiffa (1957) still seems the best chronicle of the failure of this whole program.

[29] Of course I do not deny that developments in 'pure' science may have great impact on practical technologies. But one should not forget that in between there is usually a great deal of applied 'research and development' that is necessary to produce the practical applications.

REFERENCES

Basu, D.: 1964, 'Recovery of Ancillary Information', *Sankhya* **26**, 3–16.

Birnbaum, Allan: 1962, 'On the Foundations of Statistical Inference', *J. Am. Statist. Assoc.* **57**, 269–306.

Birnbaum, Allan: 1969, 'Concepts of Statistical Evidence', in *Philosophy, Science and Method*, (S. Morgenbesser, P. Suppes, M. White, eds.), St. Martins, New York.

Birnbaum, Allen: 1971, 'A Perspective for Strengthening Scholarship in Statistics', *The Am. Statist.* 1–2.

Birnbaum, Allen: 1974, 'The Statistical Phenotype Concept, with Applications to the Skeletal Variation in the Mouse', *Genetics*, forthcoming.

Carnap, Rudolf: 1963, 'Replies and Systematic Expositions', in *The Philosophy of Rudolf Carnap*, (P. A. Schilpp, ed.), Open Court, La Salle.

Church, Alonzo: 1940, 'On the Concept of a Random Sequence', *Bull. Am. Math. Soc.* **46**, 130.

Cox, D. R.: 1958, 'Some Problems Connected with Statistical Inference', *Ann. Math. Stat.* **29**, 357–372.

de Finetti, Bruno: 1937, 'Foresight: Its Logical Laws, Its Subjective Sources', in *Studies in Subjective Probability*, (H. E. Kyburg Jr. and H. E. Smokler, eds.) Wiley, New York.

de Finetti, Bruno: 1970, *Theoria Della Probabilita*, Turin.

Edwards, A. W. F.: 1972, *Likelihood*, University Press, Cambridge.

Ellis, Brian: 1973, 'The Logic of Subjective Probability', *Brit. J. Philos. Sci.* 24, 125–152.

Feyerabend, P. K.: 1970, 'Against Method: Outline of an Anarchistic Theory of Knowledge', in *Analyses of Theories and Methods of Physics and Psychology: Minnesota Studies in the Philosophy of Science*, Vol. IV, (M. Radner and S. Winokur, eds.), University of Minnesota Press, Minneapolis.

Fisher, R. A.: 1935, *The Design of Experiments*, Oliver and Boyd, London, 1st ed.

Giere, R. N.: 1972, 'The Significance Test Controversy', *Brit. J. Philos. Sci.* 23, 170–181.

Giere, R. N.: 1973a, 'Objective Single Case Probabilities and the Foundations of Statistics', in *Logic, Methodology and Philosophy of Science IV, Proceedings of the 1971 International Congress, Bucharest*, (P. Suppes *et al.*, eds.), North Holland, Amsterdam.

Giere, R. N.: 1973b, 'History and Philosophy of Science: Intimate Relationship or Marriage of Convenience?', *Brit. J. Philos. Sci.* 24, 282–297.

Giere, R. N.: 1975, 'The Epistemological Roots of Scientific Knowledge', in *Induction, Probability, and Confirmation Theory: Minnesota Studies in the Philosophy of Science*, Vol. VI, (Grover Maxwell and Robert M. Anderson, eds.), University of Minnesota Press, Minneapolis.

Gilles, Donald: 1972, 'The Subjective Theory of Probability', *Brit. J. Philos. Sci.* 23, 138–157.

Godambe, V. P.: 1969, 'Some Aspects of the Theoretical Developments in Survey Sampling', in *New Developments in Survey Sampling*, (N. L. Johnson and H. Smith Jr., eds.), Wiley, New York.

Good, I. J.: 1965, *The Estimation of Probabilities*, MIT, Cambridge, Massachusetts:

Hacking, Ian: 1965, *Logic of Statistical Inference*, Cambridge University Press, Cambridge.

Hempel, C. G.: 1962, 'Deductive-Nomological vs. Statistical Explanation', in *Minnesota Studies in the Philosophy of Science*, Vol. III, (H. Feigl and G. Maxwell, eds.), University of Minnesota.

Jeffrey, Richard C.: 1956, 'Valuation and Acceptance of Scientific Hypotheses', *Philos. Sci.* 23, 237–246.

Kuhn, Thomas S.: 1962, *The Structure of Scientific Revolutions*, University of Chicago Press, Chicago.

Lakatos, Imre: 1970, 'Falsification and the Methodology of Scientific Research Programmes', in *Criticism and the Growth of Knowledge*, (Imre Lakatos and Alan Musgrave, eds.), Cambridge University Press, Cambridge.

Levi, Isaac: 1967, *Gambling with Truth*, Knopf, New York.

Lieberman, Bernhardt: 1971, *Contemporary Problems in Statistics*, Oxford University Press, New York.

Lindley, D. V.: 1971a, *Bayesian Statistics, A Review*, Society for Industrial and Applied Mathematics, Philadelphia.

Lindley, D. V.: 1971b, *Making Decisions*, Wiley, New York.

Luce, R. D. and Howard Raiffa: 1957, *Games and Decisions*, Wiley, New York.

Michalos, Alex: 1973, 'Rationality Between the Maximizers and the Satisficers', *Policy Sci.* 4, 229–44.

Morrison, Denton E. and Ramon E. Henkel: 1970, *The Significance Test Controversy*, Aldine Publishing Company, Chicago.

Neyman, J.: 1938, 'L'Estimation Statistique Traitée comme un Problème Classique de Probabilité', *Act. Scient. Indust.* **734**, 25–57.

Neyman, J.: 1950, *First Course in Probability and Statistics*, Henry Holt, New York:

Neyman J.: 1957, "Inductive Behavior' as a Basic Concept in Philosophy of Science', *Rev. Intern. Statist. Inst.* **25**, 7–22.

Neyman, J.: 1962, 'Two Breakthroughs in the Theory of Statistical Decision Making', *Rev. Int. Statist. Inst.* **25**, 11–27.

Peirce, C. S.: 1934–58, *Collected Papers of Charles Sanders Peirce*, 8 Vols, Harvard University, Cambridge, Massachusetts:

Popper, Karl R.: 1959, 'The Propensity Interpretation of Probability', *Brit. J. Philos. Sci.* **10**, 25–42.

Reichenbach, Hans: 1949, *The Theory of Probability*, University of California, Berkeley.

Robbins, Herbert: 1963, 'A New Approach to a Classical Statistical Decision Problem', in *Induction: Some Current Issues*, (H. Kyburg, and E. Nagel, eds.), Wesleyan University, Middletown, Connecticut: 101–110.

Salmon, Wesley C.: 1967, *The Foundations of Scientific Inference*, University of Pittsburgh, Pittsburgh.

Savage, Leonard J.: 1954, *The Foundations of Statistics*, Wiley, New York.

Savage, Leonard, J.: 1962, 'Bayesian Statistics', in *Recent Developments in Decision and Information Processes*, (R. E. Machol and P. Gray, eds.), Macmillan, New York.

Savage, Leonard J.: 1967, 'Implications of Personal Probability for Induction', *J. Philos.* **64**, 593–607.

Shimony, Abner: 1970, 'Scientific Inference', in *The Nature and Function of Scientific Theories*, (R. G. Colodny, ed.), University of Pittsburgh, Pittsburgh.

Simon, H. A.: 1957, *Models of Man, Social and Rational*, Wiley, New York.

Tukey, John W.: 1961, 'The Future of Data Analysis', *Ann. Math. Statist.* **33**, 1–67.

Venn, John: 1866, *The Logic of Chance*, London.

DISCUSSION

Commentator: Lindley: There are two points here. From a statistical viewpoint, knowing which machine is involved is equivalent to knowing the value of an ancillary statistic. Unfortunately, there are typically many ancillary statistics, and a problem that statisticians have not succeeded in solving is that of determining which, if any, ancillary to use. (Typically, different ancillaries will give different answers in terms of α and β.) Secondly, Savage and I (a convenient reference is *Bayesian Statistics: A Review* by D. V. Lindley, Philadelphia: SIAM, 1971) have shown that if the same α is chosen for each sample size, then the procedure is incoherent in the sense that you can be made to lose money for sure. If α and β are *both* fixed, then information is being discarded, and the situation is even worse.

Giere: No, your argument would require that I be indifferent between various α, β pairs – and I'm not.

Lindley: You are! I can test you.

Giere: No, because my exploratory context tells me that I want an α value above 0.05 and I want a β value of about 90% – so because of the scientific context I'm not indifferent.

Lindley: But if you fix α and β you get into real trouble, for you are simply throwing away observations.

Godambe: Would you explain again why the frequency interpretation of probability is less favoured than this interpretation?

Giere: It's because according to the frequency interpretation of error probabilities, a strongly counter-intuitive test should be preferable to a more reasonable alternative. Obviously, it is more sensible to wait until you've got to the factory and see which machine is actually operating that day. The propensity interpretation is not bothered by this objection, however, because we can make sense of the notion of a particular machine's propensity to turn out a certain quality of product.

Harper: What are you saying that couldn't be captured by using the different frequencies one appropriate to each of the machines?

Giere: Well, if you can consistently do that. Frequency theorists have attempted to talk about homogeneous references classes etc. without any conspicuous success. If you attempt to relativize to a sequence then you have to start talking about the defining characteristics of the sequence and then you are in trouble.

Kyburg: It seems to me that the two approaches are exactly on a par. Surely you can say that rule A has better propensities than rule *B*, or it has equal propensity for accepting the hypothesis when it's true as does rule *B*, but a lot of propensity for rejecting it when it's false.

Giere: Yes, you could define a propensity for the whole complex process which included the flipping of the coin to choose between the machines which operate etc. But the conceptual advantage of the propensity approach to probability theory is that it allows you to deal with the behaviour of a particular machine within the overall complex situation and ignore irrelevant sequences.

Harper: The particular case then is a random member of the composite process?

Giere: No! That route leads to disaster. Popper undermined his own propensity interpretation by taking exactly that alternative. You've got to talk about a particular case, and to relate propensity to particular physical processes the nature of which are to be specified by scientists, not statisticians.

Bunge: In common with all orthodox statisticians, Giere talks of criteria for rejecting hypotheses on the strength of data. Does not this presuppose that only hypotheses can go wrong, whereas all data are true? And is this not an empiricist prejudice? Surely one often rejects data on the strength of well confirmed theories, not just of statistical rules of thumb. What are the criteria in this case? Is statistics in a position to supply them or do we rather need (and tacitly use) the notion of the degree of truth of one statement relative to another proposition taken as a base line? And if the latter is the case, as I submit it is, would it not be advantageous to have a uniform battery of criteria for accepting or rejecting hypotheses as well as data, and moreover criteria bearing on relative partial truth?

Giere: Reply to Comments (February, 1975)

The printed version of my paper is in fact a revision which was completed shortly after the conference. It therefore already implicitly contains my immediate response to the many discussions, both formal and informal, that took place. Nevertheless, the perspective of nearly two years and the stimulus of seeing some written questions justifies a few additional comments – even on short notice. The written remarks fall into three categories. First, there is Prof. Lindley's statement that using a classical test is incoherent or wastes information, or both. Second there are several questions about the nature of physical propensities and the supposed advantages of this conception over that of simple relative frequencies. Finally there is Prof. Bunge's question about how best to deal with the fact that data are also uncertain. I will comment on each of these in order.

1. *Testing vs. Information Models of Statistical Inference*

The claim that a particular statistical procedure disregards information only makes sense relative to an agreed criterion for what counts as relevant, or usable, information. The orthodox literature contains many demonstrations that one test is more efficient than another as measured, for example, by the power against given alternative hypotheses. Such demonstrations are not controversial because there is an accepted measure of efficiency. But when a Bayesian claims that as a general rule objectivist methods discard information, he is not appealing to an accepted measure of efficiency, but to what would count as relevant information assuming a Bayesian treatment of the problem. Thus objectivists are charged with disregarding the 'information' provided by a prior probability distribution and by the observed data point. Now if there is an objective prior distribution, it will of course be used. But the objectivist statistician simply does not regard subjective opinion as constituting usable scientific information. The status of the observed data point is similar. To a Bayesian, the data point is just another set of information to be processed through Bayes' theorem. Not having prior probabilities, the objectivist cannot make the same use of the data, which is not to say that it is wholly disregarded since it is the actual data point that determines whether the test hypothesis is accepted or rejected. But this is all it does. The characteristics of the test itself are fixed independently.

In sum, the Bayesian charge that objectivist statisticians disregard information is completely question begging. It assumes his own criterion for what counts as usable information. Once this is realized, it is easy to put the shoe on the other foot. For example, the characteristics of an objective test depend on the whole sample space, i.e., on outcomes which might but do not occur. Because Bayes' theorem processes observations through the likelihood function alone, Bayesians necessarily disregard all this information about what other outcomes might have occurred. Shame on them. Clearly nothing can be resolved by such question begging accusations. Yet, in light of the fundamental nature of the dispute, one may seriously wonder whether there are any arguments that do not ultimately beg the question.

What about Lindley's claim that the use of classical objectivist tests is incoherent? Consider a test of a simple hypothesis, H, against the simple alternative, K. The operating characteristics of this test are completely determined by a single point in the α-β plane, where α is the probability of rejecting H when it is true and β the probability of rejecting K when it is true. Now in the argument to which Lindley refers, it is assumed that there is a fixed value of α, α_0 such that any point on the line $\alpha = \alpha_0$ is preferred to any α-β point not on that line. Among points on the α_0 line those with lower values of β are assumed obviously preferable. With these two assumptions we argue as follows (see Figure 1).

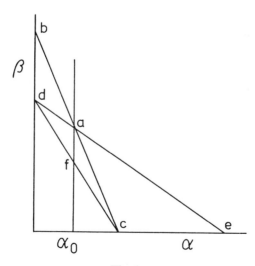

Fig. 1.

Looking first at the line $b\,a\,c$, we see that neither a nor c has a lower value for both α and β. Thus neither dominates the other, so both are admissible. But by the first assumption, test a is preferred to test c because it is on the line $\alpha=\alpha_0$. Looking at line $d\,a\,e$ we conclude similarly that a is preferred to d. Thus (in an obvious notation), $a>c$ and $a>d$. Now looking at c and d either $d>c$, or $c>d$, or $c\sim d$. Now if $d>c$, then $d>f$ since f is a mixture of d and c, and if $d>c$, d is preferred to any mixture of d and c. Similarly, if $c>d$, then $c>f$. Finally, if $c\sim d$, then $c\sim f\sim d$. Now since $a>c$ and $a>d$, it follows that $a>f$. But this contradicts the assumption that if α is fixed, the test with lower β is preferable, since this implies $f>a$. So the given assumptions do indeed generate incoherent preferences. What is one to reply?

My main response is that Lindley's assumption that tests with $\alpha=\alpha_0$ are preferred to all others does not adequately capture the proper use of statistical tests in scientific contexts. Of course mathematical statisticians prove optimality theorems by assuming a fixed α and a fixed sample size. And many experimentalists (too many, unfortunately) seem to use tests the same way – not bothering to compute the power until later, if ever. But this view of testing abstracts too much from the scientific (as opposed to mathematical) context. In any scientific inquiry, whether exploratory, confirmatory, or whatever, there will be statistical hypotheses which we must decide (provisionally, of course) either to reject as clearly false or to accept as approximately true. The point of a test is to help us make this decision. A well designed test is one that has appropriately high objective probabilities for accepting the hypothesis if it is approximately true and rejecting it if it is not. What probabilities are appropriate depends on the kind of context. But in any case there would be only a small region of the α-β plane that would include appropriate tests. Any differences within this region would be scientifically irrelevant.

Yet there is an obvious reply to all this apparently vague talk about appropriate error probabilities in various scientific contexts. Bring the parameters of the context explicitly into the decision problem. Then we will be back where we started with a space full of admissible decision strategies and no satisfactory way to choose among them. The minimax principle, for example, can easily be shown to lead to incoherent preferences by an argument analogous to that given above. Moreover by a judiciously chosen selection of pure and mixed strategies, the Bayesian will show that in using any admissible strategy we are behaving *as if* we had a particular set of prior probabilities. From which he will conclude (fallaciously, though in good operationist fashion) that we *do* have those prior beliefs and thus are really closet Bayesians after all.

Now in the first place, it is far from obvious that we could ever determine the parameters and utilities of the scientific context necessary to set up the grand decision problem the Bayesian envisions. But, more importantly, it is not clear that we have to. This is why I brought in the concept of satisficing. Even though we cannot determine what are the *optimal* error probabilities for testing a particular hypothesis, we can, it seems, decide on a test that is 'good enough' for the purposes of the scientific inquiry at hand. The weakness of this appeal is that it presupposes an objectivist conception of the purposes of scientific inquiry, thus begging important questions against Bayesians, who have a quite different idea of what science is all about. So once again we come up against the problem of finding non-question-begging ways of resolving the fundamental dispute.

2. *Propensities vs. Relative Frequencies*

Arguments to show the incoherence of objectivist strategies or the existence of unsuspected prior opinions all make use of *mixed* tests. Now there is clearly nothing underhanded in the Bayesians' appeal to mixed tests since these were first introduced by objectivist statisticians themselves. But if one could find a good objection to mixed tests, that would have important consequences for the debate. A propensity interpretation provides the basis for such an objection. It also has implications for general problems concerning ancillary statistics. And it might even provide an answer to the objection that objectivist methods do not deal with particular inferences since error probabilities have been thought to refer only to relative frequencies in sequences of tests.

The basic idea behind any propensity interpretation is that empirical probabilities are not to be identified with the characteristics of any sequence but with a particular chance set up (random experiment, stochastic process, etc.) Statistical hypotheses are then seen to be hypotheses about the propensity distribution over possible outcomes of a particular chance process. This conceptual shift may seem slight, but it is sufficient to rule out certain mixed tests and thus resolve some questions about ancillary statistics. Consider the example of Cox's two machines. If our concern is with testing hypotheses about particular processes, then the fact that the *sequence* generated by the 50-50 output of both machines has better error frequencies than those generated by the first machine taken separately is simply irrelevant so long as we are dealing with the first machine – and know it. The objectivist tradition, unfortunately, stands guilty of saying that science is concerned merely with its ratio of correct acceptances and rejections, regardless of subject matter. In fact we want to be right about particular systems or kinds of systems. Of course, as Kyburg suggests, one can consider the propensities of the whole complex process including both machines and the randomizing device. There is, however, little scientific reason to do so unless one were for some reason forced to do a test without knowing which machine was running.

As for the nature of propensities themselves, there are two main conceptions, one absolute, one relativized. In the paper and elsewhere I have argued for the absolute conception according to which propensities are physical tendencies existing in causally indeterministic processes. On this interpretation, propensities are conceived as operative on individual *trials* of indeterministic processes. Thus the objection that objectivists cannot assign error probabilities to individual inferences is overcome.

But absolute, single case propensities are apt to be regarded as mysterious, even occult. The only plausible known examples of ontologically (as opposed to merely epistemologically) indeterministic phenomena are in the quantum domain, and even this case is controversial. Moreover, although it is possible that probabilities in the macro-domains of biology, psychology and sociology are generated by propensities in underlying micro-processes, this is currently a shaky support for the attribution of propensities to macro-processes.

The force of such objections, raised repeatedly by Prof. Harper and others, has led me to take a more generous view of relativized propensities. Consider a macro-system, S, described by a set of variables, Q. Part of a propensity distribution for S would be given by the claim that relative to variables, Q, S tends to produce outcome O with frequency r. This claim may be true whether S is really deterministic or not. What we know is that the variables Q are not sufficient to determine a unique outcome, and whatever unknown variables are operating, they produce outcome O with expected frequency r. Even such a relativized conception

of propensities is sufficient to justify our ignoring the frequency characteristics of various mixed tests when we know which of the chance processes is actually operating and know its (relativized) propensity distribution. The main disadvantage of a relativized conception of propensities is that it leaves unresolved the meaningfulness of attributing propensities to individual trials. But perhaps this objection can be overcome by
some pragmatic single case rule.

3. *Uncertain Data*

With the notable exception of several philosophers such as Richard
Jeffrey and Isaac Levi, most writers on the foundations of statistics,
myself included, have taken data as given. Prof. Bunge objects that this is
unrealistic and that any comprehensive account of scientific inference
would also include a means of evaluating data in light of hypotheses. I
agree completely. But I disagree that this should be done by introducing
a notion of degree of truth that would cover both singular data statements
and statistical hypotheses. I would hope that one could develop criteria
for the acceptance and rejection of data statements – criteria that would
take account of hypotheses already accepted. This would provide the
methodological unity Prof. Bunge desires, though on a different basis.
But this program has yet to be carried out. In the mean time, there are
many pressing problems even assuming the data as given. Perhaps this is
an empiricist prejudice, but I do not think the solution to these latter
problems will be much affected by any criteria for dealing with uncertain
data.

PHILLIPS MEMORIAL
LIBRARY
PROVIDENCE COLLEGE

V. P. GODAMBE AND M. E. THOMPSON

PHILOSOPHY OF SURVEY-SAMPLING PRACTICE

The 'induction' which philosophers have dealt with over the centuries, particularly since Hume, is illustrated popularly by the assertion, "Since the sun has risen every day over several thousand days (assuming the historical record), it will rise tomorrow". To find a general rationale behind this assertion is considered by philosophers to be the 'problem of induction'. Now philosophers of probability such as Carnap and Reichenbach have attempted to relate this 'problem of induction' with 'statistical inference'. However, in so doing they completely overlooked a central feature of 'statistical inference', namely artificial randomisation, or randomisation introduced by some suitable artifice such as random number tables and the like. Thus, in contrast to the philosopher asking, "What is the probability of the sun rising tomorrow?" the statistician was raising the problem: "Suppose balls are drawn at random (with suitable artifice of randomisation) from a bag containing one thousand black and white balls in unknown proportion; what is the probability of the nth ball being white, given that the previous $n-1$ random draws have resulted in white balls?" (Kempthorne (1971) has made a somewhat similar point). This latter problem is obviously scientifically realistic while the former one is not. This scientifically realistic problem of 'statistical inference' is generally not understood by philosophers of probability because of their rigid adherence to the framework provided by the 'problem of induction' referred to above. (Although this last remark is not quite true of hypothetico-deductive philosophers like Popper, their failure to come to grips with the scientific problem of statistical inference is equally appalling.)

Now it looks as though from the beginning statistical inference basically was supposed to have been applicable to the exclusive situation when a random sample (that is, one drawn with appropriate randomisation) from a *well specified* population was available. The fundamental necessity of this 'specification' was emphasized by Fisher in his writings. However, it seems that the results of the general statistical theory founded and developed by Fisher, Neyman and others are applicable essentially to

Harper and Hooker (eds.), Foundations of Probability Theory, Statistical Inference, and Statistical Theories of Science, Vol. II, 103–123. All Rights Reserved. Copyright © 1976 by D. Reidel Publishing Company, Dordrecht-Holland.

situations where this 'specification' is still of a hypothetical population, for instance, that of infinitely many tosses of a given coin. Curiously enough, these results are either inapplicable or meaningless when the specification is ideally achieved, namely when the population is an actual (finite) aggregate of individuals, and the value of the variate under study is unambiguously defined for each individual. For instance the population may consist of households in a given city, with the variate under study being household income over a specified time period. It is well known that the optimum properties of the sample mean established in relation to hypothetical populations are no longer valid for actual populations (Godambe, 1955; Godambe and Joshi, 1965). The realization of this has led to the recent new developments in survey-sampling (Johnson and Smith, 1969).

Thus philosophers in their overzealousness to find a rationale for 'induction' completely ignored or misunderstood artificial randomisation; and the early pioneers of general statistical theory, in spite of their valuable contribution to the concept of randomisation, mostly confused the issues underlying artificial randomisation by their failure to see clearly that artificial randomisation presupposes an actual (finite) population of labelled individuals, in contrast to hypothetical populations (Godambe, 1969). Such actual populations of labelled individuals are primarily dealt with in survey-sampling. As in the example above, the population might be of housholds in a city which are listed in some county office. One may be interested in estimating the average income in the city, and one may do this by drawing a random sample from the list of households and actually collecting the information about income of the households included in the sample. Not surprisingly, practitioners of survey-sampling have made the most important practical contributions to the concept of randomisation. The concept has been enriched far beyond the naive and unharnessable 'randomness' of philosophers and many statisticians ignorant of survey-sampling. In particular, the enrichment incorporated in the idea of unequal probability sampling designs has had great intuitive appeal for most practitioners, although its logic and theoretical significance have often remained obscure.

A theoretical argument that essentially goes against all artificial randomisation can be based on the likelihood principle (Barnard *et al.*, 1962) or the conditionality principle (Cox, 1959; Birnbaum, 1962, 1969). A direct upshot of these principles is that after the sample is drawn, the inference must be independent of the mode of randomisation that has been used to draw the sample (Godambe, 1966). Since Bayes theorem, granting a valid prior distribution, implies among other things the likelihood and conditionality principles, Bayesian statisticians have difficulty interpreting randomisation strictly within the Bayesian framework. Ericson (1969) has proposed a justification for 'randomisation' as a means to validate his assumption of an exchangeable prior distribution for a survey population. However, even this justification, it seems, can at best deal with situations up to stratified random sampling, leaving outside its scope most of the richer and more sophisticated modes of randomisation provided by unequal probability sampling.

The conditionality and likelihood principles can obviously provide no meaningful interpretation for the 'frequency distribution generated by randomisation' (Godambe, 1969a; Godambe and Thompson, 1971). Yet the use of such frequency distributions, in appropriate cases, has immediate intuitive appeal for many practitioners. Indeed, this frequency distribution generated by randomisation was the central object of study among early investigators of randomisation in the thirties, notably Fisher (1935, 1936), Neyman (1935), Eden and Yates (1933), Welch (1937) and Pitman (1937). It is very unfortunate that these pre-war studies of randomisation, which were mainly in relation to design of experiments, did not reach their logical conclusions. (This we can say even after acknowledging several subsequent contributions, notably those of Kempthorne (1955), Box and Andersen (1955).) These conclusions would have played a crucial role when, in the postwar period, the richer modes of randomisation referred to above were being employed by survey practitioners. Possibly the confusion prevailing in the literature on 'sampling theory' in the forties, fifties and even later could have been avoided.

The interpretation of randomisation given subsequently is essentially along the lines of argument set forth in Godambe (1969) and Godambe and Thompson (1971). It is inseparably tied to the 'frequency distribution generated by randomisation' referred to above. Briefly, if ξ is 'assumed' as a prior distribution, then given the data, Bayes theorem provides the necessary inference. Up to this, no randomisation is relevant. But consider the possibility that the assumption that 'ξ is the prior distribution' is effectively wrong. This possibility, if the prior distribution is in some sense objective, clearly exists. Even a subjectivist could think of this possibility (though with some difficulty) as a possible mistake in calculations. Now only when the validity of the 'assumed' prior is in question does the frequency distribution generated by randomisation become meaningful. This frequency distribution, for an appropriate mode of randomisation, can to some extent provide validation of Bayes inference based on a wrong prior distribution. This of course is too sharp a description (adopted for convenience of communication); and in real situations there is considerable interaction of priors and sampling frequencies, as will be clear from later discussion. Nevertheless, acceptance or rejection of this interpretation would depend on acceptance or rejection of the following *philosophy* or *philosophical view* on the part of the survey practitioner:

> Be bold enough to make intelligent 'assumptions' based on prior knowledge concerning the actual (finite) population under study. (These 'assumptions' may be made in terms of prior distributions or otherwise.) Draw 'inferences' on the basis of these 'assumptions' and the 'sample data'. But be wise enough to draw the sample with an appropriate mode of randomisation, so that in case the 'assumptions' go wrong, the 'inferences' are protected, to the extent possible, by way of their 'frequency interpretation' (obtained through the frequency distribution generated by the randomisation).

This also seems to be the underlying spirit in the early writings of Fisher and others mentioned previously. *It is important to note that the 'frequency' referred to in the above paragraph becomes the 'statistical probability (to be used for inferences) in the drawing of a single sample' only when the underlying 'assumptions' are no longer available.* This requirement is analogous to the well known condition of 'non-existence of relevant subsets', laid down by Fisher to define the probability of an event.

The choice of appropriate modes of randomisation in sampling to protect probability statements is illustrated by the following simple examples. The finite population in these examples will consist of N units, labelled $1, ..., N$. With unit i will be associated a variate value x_i. The N-dimensional vector $\mathbf{x} = (x_1, ..., x_N)$ is called the population vector; and a prior distribution ξ is a probability distribution for $\mathbf{x}$ from which the statistician assumes the population vector to be sampled to have come.

EXAMPLE 1: ('Quality control'). Suppose it is *given* that x consists of M zeros and $N-M$ ones in *some order*, and the prior distribution ξ is such that *every order* has the same probability. Under ξ, whatever way a unit is drawn, the prior probability of obtaining a zero is M/N. However, if a unit is drawn at random (with a suitable artifice of randomisation) from $\{1, ..., N\}$, so that each unit is equally likely to be drawn, the probability of obtaining a zero is M/N regardless of the order of x. This later probability statement, based on the frequency distribution generated by artifical randomisation, is numerically valid under the assumption of ξ or without the assumption; and *without the assumption* this probability statement has obvious inferential meaning, which the earlier probability statement has not.

EXAMPLE 2: Suppose that $N=N_1+N_2$, and it is *given* that the first N_1 coordinates of x contain M_1 zeros, while the last N_2 contain M_2 zeros. Let the prior distribution ξ be such that every x consisting of zeros and ones and satisfying these conditions has the same probability. Under ξ, whatever way a unit is drawn, the prior probability of obtaining a zero is M_1/N_1, given that the unit drawn belongs to the first N_1 units, and the probability is M_2/N_2, given that it belongs to the last N_2. This time if a unit is drawn at random from $\{1, ..., N\}$, M_1/N_1 is also the conditional (frequency) probability of a zero, given that the unit is one of $\{1, ..., N_1\}$. Thus without the assumption of ξ, given randomisation and that the unit is drawn from $\{1, ..., N_1\}$, the prior probability M_1/N_1 is again supported in a conditional frequency sense.

EXAMPLE 3: (Unequal probability sampling). Suppose $N=3$ and it is *given* that x, is either $(0, 1, 1)$ or $(1, 0, 0)$. Let the prior distribution ξ be such that x is $(0, 1, 1)$ or $(1, 0, 0)$, each with probability $\frac{1}{2}$. Here whatever way a unit is drawn, the prior probability of obtaining a zero is $\frac{1}{2}$. But if prior probabilities, (ξ), were not available for the two possible values of x, namely $(0, 1, 1)$ and $(1, 0, 0)$, a probability of $\frac{1}{2}$ in the frequency sense could be obtained from the (unequal probability) sampling design which draws units 1, 2, 3 with probabilities $\frac{1}{2}, \frac{1}{4}, \frac{1}{4}$ respectively.

Let us turn now to closer consideration of the statistical problem of estimation. Many situations in survey-sampling, like the household income example mentioned above, involve estimation of the 'population total' $T=\sum_1^N x_i$, or equivalently the 'population mean' T/N. In several important ways, some of which will now be discussed, randomisation can be used to validate estimation procedures based on prior knowledge of the population.

In standard notation, a 'sample' s is a subset of the population $\{1, ..., N\}$, and a 'sampling design' is a probability function $p(s)$ on the set of all possible samples. The 'inclusion probability' π_i of unit i is $\sum_{s \ni i} p(s)$. The 'sample size' $n(s)$ is the number of units belonging to the sample s, and a 'fixed size design of size n' is one for which $p(s)>0$ implies $n(s)=n$. Since the outcome of a sampling experiment consists of the sample drawn and the variate values observed, it may be written as $(s, x_i: i \in s)$. An estimator of T, which is a function of the outcome, will be written as $e=e(s, x_i: i \in s)$. Expectations with respect to a prior distribution ξ and with respect to the probability function p will be denoted respectively by $\mathscr{E}$ and E. In the terminology of Royall (1970), an estimator e of T is said to be ξ-unbiased if

$$\mathscr{E}(e - T) = \int \left[e(s, x_i: i \in s) - \sum_1^N x_i \right] d\xi(x) = 0$$

for every s, and p-unbiased if

$$E(e - T) = \sum_s \left[e(s, x_i: i \in s) - \sum_1^N x_i \right] p(s) = 0$$

for every possible x. Different estimators and designs are usually compared on the basis of distributional properties of the 'squared error' $(e-T)^2$.

When a valid prior distribution ξ is granted, then no matter what sampling design p is used, inferences about T given the data are most naturally based on the posterior distribution of T given $(s, x_i: i \in s)$, and this distribution does not involve p, (Godambe (1966). Since the posterior expected squared error $\mathscr{E}[(e-T)^2 \mid s, x_i: i \in s]$ of an estimator is minimized by taking e to be the posterior expectation $e^*=\mathscr{E}(T \mid s, x_i: i \in s)=$ $=\sum_{i \in s} x_i + \mathscr{E}(\sum_{i \notin s} x_i \mid s, x_i: i \in s)$, the function e^* is called the 'Bayes estimator' of T. It is interesting to note that if the prior ξ is such that the coordinates of x are independently distributed, $e^* - T$ is distributed a posteriori independently of $(x_i: i \in s)$, so that for a *given s*, $\mathscr{E}[(e^*-T)^2 \mid s,$ $x_i: i \in s]=$the prior expected squared error $\mathscr{E}(e^*-T)^2$. However, it is clear that for the minimization of either the posterior or the prior expected squared error, the appropriate design of a given size is the non-randomised (purposive) design which chooses with probability 1 the allowed sample s with smallest $\mathscr{E}(e^*-T)^2$.

The Bayes estimator e^* is evidently ξ-unbiased; but since its value generally depends on the particular ξ assumed, its usefulness is limited primarily to situations in which a single prior can be adopted with assurance. (The reason for the qualification is apparent below in connection with the difference estimator.) A more commonly encountered situation in survey-sampling is one in which the values of some auxiliary variate $\mathbf{y}=$ $=(y_1,...,y_N)$ are known, and prior knowledge is expressed rather broadly in terms of these. For example, one may have reason to believe that the x_i are independently distributed with $\mathscr{E}x_i=\beta y_i$, and variances $\mathscr{V}x_i=$ $=\sigma^2 v(y_i)$, where v is a known function but β and σ^2 are unspecified. This is an expression of prior knowledge in terms of a wide *class* $C_{v,\mathbf{y}}$ of priors ξ (Godambe, 1968), and the Bayes estimator $e^*=\sum_{i\in s} x_i + \beta\sum_{i\notin s} y_i$ is of little use, β being unspecified. Interestingly, however, there are still meaningful optimality criteria for estimators based on the ξ distributions alone; for as Royall (1970) has shown, there exist estimators which are ξ-unbiased for all prior distributions ξ in the class $C_{v,\mathbf{y}}$, and if only homogeneous linear estimators of the form $\sum_{i\in s} b(s,i) x_i$ are allowed, there is one sample s and one estimator which is ξ-unbiased and minimizes $\mathscr{E}(e-T)^2$ for all $\xi\in C_{v,\mathbf{y}}$. This optimal homogeneous linear estimator, for which $b(s,i)=1+(\sum_{i\notin s} y_i)[y_i/v(y_i)]/[\sum_{i\in s} (y_i^2/v(y_i))]$, will be called Royall's estimator. However, restriction to homogeneous linear estimators has no obvious justification.

Thus the *recognition* that the assumption of a unique prior distribution ξ is mostly unrealistic has put us in a problematic situation, namely that of non-existence of a unique Bayes estimator for all $\xi\in C_{v,\mathbf{y}}$, as stated in the preceding paragraph. The solution of the problem given below is in terms of artificial randomisation, i.e. in terms of sampling designs. Curiously enough the solution depends upon the further *recognition* of the fact that even a class of prior distributions so wide as $C_{v,\mathbf{y}}$ may fail to reflect the true relationship (if any) between $\mathbf{x}$ and $\mathbf{y}$. In presenting the solution we pursue a line of argument closely related to that presented in Godambe (1969) and Godambe and Thompson (1971).

If ξ is uniquely specified and e^* is the Bayes estimate for T with respect to ξ, then e^* is $\mathscr{E}(T\,|\,s, x_i: i\in s)$, namely the expectation of T *after the* *data.* Writing this as $\mathscr{E}((e^* - T) \mid s, x_i: i \in s) = 0$ and integrating out $x_i: i \in s$ gives us $\mathscr{E}(e^* - T) = 0$ for every s. Now suppose ξ is not unique, but ranges through $C_{v, y}$ say. Suppose, however, that the estimator e (defined independently of ξ) is such that for all $\xi \in C_{v, y}$ and all possible s, $\mathscr{E}(e - T) = 0$. (Then extending the formalism of Godambe and Thompson (1971) we could define $e - T$ to be a *pivot around zero*.). We now ask the

QUESTION: Is there a sense in which it is reasonable to regard e as the expectation of T, after the data? In other words, under what *conditions* can we *invert* the *pivotal equation*, $\mathscr{E}(e - T) = 0$, after the data?

Note that when the prior $\xi \in C_{v, y}$ is uniquely specified and e^* is the Bayes estimate for T with respect to ξ, the distribution of $e^* - T$ (the 'Bayes' pivot around zero) given s, $x_i: i \in s$ determines the distribution of T after the data. The moments $\mathscr{E}((e^* - T)^k \mid s, x_i: i \in s) = \mathscr{E}((e^* - T)^k \mid s) = \mathscr{E}((e^* - T)^k)$ are the central moments of T after the data. When, however, ξ ranges through $C_{v, y}$, and an estimator e exists such that for all $\xi \in C_{v, y}$ and all possible s, $\mathscr{E}(e - T) = 0$, we may further ask what inferential role should be played by the moments $\mathscr{E}(e - T)^k$.

There is one sense in which these moments $\mathscr{E}(e - T)^k$ can be regarded as moments of T after the data: The principles of likelihood and conditionality (Godambe, 1966), both *imply* that after s is drawn the inference should be independent of the mode of randomisation (sampling design) adopted to draw s. This *implication* is regardless of whether the prior ξ is unique or whether it at all exists. Thus when (as in the present case for reasons discussed below) conditioning on the entire data s, $x_i: i \in s$ is *practically not feasible*, we may condition the inference on 's'alone. Under the situation $\mathscr{E}(e - T)^k$ can be considered as moments *after the data*. Now if E is the expectation due to randomisation i.e. sampling design, *before* the *data* the moments of $e - T$ are given by $E(e - T)^k$, assuming $E(e - T) = 0$. Following the logic of Godambe and Thompson (1971) we now suggest the following answer to the QUESTION raised in the preceding paragraph: 'the pivotal equation $\mathscr{E}(e - T) = 0$ could be inverted if the $e - T$ preserves the same distribution before and after the data'. Or we may say '*if $e - t$ preserves the same moments before and after the data*'. Since unlike Godambe and Thompson (1971), here we do not impose any structural assumptions on the population vector **x**, we might just equate the first two moments before the data with the corresponding ones after the data i.e.

(C') $E(e - T) = \mathscr{E}(e - T),$ $(= 0)$
$$E(e - T)^2 = \mathscr{E}(e - T)^2$$

for all $\xi \in C_{v,y}$ and all s. Further of all the 'pivots around zero' $e - T$ satisfying the conditions (C') above, the one having smallest $E(e-T)^2$, $\mathscr{E}(e-T)^2$ can in a sense be said to *best* preserve the *sameness* of the entire distribution before and after the data. (In searching for the *best pivot* around zero, while preserving the condition $E(e-T)=0$, one might like to replace in the above conditions (C'), the expectations E, *before* the *data* by $\mathscr{E}E$. Then instead of (C') we have

(C) $\mathscr{E}E(e - T) = \mathscr{E}(e - T),$
$$\mathscr{E}E(e - T)^2 = \mathscr{E}(e - T)^2,$$

for all $\xi \in C_{v,y}$ and all s. Evidently conditions (C) are weaker than conditions (C')). Subsequently we shall establish *approximate bestness* of certain commonly used pivots around zero, $e-T$. The difficulties involved would suggest that if *after* the *data* expectations were conditioned on the entire data $(s, x_i: i \in s)$, instead of 's' alone as in (C) or (C'), the task would have been *practically impossible*.

A simple minded conclusion of the discussion in the above paragraph is that among the inferences one can make in terms of $\mathscr{E}(e-T)$ and $\mathscr{E}(e-T)^2$ on the *assumption* $\xi \in C_{v,y}$, which can be replaced by inferences in terms of $E(e-T)$ and $E(e-T)^2$ if the *assumption* is *invalid*, the most accurate ones are provided by the *best pivot* satisfying (C) or (C') in the preceding paragraph.

We shall now discuss examples of such inferences based on three commonly used estimators, namely the Horvitz-Thompson estimator, the difference estimator and the ratio estimator. (For details see Appendix.)

In the first case, suppose that the sampling design p^* can be chosen so that the sample size n is fixed and the inclusion probabilities π_i are proportional to the y_i. Then the Horvitz-Thompson estimator (of T)

$$e_{HT} = \sum_{i \in s} \frac{x_i}{\pi_i}$$

is both p^*-unbiased and ξ-unbiased for every ξ in $C_{v,y}$. If $v(y_i)=y_i^2$, then for any $\xi \in C_{v,y}$, the estimator e_{HT} and the design p^* jointly minimize the expected mean squared error $\mathscr{E}E(e-T)^2 = E\mathscr{E}(e-T)^2$ among all designs

p of size n and p-unbiased estimators e (Godambe and Joshi, 1965). Now it can easily be shown that for $\xi \in C_{v, y}$ with $v(y_i) = y_i^2$

$$\mathscr{E}(e_{\mathrm{HT}} - T)^2 = \frac{\sigma^2 \left[\sum_1^N y_i \right]^2}{n} - \frac{2\sigma^2}{n} \left[\sum_1^{N} y_i \right] \left[\sum_{i \in s} y_i \right] + \sigma^2 \sum_1^{N} y_i^2,$$

which for large N, assuming some regularity conditions on $\mathbf{y}$, is approximately the same as

$$\mathscr{E}E(e_{\mathrm{HT}} - T)^2 = \frac{\sigma^2 \left[\sum_1^N y_i \right]^2}{n} - \sigma^2 \sum_1^N y_i^2 .$$

Also for large populations under suitable conditions arguments involving laws of large numbers may be invoked to show that $E(e_{\mathrm{HT}} - T)^2$ is a good approximation to $\mathscr{E}E(e_{\mathrm{HT}} - T)^2$ and thus to $\mathscr{E}(e_{\mathrm{HT}} - T)^2$ when $\xi \in C_{v, y}$. It follows $e_{\mathrm{HT}} - T$ satisfies (for large N) conditions (C) and (C') approximately, and minimizes $\mathscr{E}E(e_{\mathrm{HT}} - T)^2$ among pivots around zero which satisfy these conditions.

Now the assumption $\xi \in C_{v, y}$ being invalid (without an alternative to replace it) means that we are unable to formulate any possible valid prior distribution. In other words this means that we cannot relate the information provided by any two units and the corresponding x-values, say (j, x_j) and $(j', x_{j'})$, in the sample. Under this condition it is very plausible to choose an estimator $e(s, x_i : i \in s)$ such that *the contribution of any $(j, x_j), j$ in s, to e should be independent of every other $(j', x_{j'}), j'$ in s.* This has an obvious mathematical formulation: Let for any sample s, $(x_i : i \in s)$ and $(x_i' : i \in s)$ be such that $x_i = x_i'$ excepting when $i = j$, i.e. $x_j \neq x_j'$, $i, j \in s$. Then we want e to satisfy

(i) $e(s, x_i' : i \in s) - e(s, x_i : i \in s) = h(j, x_j, x_j' - x_j).$

That is, the difference between the two values of e obtained by changing just one x-value in s (in (i) above it is x_j) is a function h with three arguments namely the unit label, the corresponding x-value and the increment in the x-value. Further, if for any sample s, such that $x_i = 0$, for all $i \in s$; then

(ii) $e(s, x_i : i \in s) = 0.$

It is easy to show that any estimator e satisfying (i) and (ii) above is given by

$$e(s, x_i: i \in s) = \sum_{i \in s} \alpha(i, x_i),$$

where α is some function with two arguments namely i and x_i. Further if unbiasedness is assumed $\alpha(i, x_i) = x_i/\pi_i$. Thus conditions (i), (ii) above together with the condition of unbiasedness uniquely yield the HT-estimator. (The proof of this will be published elsewhere).

As a second example, suppose that the sampling design p_0 is simple random sampling without replacement (size n), in which any s consisting of n units is equally likely to be drawn. The difference estimator

$$e_D = \sum_1^N y_i + \frac{N}{n} \sum_{i \in s} (x_i - y_i)$$

is p_0-unbiased. It is also ξ-unbiased for any ξ belonging to the class $D_{v,y}$ of priors for which the x_i are independent, $\mathscr{E}x_i = y_i + \alpha$ and $\mathscr{V}(x_i) = = \sigma^2 v(y_i)$, α and σ^2 unspecified. Moreover, it can be shown that when $v(y_i) = $ a constant v, then for any $\xi \in D_{v,y}$ the estimator e_D together with p_0 jointly minimize $\mathscr{E}E(e-T)^2 = E\mathscr{E}(e-T)^2$ in competition with all size n sampling designs p and p-unbiased estimators e. Conditions (C) are satisfied by the pivot $e_D - T$ when $v(y_i) = v$ because $\mathscr{E}(e_D-T)^2$ is actually equal to $\mathscr{E}E(e_D-T)^2$ for each s of size n. Further since $\mathscr{E}(e_D-T)^2$ is well approximated by $E(e_D-T)^2$ with high (ξ) probability for large N, conditions (C') are satisfied approximately.

Finally, a very commonly used estimator is the ratio estimator

$$e_R = \sum_1^N y_i \left[\frac{\sum\limits_{i \in s} x_i}{\sum\limits_{i \in s} y_i} \right].$$

It is ξ-unbiased for any $\xi \in C_{v,y}$, and p-unbiased for the design having $p(s)$ proportional to $\sum_{i \in s} y_i$ whenever $n(s) = n$. Unlike e_{HT} and e_D, it has no known exact optimality property as a member of the class of p-unbiased estimators. Furthermore, for $\xi \in C_{v,y}$ we have

$$\mathscr{E}(e_R - T)^2 = \frac{y^2 \sigma^2 v_s}{y_s^2} - \frac{y 2 \sigma^2 v_s}{y_s} + \sigma^2 \sum_1^N v(y_i)$$

where $y = \sum_1^N y_i$, $y_s = \sum_{i \in s} y_i$ and $v_s = \sum_{i \in s} v(y_i)$. Whenever this depends markedly on s, there must be an essential difference between $\mathscr{E}(e_R - T)^2$ and $E(e_R - T)^2$. Except when $y_i \approx$ const. y, conditions (C') cannot be satisfied for the pivot $e_R - T$. Note that, again unlike e_{HT} and e_D, e_R is a 'Royall's estimator' for a certain $C_{v, y}$, namely the one for which $v(y_i) = = y_i$, and its great efficiency in many practical situations when the assumption of $C_{v, y}$ is valid is accounted for by this fact. However, unless $y_i \approx$ const. y, statements made in terms of $\mathscr{E}(e_R - T)^2$ will not be capable of validation by artificial randomisation in terms of $E(e_R - T)^2$ in case the assumption of $C_{v, y}$ is invalid.

University of Waterloo

REFERENCES

Barnard, G. A., Jenkins, G. M., and Winsten, C. B.: 1962, 'Likelihood Inference and Time Series (with discussion)', *J. R. Statist. Soc., Ser. A* **125**, 321–72.

Birnbaum, A.: 1962, 'On the Foundations of Statistical Inference (with discussion)', *J.A.S.A.* **57**, 269–326.

Birnbaum, A.: 1969, 'Concepts of Statistical Evidence', in S. Morgenbesser, P. Suppes and M. White, (eds.), *Philosophy, Science, and Method: Essays in Honor of Ernest Nagel*, St. Martin's Press, New York.

Box, G. E. P. and Andersen, S. L.: 1955, 'Permutation Theory in the Derivation of Robust Criteria and the Study of Departures from Assumption (with discussion)', *J. R. Statist. Soc., Ser. B* **17**, 1–34.

Cox, D. R.: 1959, 'Some Problems Connected with Statistical Inference', *Ann. Math. Statist.* **29**, 357–372.

Eden, T. and Yates, F.: 1933, 'On the Validity of Fisher's *z* Test...', *J. Agric. Sci.* **23**, 6–16.

Ericson, W. A.: 1969, 'Subjective Bayesian Models in Sampling Finite Populations, I', *J. R. Statist. Soc., Ser. B* **31**, 195–234.

Fisher, R. A.: 1935, *The Design of Experiments*, Oliver and Boyd, Edinburgh.

Fisher, R. A.: 1936, '"The Coefficient of Racial Likeness" and the Future of Craniometry', *J. Roy. Anthrop. Inst.* **66**, 57–63.

Fisher, R. A.: 1956, *Statistical Methods and Scientific Inference*, Oliver and Boyd, Edinburgh.

Godambe, V. P.: 1955, 'A Unified Theory of Sampling from Finite Populations', *J. R. Statist. Soc., Ser. B.* **17**, 268–278.

Godambe, V. P.: 1966, 'A New Approach to Sampling from Finite Populations, I', *J. R. Statist. Soc., Ser. B* **28**, 310–319.

Godambe, V. P.: 1968, 'Bayesian Sufficiency in Survey-Sampling', *Ann. Inst. Statist. Math.* **20**, 363–373.

Godambe, V. P.: 1969, 'Some Aspects of the Theoretical Developments in Survey-Sampling', in N. L. Johnson and H. Smith, (eds.), *New Developments in Survey-Sampling*, Wiley Interscience, New York, 27–53.

Godambe, V. P.: 1969a, 'A Fiducial Argument with Application to Survey-Sampling', *J. R. Statist. Soc., Ser. B* **31**, 246–260.

Godambe, V. P. and Joshi, V. M.: 1965, 'Admissibility and Bayes Estimation in Sampling from Finite Populations, I', *Ann. Math. Statist.* **36**, 1707–1722.

Godambe, V. P. and Thompson, M. E.: 1971, 'Bayes, Fiducial and Frequency Aspects of Statistical Inference in Regression Analysis in Survey-Sampling (with discussion)', *J. R. Statist. Soc., Ser. B* **33**, 361–390.

Johnson, N. L. and Smith, H., (eds.): 1969, *New Developments in Survey Sampling*, Wiley Interscience, New York.

Kempthorne, O.: 1955, 'The Randomization Theory of Experimental Inference', *J. Amer. Stat. Ass.* **50**, 946–967.

Kempthorne, O.: 1971, 'Probability, Statistics, and the Knowledge Business', in V. P. Godambe and D. A. Sprott, (eds.), *Foundations of Statistical Inference*, Holt, Rinehart and Winston of Canada, Toronto, 470–492.

Neyman, J., with the cooperation of K. Iwaskiewicz and St. Kolodoziejczyk: 1935, Statistical Problems in Agricultural Experimentation, *Supplement to the J. R. Statist. Soc.* **2**, 107–154.

Pitman, E. J. G.: 1937, 'Significance Tests which can be Applied to Samples from any Populations, III', *Biometrika* **29**, 322–35.

Royall, R. M.: 1970, 'On Finite Population Sampling Theory under Certain Linear Regression Models', *Biometrika* **57**, 377–387.

Welch, B. L.: 1937, 'On the z Test in Randomized Blocks and Latin Squares', *Biometrika* **29**, 21–52.

APPENDIX

1. THE LIMITING BEHAVIOUR OF THE SAMPLING VARIANCE OF e_{HT}

Let us regard the auxiliary vector $(y_1, \ldots, y_N)$ as the initial segment of an infinite sequence $y_1, y_2, \ldots$. We denote by Y_N the sum $\sum_{i=1}^{N} y_i$. If the inclusion probabilities π_i are chosen proportional to the y_i for the population $\{1, \ldots, N\}$, then π_i must equal $n y_i / Y_N$, in view of the well known fact that $\sum_{i=1}^{N} \pi_i = n$ for a fixed size design of size n (Godambe, 1955). Let us use the notation π_{iN} for π_i, to indicate its dependence on N.

Now for $\xi \in C_{v,y}$ with $v(y_i) = y_i^2$, we have $\mathscr{E} x_i = \beta y_i$ and $\mathscr{V} x_i = \sigma^2 y_i^2$, so that

$$(A1) \quad \mathscr{E}(e_{HT} - T)^2 = \mathscr{V}\left[\sum_{i \in s} x_i \left[\frac{1}{\pi_{iN}} - 1\right] - \sum_{i \notin s} x_i\right] =$$

$$= \sum_{i \in s} \left((1 - \pi_{iN})^2 \sigma^2 y_1^2 / \pi_{iN}^2\right) + \sum_{i \notin s} \sigma^2 y_i^2 =$$

$$= \sigma^2 \left\{ (Y_N^2/n) - 2(y_s Y_N/n) + \sum_{i=1}^{N} y_i^2 \right\} =$$

$$= (\sigma^2 Y_N^2/n) \left\{ 1 - 2(y_s/Y_N) + n \sum_{i=1}^{N} y_i^2/Y_N^2 \right\},$$

where $y_s = \sum_{i \in s} y_i$ and $\mathscr{V}$ denotes variance with respect to ξ. Moreover, from (A1)

(A2)　　$E\mathscr{E}(e_{HT} - T)^2 = (\sigma^2 Y_N^2/n) \left\{ 1 - n \sum_{i=1}^{N} y_i^2/Y_N^2 \right\},$

since

$$E y_s = \sum_{i=1}^{N} y_i \pi_{iN} = n \sum_{i=1}^{N} y_i^2/Y_N.$$

Thus as $N \to \infty$, $\mathscr{E}(e_{HT} - T)^2$ and $E\mathscr{E}(e_{HT} - T)^2$ are asymptotically equivalent (for each s) provided $Y_N \to \infty$ and

$$E(y_s/Y_N) = n \sum_{1}^{N} y_i^2/Y_N^2 \to 0.$$

This would happen if for example $0 < r \leqslant y_i < R < \infty$ for all i or if the y_i were independent observations on a positive random variable with finite variance.

Coming to the sampling variance of e_{HT}, we have

$$\frac{n}{N^2} E(e_{HT} - T)^2 = \frac{n}{N^2} \sum_{i=1}^{N} \left[\frac{x_i}{\pi_{iN}} \right]^2 \pi_{iN} +$$

(A3)　　$$+ \frac{n}{N^2} \sum_{i=1}^{N} \sum_{\substack{j=1 \\ i \neq j}}^{N} \frac{x_i}{\pi_{iN}} \frac{x_j}{\pi_{jN}} \pi_{ijN} - \frac{n}{N^2} T^2,$$

π_{ijN} being the joint inclusion probability of i and j.

The first term of (A3) can be written $(Y_N/N^2) \sum_{i=1}^{N} x_i^2/y_i$ with mean value $(Y_N^2/N^2)(\sigma^2 + \beta^2)$. It can be shown that the ratio of this term to its mean, namely $(\sum_{i=1}^{N} (x_i/y_i)^2 y_i/((\sigma^2 + \beta^2)Y_N))$, approaches 1 in ξ-probability as $N \to \infty$, provided

(i) $Y_N \to \infty$,

(ii)   the random variables $X_i = (x_i/y_i)^2 - (\sigma^2 + \beta^2)$ are uniformly integrable, and

(iii) $\sum_{i=1}^{N} y_i^2/Y_N^2 \to 0$.

By way of proof, we note that since $\sum_{i=1}^{N} y_i^2/Y_N^2 \to 0$ as $N \to \infty$ there exist numbers $a_i \to \infty$ such that $\sum_{i=1}^{N} y_i^2 a_i^2/Y_N^2 \to 0$. Let

$$X_i' = X_i \quad \text{if} \quad |X_i| \leqslant a_i$$
$$= 0 \quad \text{if} \quad |X_i| > a_i$$

and

$$X_i'' = X_i - X_i'.$$

Since the quantity being considered is $1 + \sum_{i=1}^{N} X_i y_i/((\sigma^2 + \beta^2) Y_N)$, we want to show that

$$\sum_{i=1}^{N} X_i y_i/Y_N = \sum_{i=1}^{N} (X_i' y_i + X_i'' y_i)/Y_N \to 0$$

in probability as $N \to \infty$. But the variance of the first sum is less than or equal to $\sum_{i=1}^{N} y_i^2 a_i^2/Y_N^2$, which approaches 0 in probability. The expectation of the absolute value of the second sum is less than or equal to $\sum_{i=1}^{N} y_i \mathscr{E}|X_i''|/Y_N$. Since $a_i \to \infty$, condition (ii) implies that $\mathscr{E}|X_i''| \to 0$ as $i \to \infty$. Thus with the addition of (i), $\sum_{i=1}^{N} y_i \mathscr{E}|X_i''|/Y_N \to 0$ as $N \to \infty$. The proof is now complete.

The ratio of the second term to its mean is

$$\sum_{\substack{i=1 \\ i \neq j}}^{N} \sum_{j=1}^{N} (x_i x_j/y_i y_j) \pi_{ijN}/(\beta^2 n(n-1)).$$

The second term is thus asymptotically equivalent to its mean in ξ-probability if

$$\text{Var}\left[\sum_{\substack{i=1 \\ i=j}}^{N} \sum_{j=1}^{N} (x_i x_j/y_i y_j) \pi_{ijN}/n(n-1)\right] = (1/n(n-1))^2 \times$$

$$\times \left[\sum_{i=1}^{N} \sum_{j=1}^{N} \sigma^4 \pi_{ijN}^2 + 2\beta^2 \sigma^2 \sum_{j=1}^{N} (n-1)^2 \pi_{jN}^2\right] \to 0.$$

This will be true if condition (iii) above holds, namely if

$$\sum_{i=1}^{N} y_i^2/Y_N^2 \to 0.$$

For

$$\sum_{j=1}^{N} \pi_{jN}^2/n^2 = \sum_{j=1}^{N} y_j^2/Y_N^2,$$

and

$$\sum_{\substack{i=1 \\ i \neq j}}^{N} \sum_{j=1}^{N} \pi_{ijN}^2 \leqslant \sum_{i=1}^{N} \left(\sum_{j \neq i} \pi_{ijN} \right)^2 = (n-1)^2 \sum_{i=1}^{N} \pi_{iN}^2.$$

Finally, as the last term of (A3) we have $(n^{1/2}T/N)^2$; the ratio of $n^{1/2}T/N$ to its mean $n^{1/2}\beta Y_N/N$ has variance $\sigma^2 \sum_{i=1}^{N} y_i^2/(\beta^2 Y_N)$, and therefore condition (iii) above implies that $n(T/N)^2$ is asymptotically equivalent to $n\beta^2 Y_N^2/N^2$ as $N \to \infty$.

We may finally conclude, then, that whenever (i), (ii), and (iii) hold, $E(e_{HT} - T)^2$ and $\mathscr{E}(e_{HT} - T)^2$ are asymptotically equivalent in probability with respect to ξ.

2. THE USE OF e_{HT} IN SUPPORTING ξ-BASED INFERENCE

More generally than in §1, let us consider an estimator of the form

$$e(s, \mathbf{x}) = C_N \sum_{i \in s} x_i/n\gamma_i$$

where $\gamma_1, \gamma_2, \ldots$ are positive numbers and $C_N = \sum_{i=1}^{N} \gamma_i$. Let the (fixed size) sampling design have inclusion probabilities π_{iN} and joint inclusion probabilities π_{ijN}. If the inclusion probabilities are chosen proportional to the γ_i and the design has size n, then π_{iN} must equal $n\gamma_i/C_N$.

Now for $\xi \in C_{v, \mathbf{y}}$ with $v(y_i) = y_i^2$,

$$\mathscr{E}(e - T) = \frac{C_N \beta}{n} \left[\sum_{i \in s} (y_i/\gamma_i) - nY_N/C_N \right].$$

Thus e is ξ-unbiased only when $\sum_{i \in s} (y_i/\gamma_i) = nY_N/C_N$ for each s associated with a positive sampling probability.

Also,

$$\mathscr{E}(e-T)^2 = \mathscr{E}\left[\sum_{i\in s} x_i\left(\frac{C_N}{n\gamma_i}-1\right) - \sum_{i\notin s} x_i\right]^2 =$$

$$= \sigma^2\left\{\frac{C_N^2}{n^2}\sum_{i\in s}\frac{y_i^2}{\gamma_i^2} - \frac{2C_N}{n}\sum_{i\in s}\frac{y_i^2}{\gamma_i} + \sum_{i=1}^{N} y_i^2\right\} +$$

$$+ \frac{C_N^2\beta^2}{n^2}\left[\sum_{i\in s}\frac{y_i}{\gamma_i} - \frac{nY_N}{C_N}\right]^2.$$

This is approximately independent of s for large N if $\gamma_i = y_i$ for each i, and is then equivalent to $\sigma^2 C_N^2/n$.

On the other hand, the sampling bias is

$$E(e-T) = \frac{C_N}{n}\sum_{i=1}^{N}\frac{x_i\pi_{iN}}{\gamma_i} - T,$$

and is identically 0 if $\pi_{iN} = n\gamma_i/C_N$. The mean square error is

$$E(e-T)^2 = \sum_{i=1}^{N} x_i^2\left[\frac{C_N^2\pi_{iN}}{n^2\gamma_i^2} - \frac{2C_N}{n}\frac{\pi_{iN}}{\gamma_i} + 1\right] +$$

$$+ 2\sum_{i=1}^{N}\sum_{\substack{j=1\\i\neq j}}^{N} x_i x_j\left\{\frac{C_N^2\pi_{ijN}}{n^2\gamma_i\gamma_j} - \frac{C_N\pi_{jn}}{n\gamma_i} - \frac{C_N\pi_{iN}}{n\gamma_j} + 1\right\}.$$

Under relatively mild conditions on the γ_i, the y_i and the π_{iN} these quantities may be expected to be 'asymptotically equivalent' to their ξ-expected values

$$\mathscr{E}E(e-T) = 0$$

and

(A4) $$\mathscr{E}(e-T)^2 = \sigma^2\sum_{i=1}^{N}\left[\frac{C_N^2}{n^2}\frac{y_i^2}{\gamma_i^2}\pi_{iN} - \frac{2C_N}{n}\frac{y_i^2}{\gamma_i}\pi_{iN} + y_i^2\right] +$$

$$+ \beta^2\frac{C_N^2}{n^2}E\left[\sum_{i\in s}\frac{y_i}{\gamma_i} - \frac{nY_N}{C_N}\right]^2$$

as $N\to\infty$. That is, we may expect that

$$E(e - T)/T \to 0$$

and

$$E(e - T)^2/\mathscr{E}E(e - T)^2 \to 1$$

in probability as $N \to \infty$.

Now if the relevant class of priors is known to be $C_{v,\mathbf{y}}$, the γ_i may be chosen proportional to the known values of y_i. In this case in (A4) $\mathscr{E}E(e-T)^2$ is equivalent to $\sigma^2 C_N^2/n$ as $N \to \infty$, for a *wide variety* of choices of the inclusion probabilities π_{iN}. Thus a large number of randomized sampling designs will make the p-based inference concerning T agree with the ξ-based inference; there is no apparent necessity to choose the π_{iN} proportional to the γ_i unless exact p-unbiasedness is desired. But suppose on the other hand that the values of y_i are not known or are faultily determined, so that in the resulting situation the γ_i are not proportional to the true y_i. The use of the ξ-distribution of e under the false assumption that the γ_i and y_i are proportional may be quite erroneous. In this case, there is an appreciable advantage in choosing π_{iN} proportional to the γ_i, to make $E(e-T)=0$, and in using estimates of $E(e-T)^2$ to make interval estimates of T. Note that

$$\mathscr{E}E(e - T)^2 = \sigma^2 \frac{C_N}{n} \sum_{i=1}^{N} \frac{y_i^2}{\gamma_i} - \sigma^2 \sum_{i=1}^{N} y_i^2 + \\ + \beta^2 \frac{C_N^2}{n^2} E\left[\sum_{i \in s} \frac{y_i}{\gamma_i} - \frac{nY_N}{C_N}\right]^2,$$

which is greater than its value with $\gamma_i = y_i$, and is not equivalent to that value asymptotically in general. We may therefore say, when $\pi_{iN} \propto \gamma_i$, that for a large population with prior $\xi \in C_{v,\mathbf{y}}$ the H.T. estimator $C_N \sum_{i \in s} x_i/n\gamma_i = e$ has less than optimal efficiency, but that choosing a matching design ensures the accuracy of the estimation.

3. The limiting behaviour of the sampling variance of e_D

The difference estimator can be written

$$e_D = \sum_{1}^{N} y_i + \frac{N}{n} \sum_{i \in s} z_i,$$

where $z_i = x_i - y_i$, and it is a p_0- and ξ-unbiased estimator of

$$T = \sum_1^N y_i + \sum_1^N z_i.$$

Thus

$$\mathscr{E}(e_D - T)^2 = \mathscr{E}\left[N\sum_{i \in s} z_i/n - \sum_1^N z_i\right]^2.$$

For $\xi \in D_{v,y}$ when $v(y_i) = $ constant v,

$$\mathscr{E}(e_D - T)^2 = \sigma^2\left[\frac{N}{n} - 1\right]^2 n + \sigma^2(N - n) = \frac{N(N - n)}{n}\sigma^2,$$

which is independent of s and therefore equal to $E\mathscr{E}(e_D - T)^2$. The sampling variance $E(e_D - T)^2 = \mathrm{Var}(N\sum_{i \in s} z_i/n)$, and since $\xi \in D_{v,y}$ implies a prior in $C_{v,1}$ for $\mathbf{z}$, the results of the previous section guarantee that $E(e_D - T)^2$ is asymptotically equivalent to $N(N - n)\sigma^2/n$ as $N \to \infty$, provided the random variables z_i^2 are uniformly integrable.

4. THE MEAN SQUARE ERRORS OF e_R

When $\xi \in C_{v,y}$,

$$\mathscr{E}(e_R - T)^2 = \mathscr{V}\left[\sum_{i \in s} x_i\left[\frac{Y_N}{y_s} - 1\right] - \sum_{i \notin s} x_i\right] =$$

$$= \left[\frac{Y_N}{y_s} - 1\right]^2 \sigma^2 v_s + \sigma^2\left[\sum_1^N v(y_i) - v_s\right] =$$

$$= \frac{Y_N^2}{y_s^2}\sigma^2 v_s - \frac{2Y_N}{y_s}\sigma^2 v_s + \sigma^2\sum_1^N v(y_i)$$

where

$$Y_N = \sum_1^N y_i,\ y_s = \sum_{i \in s} y_i,\ v_s = \sum_{i \in s} v(y_i).$$

The dominant term is the first one if $Y_N \to \infty$ and $\sum_{i=1}^N v(y_i)/Y_N^2 \to 0$ as $N \to \infty$. This first term will depend on s unless v_s is proportional to y_s^2, a situation which cannot generally happen when the y_i take different values and all s of size n are taken into account. Thus the sampling mean square error cannot be equivalent to $\mathscr{E}(e_R - T)^2$ for all s.

However, suppose the design adopted is simple random sampling without replacement and n draws. Then if n is large enough that y_s has a negligible coefficient of variation, we have

$$
(A5) \qquad E\left(e_R - T\right)^2 \approx \frac{N\left(N - n\right)}{N - 1} \frac{1}{n} \sum_{i=1}^{N} \left[x_i - \frac{T}{Y_N} y_i \right]^2
$$

and

$$
(A6) \qquad \mathscr{E} E\left(e_R - T\right)^2 \approx \left[\frac{N}{n} - 1\right] \sigma^2 \sum_{i=1}^{N} v\left(y_i\right).
$$

If N becomes large while the coefficient of variation of y_s remains negligible, (A5) and (A6) will be equivalent (for suitable values of $v(y_i)$), and comparable in value to $\mathscr{E}\left(e_R - T\right)^2$.

ACKNOWLEDGEMENT

We would like to thank Angela Lange and Jill Wiens for their excellent typing.

DISCUSSION

Commentator Lindley: According to the Bayesian viewpoint, a probability statement is made on the basis of the knowledge to hand at the time the statement is made (the conditioning event) and it is incorrect to call such probabilities incorrect because they change when the conditioning event changes. Of course a function may alter its value when one of its arguments changes. Indeed, I can see no sense in which a single probability statement can be said to be incorrect. Right or wrong are terms that can only be applied to the relationships between probabilities. These are correct if, and only if, they obey the rules of the probability calculus.

Godambe: If prior probabilities are frequency probabilities they can be wrong, you can wrongly assume a prior.

Good: It is universally true of all objectivistic methods in statistics that the only way they can be precise is by the suppression of information. This fact itself is usually suppressed – what one tacitly assumes, apparently, is that the statistician has a 'statistician's stooge' who arranges to randomize the experiment question and not to tell the statistician the precise details of the method of randomization. (In fact, a good statistician would insist that the stooge be shot immediately he attempted to reveal such information.) Only then can the objectivist obtain precise results, but of course he has suppressed information that can be important.

Godambe: No response tendered (editors)

I. J. GOOD

THE BAYESIAN INFLUENCE,
OR HOW TO SWEEP SUBJECTIVISM
UNDER THE CARPET

ABSTRACT. On several previous occasions I have argued the need for a Bayes/non-Bayes compromise which I regard as an application of the 'Type II' principle of rationality. By this is meant the maximization of expected utility when the labour and costs of the calculations are taken into account. Building on this theme, the present work indicates how some apparently objective statistical techniques emerge logically from subjective soil, and can be further improved if their subjective logical origins (if not always historical origins) are not ignored. There should in my opinion be a constant interplay between the subjective and objective points of view, and not a polarization separating them.

Among the topics discussed are, two types of rationality, 27 'Priggish Principles', 46656 varieties of Bayesians, the Black Box theory, consistency, the unobviousness of the obvious, probabilities of events that have never occurred (namely all events), the Device of Imaginary Results, graphing the likelihoods, the hierarchy of types of probability, Type II maximum likelihood and likelihood ratio, the statistician's utilities versus the client's, the experimenter's intentions, quasi-utilities, tail-area probabilities, what is 'more extreme'?, 'deciding in advance', the harmonic mean rule of thumb for significance tests in parallel, density estimation and roughness penalties, evolving probability and pseudorandom numbers and a connection with statistical mechanics.

I. PREFACE

Owing to a delay in Richard Jeffrey's plane, I was invited, at a few minutes' notice, to give the first talk in the conference, instead of the last one, so that the spoken version was not fully prepared. It contained a few ambiguities that I might otherwise have avoided, and perhaps a little of the discussion was fortunately more heated than it would otherwise have been. I hope this printed version is more lucid although I have tried to preserve some of the informal tone of the spoken version. I should like to thank the discussants for helping me to achieve improved clarity, although some of the clarifications were already in the preliminary typed version, and some were even in the spoken version but might not have been enunciated clearly enough judging by the cassette recording.

There is one respect in which the title of this paper is deliberately ambiguous: it is not clear whether it refers to the *historical* or to the *logical* influence of 'Bayesian' arguments. In fact it refers to both, but

Harper and Hooker (eds.), Foundations of Probability Theory, Statistical Inference, and Statistical Theories of Science, Vol. II, 125–174. *All Rights Reserved. Copyright © 1976 by D. Reidel Publishing Company, Dordrecht-Holland.*

with more emphasis on the logical influence. Logical aspects are more fundamental to a science or philosophy than are the historical ones, although they each shed light on the other. The logical development is a candidate for being the historical development on *another* planet.

I have taken the expression the 'Bayesian influence' from a series of lectures in mimeographed form (Good, 1971a). In a way I am fighting a battle that has already been won to a large extent. For example, the excellent statisticians L. J. Savage, D. V. Lindley, G. E. P. Box (R. A. Fisher's son-in-law) and J. Cornfield were converted to the Bayesian fold years ago. For some years after World War II, I stood almost alone at meetings of the Royal Statistical Society in crusading for a Bayesian point of view. Many of the discussions are reported in the *Journal*, *series B*, but the most detailed and sometimes heated ones were held privately after the formal meetings in dinners at Berterolli's restaurant and elsewhere, especially with Anscombe, Barnard, Bartlett, Daniels, and Lindley. These protracted discussions were historically important but have never been mentioned in print before as far as I know. There is an unjustifiable convention in the writing of the history of science that science communication occurs only through the printed word.

It is difficult for me to tone down the missionary zeal acquired in youth, but perhaps the good battle is justified since there are still many heathens.

II. INTRODUCTION

On many previous occasions, and especially at the Waterloo conference of 1970, I have argued the desirability of a Bayes/non-Bayes compromise which, from one Bayesian point of view, can be regarded as the use of a 'Type II' principle of rationality. By this is meant the maximization of expected utility when the labour and costs of calculations and thinking are taken into account. Building on this theme, the present paper will indicate how some apparently objective statistical techniques emerge logically from subjective soil, and can be further improved by taking into account their logical, if not always historical, subjective origins. There should be in my opinion a constant interplay between the subjective and objective points of view and not a polarization separating them.

Sometimes an orthodox statistician will say of one of his techniques that it has 'intuitive appeal'. This is I believe always a guarded way of saying that it has an informal approximate Bayesian justification.

Partly as a matter of faith, I believe that *all* sensible statistical procedures can be derived as approximations to Bayesian procedures. As I have said on previous occasions, "To the Bayesian all things are Bayesian".

Cookbook statisticians, taught by non-Bayesians, sometimes give the impression to *their* students that cookbooks are enough for all practical purposes. Any one who has been concerned with complex data analysis knows that they are wrong: that subjective judgment of probabilities cannot usually be avoided, even if this judgment can later be used for constructing apparently non-Bayesian procedures in the approved sweeping-under-the-carpet manner.

(a) WHAT IS SWEPT UNDER THE CARPET?

I shall refer to 'sweeping under the carpet' several times, so I shall use the abbreviations UTC and SUTC (laughter). One part of this paper deals with what is swept under the carpet, and another part contains some examples of the SUTC process. (The past tense, etc., will be covered by the same abbreviation.)

Let us then consider what it is that is swept under the carpet. Maurice Bartlett once remarked, in a discussion at a Research Section meeting of the Royal Statistical Society, that the word 'Bayesian' is ambiguous, that there are many varieties of Bayesians, and he mentioned for example, "Savage Bayesians and Good Bayesians" (laughter), and in a letter in the *American Statistician* I classified 46656 varieties (Good, 1971b). There are perhaps not that number of practicing Bayesian statisticians, but the number comes to 46656 when you cross-classify the Bayesians in a specific manner by eleven facets. Some of the categories are perhaps logically empty but the point I was making was that there is a large variety of possible interpretations and some of the arguments that one hears against *the* Bayesian position are valid only against *some* Bayesian positions. As so often in controversies 'it depends what you mean'. The particular form of Bayesian position that I adopt might be called non-Bayesian by some people and naturally it is my own views that I would like most to discuss. I speak for some of the Bayesians all the time and for all the Bayesians some of the time. In the spoken version of this paper I named my position after 'the Tibetan Lama K. Caj Doog', and I called my position 'Doogian'. Although the joke wears thin, it is convenient to have a name for this viewpoint, but 'Bayesian' is misleading, and 'Goodian' or 'Good' is absurd, so I shall continue with the joke even in print. (See also Smith, 1961, p. 18, line minus 15, word minus 2.)

Doogianism is mainly a mixture of the views of a few of my eminent pre-1940 predecessors. Many parts of it are therefore not original, but, taken as a whole I think it has some originality; and at any rate it is convenient here to have a name for it. It is intended to be a general philosophy for reasoning and for rationality in action and not just for statistics. It is a philosophy that applies to all activity, to statistics, to economics, to the practice and philosophy of science, to ordinary behaviour, and, for example, to chess-playing. Of course each of these fields of study or activity has its own specialized problems, but, just as the theories of each of them should be consistent with ordinary logic, they should in my opinion be consistent also with the theory of rationality as presented here and in my previous publications, a theory that is a useful and practically necessary extension of ordinary logic. Ordinary Aristotelean logic has often been taught in the past in the hope that it would lead to an improvement in the effective intelligence of students. Perhaps the Doogian theory of rationality would be more useful for this purpose, while at the same time leading to greater unity of the various branches of knowledge and therefore perhaps to a decrease in Philistinism.

At the Waterloo conference (Good, 1970/71a), I listed 27 Priggish Principles that summarise the Doogian philosophy, and perhaps the reader will consult the Proceedings and some of its bibliography for a more complete picture, and for historical information. Here it would take too long to work systematically through all 27 principles and instead I shall concentrate on the eleven facets of the Bayesian Varieties in the hope that this will give a fairly clear picture. I do not claim that any of these principles were 'discovered last week' (to quote Oscar Kempthorne's off-the-cuff contribution to the spoken discussion), in fact I have developed, acquired or published them over a period of decades, and most of them were used by others before 1940, in one form or another, and with various degrees of bakedness or emphasis. The main merit that I claim for the Doogian philosophy is that it codifies and exemplifies an adequately complete and simple theory of rationality, complete in the sense that it is I believe not subject to the criticisms that are usually directed at other forms of Bayesianism, and simple in the sense that it attains realism with the minimum of machinery. To pun somewhat, it is 'minimal sufficient'.

(b) RATIONALITY, PROBABILITY, AND THE BLACK BOX THEORY

In some philosophies of rationality, a rational man is defined as one whose judgments of probabilities, utilities, and of functions of these, are all both consistent and sharp or precise. Rational men do not exist, but the concept is useful in the same way as the concept of a reasonable man in legal theory. A rational man can be regarded as an ideal to hold in mind when we ourselves wish to be rational. It is sometimes objected that rationality as defined here depends on betting behaviour, and people sometimes claim they do not bet. But since their every decision is a bet I regard this objection as unsound: besides they could in principle be forced to bet in the usual monetary sense. It seems absurd to me to suppose that the *rational* judgment of probabilities would normally depend on whether you were forced to bet rather than betting by free choice.

There are of course people who argue (rationally?) against rationality, but presumably they would agree that it is sometimes desirable. For example, they would usually prefer that their doctor should make rational decisions, and, when they were fighting a legal case in which they were sure that the evidence 'proved' their case, they would presumably want the judge to be rational. I believe that the dislike of rationality is often merely a dishonest way of maintaining an indefensible position. Irrationality is intellectual violence against which the pacifism of rationality may or may not be an adequate weapon.

In practice one's judgments are not sharp, so that to use the most familiar axioms it is necessary to work with judgments of inequalities. For example, these might be judgments of inequalities between probabilities, between utilities, expected utilities, weights of evidence (in a sense to be defined and apparently first proposed by Charles Saunders Peirce, 1878), or any other convenient function of probabilities and utilities. We thus arrive at a theory that can be regarded as a combination of the theories espoused by F. P. Ramsey (1926/31/50/64), who produced a theory of precise subjective probability and utility, and of J. M. Keynes (1921), who emphasized the importance of inequalities (partial ordering) but regarded logical probability or credibility as the fundamental concept, at least until he wrote his obituary on Ramsey (Keynes, 1933).

To summarise then, the theory I have adopted since about 1938 is a theory of subjective (personal) probability and utility in which the judgments take the form of inequalities (but see Section III (iii) below). This theory can be formulated as the following 'black box' theory.

A black box theory is the ideal form for *any* nearly perfected scientific or philosophical theory based on a system of axioms. One can think of the mathematical structure that can be built on the axioms as a black box or machine into which judgments can be plugged and out of which you can extract 'discernments' and then feed them back into the box. The entire set of judgments and discernments at any one time constitutes the 'body of beliefs' and the purpose of the theory is to enlarge the body of beliefs and to detect inconsistencies in it. When the inconsistencies are judged to be important they are to be removed by more mature judgments. The *axioms* are inside the black box, the *rules* of application are defined by the input and output, and in addition there are an unlimited number of *suggestions* or psychological devices for helping you to make judgments.

In the particular case of a theory of subjective probability, the most typical form of judgment is an inequality of the form

$$P'(E \mid F) \geqslant P'(G \mid H),$$

where P' denotes a 'degree of belief' or 'intensity of conviction', but is not intended necessarily to have a numerical value. It is the *inequality* between two degrees of belief that is supposed to be meaningful. When this inequality enters the black box, the primes are supposed to be dropped and $P(E \mid F)$ etc. are intended now to be represented by precise numbers (which you do not necessarily know). This device of imagining that the black box deals with numbers makes the formalism of the theory almost as simple as if we had assumed that the degrees of belief were themselves numerical. To extend this theory to rationality, we need merely to let judgments of preferences to be allowed also, and to append the 'principle of rationality', the recommendation to maximize expected utility. (Good, 1947/50, 1951/52, 1960/62.)

The axioms, which are expressed in conditional form, have a familiar appearance (but see the reference to 'evolving probability' below), and I shall not state them here.

There is emphasis on human judgment in this theory, based on a respect for the human brain. Even infrahuman brains are remarkable instruments that cannot yet be replaced by machines, and it seems unlikely to me that decision-making in general, and statistics in particular, can become independent of the human brain until the first ultraintelligent machine is built. Harold Jeffreys once remarked that the brain may not be a perfect reasoning machine but is the only one available. That is still true, though a little less so than when Jeffreys first said it. It has been operating for millions of years in billions of individuals and it has developed a certain amount of magic. On the other hand I believe that some formalizing is useful and that the ultraintelligent machine will also use a subjectivistic theory as an aid to its reasoning.

So there is this respect for judgment in the theory. But there is also respect for logic. Judgments and logic must be combined, because, although the human brain is clever at perceiving facts, it is also cunning in the rationalisation of falsity for the sake of its equilibrium. You *can* make bad judgments so you need a black box to check your subjectivism and to make it more objective. That then is the purpose of a subjective theory; to increase the objectivity of your judgments, to check them for consistency, to detect the inconsistencies and to remove them. Those who want their subjective judgments to be free and untrammeled by axioms regard themselves as objectivists: paradoxically, it is the subjectivists who are prepared to discipline their own judgments!

For a long time I have regarded this theory as almost intuitively obvious, partly perhaps because I have used it in many applications without inconsistencies arising, and I know of no other theory that I could personally adopt. It is the one and only True Religion. My main interest has been in developing and applying the theory, rather than finding *a priori* justification for it. But such a justification has been found by C. A. B. Smith (1961), based on a justification by Ramsey (1926/31/50/64), de Finetti (1937/64), and L. J. Savage (1954) of a slightly different theory (in which sharp values of the probabilities and utilities are assumed). These justifications show that, on the assumption of certain compelling desiderata, a rational person will hold beliefs, and preferences, *as if* he had a system of subjective probabilities and utilities satisfying a familiar set of axioms. He might as well be explicit about it: after all it is doubtful whether any of our knowledge is better than of the 'as if' variety.

Another class of justifications, in which utilities are not mentioned, is exemplified by Bernstein (1921/22), Koopman (1940a, b), and R. T. Cox (1946, 1961). (See also Good, 1947/50, pp. 105–6, and 1962a.) A less convincing, but simpler justification is that the product and addition axioms are forced (up to a monotonic transformation of probabilities) when considering ideal games of chance, and it would be *surprising* if the same axioms did not apply more generally. Even to deny this seems to me to show poor or biased judgment.

Since the degrees of belief, concerning events over which he has no control, of a person with ideally good judgment, should surely not depend on whether he intends to use his beliefs in any specific manner, it seems desirable to have justifications that do not mention preferences or utilities. But utilities necessarily come in whenever the beliefs are to be used in a practical problem involving action.

(c) CONSISTENCY AND THE UNOBVIOUSNESS OF THE OBVIOUS

Everybody knows, all scientists know, all mathematicians know, all players of chess know, that from a small number of sharp axioms you can develop a very rich theory in which the results are by no means obvious, even though they are, in a technical sense, tautological. This is an exciting feature of mathematics, especially since it suggests that it might be a feature of the universe itself. Thus the completion of the basic axioms of probability by Fermat and Pascal led to many interesting results, and the further near completion of the axioms for the *mathematical* theory of probability by Kolmogorov led to an even greater expansion of mathematical theory.

Mathematicians are I think often somewhat less aware that a system of rules and suggestions of an axiomatic system can also stimulate useful technical advances. The effect can be not necessarily nor even primarily to produce theorems of great logical depth, but sometimes more important to produce or to emphasize attitudes and techniques for reasoning and for statistics that seem obvious enough after they are suggested, but continue to be overlooked even then.

One reason why many theoretical statisticians prefer to prove mathematical theorems rather than to emphasize logical issues is that a theorem has a better chance of being indisputably novel. The person who proves the theorem can claim all the credit. But with general principles, however important, it is usually possible to find something in the past that to some extent foreshadows it. There is usually something old under the sun. Natural Selection was stated by Aristotle, but Darwin is not denied credit, for most scientists were still overlooking this almost obvious principle.

Let me mention a personal example of how the obvious can be overlooked.

In 1953, I was interested in estimating the physical probabilities for large contingency tables (using essentially a log-linear model) when the entries were very small, including zero. In the first draft of the write-up I wrote that I was concerned with the estimation of *probabilities of events that had never occurred before* (Good, 1953/56). Apparently this *concept* was itself an example of such an event, as far as the referee was concerned, because he felt it was too provocative, and I deleted it in deference to him. Yet this apparently 'pioneering' remark is obvious: every event in life is unique, and every real-life probability that we estimate in practice is that of an event that has never occurred before, provided that we do enough cross-classification. Yet there are many 'frequentists' who still sweep this fact UTC.

A statistical problem where this point arises all the time is in the estimation of physical probabilities corresponding to the cells of multi-dimensional contingency tables. Many cells will be empty for say a 2^{20} table. A Bayesian proposal for this problem was made in Good (1965b, p. 75), and I am hoping to get a student to look into it; and to compare it with the use of log-linear models which have been applied to this problem during the last few years. One example of the use of a log-linear model is, after taking logarithms of the relative frequencies, to apply a method of smoothing mentioned by Good (1958a) in relation to factorial experiments: namely to treat non-significant interactions as zero (or of course they could be 'flattened' Bayesianwise instead for slightly greater accuracy).

Yet another problem where the probabilities of events that have never occurred before are of interest is the species sampling problem. One of its aspects is the estimation of the probability that the next animal or word sampled will be one that has not previously occurred. The answer turns out to be approximately equal to n_1/N, where n_1 is the number of species that have so far occurred just once, and N is the total sample size: see

Good (1953), and Good and Toulmin (1956); this work originated with an idea of Turing's (1940) which anticipated the empirical Bayes method in a special case. (See also Robbins, 1968). The method can be regarded as non-Bayesian but with a Bayesian influence underlying it. More generally, the probability that the next animal will be one that has so far been represented r times is approximately $(r+1) n_{r+1}/N$, where n_r is the 'frequency of the frequency r', that is, the number of species each of which has already been represented r times. (In practice it is necessary to smooth the n_r's when applying this formula, to get adequate results, when $r > 1$.) I shall here give a new proof of this result. Denote the event of obtaining such an animal by E_r. Since the order in which the N animals were sampled is assumed to be irrelevant (a Bayesian-type assumption of permutability), the required probability can be estimated by the probability that E_r would have occurred on the last occasion an animal was sampled if a random permutation were applied to the order in which the N animals were sampled. But E_r would have occurred if the last animal had belonged to a species represented $r+1$ times *altogether*. This gives the result, except that for greater accuracy we should remember that we are talking about the $(N+1)$st trial, so that a more accurate result is $(r+1) \mathscr{E}_{N+1}(n_{r+1})/(N+1)$. Hence the expected physical probability q_r corresponding to those n_r species that have so far occurred r times is

$$\mathscr{E}(q_r) = \frac{r+1}{N+1} \cdot \frac{\mathscr{E}_{N+1}(n_{r+1})}{\mathscr{E}_N(n_r)}.$$

This is formula (15) of Good (1953) which was obtained by a more Bayesian argument. The 'variance' of q_r was also derived in that paper, and a 'frequency' proof of it would be more difficult. There is an interplay here between Bayesian and frequency ideas.

One aspect of Doogianism which dates back at least to F. P. Ramsey (1926/31/50/64) is the emphasis on *consistency*: for example, the axioms of probability can provide only *relationships* between probabilities and cannot manufacture a probability out of nothing. Therefore there must be a hidden assumption in Fisher's fiducial argument. [This assumption is pinpointed in Good (1970/71a), p. 139, line minus 4. The *reason* Fisher overlooked this is also explained there.]

The idea of consistency seems weak enough, but it has the following immediate consequence which *is* often overlooked.

Owing to the adjectives 'initial' and 'final' or 'prior' and 'posterior', it is usually assumed that initial probabilities must be assumed before final ones can be calculated. But there is nothing in the theory to prevent the implication being in the reverse direction: we can make judgments of initial probabilities and infer final ones, or we can equally make judgments of final ones and infer initial ones by *Bayes' theorem in reverse*. Moreover this can be done corresponding to entirely *imaginary* observations. This is what I mean by the Device of Imaginary Results (for the judging of initial probabilities). (See, for example, Good, 1947/50, Index). I found this device extremely useful in connection with the choice of a prior for multinomial estimation and significance problems (Good, 1965/67) and I believe the device will be found to be of the utmost value in future Bayesian statistics. Hypothetical experiments have been familiar for a long time in physics, and in the arguments that led Ramsey to the axioms of subjective probability, but the use of Bayes's theorem in reverse is less familiar. "Ye priors shall be known by their posteriors" (Good, 1970/71a, p. 124). Even the slightly more obvious technique of imaginary bets is still disdained by many decision makers who like to say "That possibility is purely hypothetical". Anyone who disdains the hypothetical is a philistine.

III. THE ELEVENFOLD PATH OF DOOGIANISM

As I said before, I should now like to take up the 46656 varieties of Bayesians, in other words the eleven facets for their categorization. I would have discussed the 27-fold path of Doogianism if there had been space enough.

(i) RATIONALITY OF TYPES I AND II

I have already referred to the first facet. Rationality of Type I is the recommendation to maximize expected utility, and Type II is the same except that it allows for the cost of theorizing. It means that in any practical situation you have to decide when to stop thinking. You can't allow the current to go on circulating round and round the black box or the cranium forever. You would like to reach a sufficient maturity of judgments, but you have eventually to reach some conclusion or to make some decision and so you must be prepared to sacrifice strict logical consistency. At best you can achieve consistency as far as you have so far seen (Good, 1947/50, p. 49). There is a time element, as in chess, and this is realistic of most practice. It might not appeal to some of you who love ordinary logic, but it is a mirror of the true situation.

It may help to convince some readers if I recall a remark of Poincaré's that some antinomies in ordinary (non-probabilistic) logic can be resolved by bringing in a time element.

The notion of Type II rationality, which I believe resolves a great many of the controversies between the orthodox and Bayesian points of view, also involves a shifting of your probabilities. The subjective probabilities shift as a consequence of thinking. I am not referring to the shifting of probabilities in the light of new empirical evidence here; I am referring to the shifting of probabilities in the light of pure thought. A banal example is that if you are asked to bet or to estimate the probability that the millionth digit of π is a 7, you would sensibly regard the betting probability as 0.1 even though you know that the logical probability is either 0 or 1. We could do the calculations; as soon as we finish doing the calculations, if the machine hasn't gone wrong by that time we would know that the 'true probability', so to speak, is 0 or 1. If it is too expensive to do the calculations, for the time being you would assume the probability to be 0.1. That is a typical situation although it sounds too special. But it is not special. For example, the pure mathematician constantly needs to estimate probabilities of theorems in a rough and ready way. The conscious recognition of Type II rationality, or not, constitutes the two aspects of the first facet.

Another name for the principle of Type II rationality might be the *Principle of Non-dogmatism*.

(ii) KINDS OF JUDGMENT

Inequalities between probabilities and between expected utilities are perhaps the most standard type of judgment, but other kinds are possible. Because of my respect for the human mind, I believe that one should allow any kind of judgments that are relevant. One kind that I believe will ultimately be regarded as vying in importance with the two just mentioned is a judgment of 'weights of evidence' (defined later) a term introduced by Charles Saunders Peirce (1878) although I did now know this when I wrote my 1950 book.

Some readers will say, 'There goes Good again, plugging weights of evidence'. In my defence I make the following claim: The only hope for the future of democracy (pardon the ambiguity) is that the masses should learn how to think rationally. This means that they must sooner or later make conscious judgments of probabilities and utilities. It is only slightly less valuable that they should be capable of estimating weights of evidence. It will encourage a revival of reasoning if statisticians adopt Peirce's appealing terminology which historically preceded the talk of the closely related and linguistically clumsy 'log-likelihood ratio' by over forty years.

One implication of the 'suggestion' that all types of judgments can be used is to encourage you to compare your 'overall' judgments with your detailed ones; for example, a judgment by a doctor that it is better to operate than to apply medical treatment (on the grounds perhaps that this would be standard practice in the given circumstances) can be 'played off' against separate judgments of the probabilities and utilities of the outcomes of the various treatments.

(iii) PRECISION OF JUDGMENTS

Most theories of subjective probability deal with numerically precise probabilities. These would be entirely appropriate if you could always state the lowest odds that you would be prepared to accept in a gamble, but in practice there is usually a degree of vagueness. Hence I assume that subjective probabilities are only partially ordered. In this I follow Keynes and Koopman, for example, except that Keynes dealt primarily with logical probabilities, and Koopman with 'intuitive' ones (which means either logical or subjective). F. P. Ramsey (1926/31/50/64) dealt with subjective probabilities, but 'sharp' ones, as mentioned before.

A theory of 'partial ordering' (inequality judgments) for probabilities is a compromise between Bayesian and non-Bayesian ideas. For if a probability is judged merely to lie between 0 and 1, this is equivalent to making no judgment about it at all. The vaguer the probabilities the closer is this Bayesian viewpoint to a non-Bayesian one.

Often, in the interests of simplicity, I assume sharp probabilities, as an approximation, in accordance with Type II rationality.

(iv) ECLECTICISM

Many Bayesians take the extreme point of view that Bayesian methods should always be used in statistics. My view is that non-Bayesian methods are acceptable *provided that they are not seen to contradict your honest judgments, when combined with the axioms of rationality.* This facet number (iv) is an application of Type II rationality. I believe it is sometimes, but not by any means always, easier to use 'orthodox' (non-Bayesian) methods, and that they are often *good enough.* It is always an application of Type II rationality to say that a method is good enough.

(v) SHOULD UTILITIES BE BROUGHT IN FROM THE START IN THE DEVELOPMENT OF THE THEORY?

I have already stated my preference for trying to build up the theory of subjective probability without reference to utilities and to bring in utilities later. The way the axioms are introduced is not of great practical importance, provided that the same axioms are reached in the end, but it is of philosophical interest. Also there is practical interest in seeing how far one can go without making use of utilities, because one might wish to be an 'armchair philosopher' or 'fun scientist' who is more concerned with discovering facts about Nature than in applying them. ('Fun scientist' is not intended to be a derogatory expression.) Thus, for example, R. A. Fisher and Harold Jeffreys never used ordinary utilities in their statistical work as far as I know (and when Jeffreys chaired the meeting in Cambridge when I presented my paper, Good (1951/52), he stated that he had never been concerned with economic problems in his work on probability). See also the following remarks concerned with quasiutilities.

(vi) QUASIUTILITIES

Just as some schools of Bayesians regard subjective probabilities as having sharp (precise) values, some assume that utilities are also sharp. The Doogian believes that this is often not so. It is not merely that utility inequality judgments of course vary from one person to another, but that utilities for individuals can also often be judged by them only to lie in wide intervals. It consequently becomes useful and convenient to make use of substitutes for utility which may be called *quasiutilities* or *pseudo-utilities.* Examples and applications of quasiutilities will be considered later in this paper. The conscious recognition or otherwise of quasi-utilities constitutes the sixth facet.

(vii) PHYSICAL PROBABILITY

Different Bayesians have different attitudes to the question of physical probability. de Finetti regards it as a concept that can be defined in terms of subjective probability, and does not attribute any other 'real existence' to it. My view (or that of my alter ego) is that it seems reasonable to suppose that physical probabilities do exist, but that they can be measured only by means of a theory of subjective probability. For a fuller discussion of this point see de Finetti (1968/70) and Good (1968/70). The question of the real existence of physical probabilities relates to the problem of determinism versus indeterminism and I shall have something more to say on this.

When I refer to physical probability I do not assume the long-run frequency definition: physical probability can be applied just as well to unique circumstances. Popper suggested the word 'propensity' for it, which I think is a good term, although I think the suggestion of a word cannot by itself be regarded as the propounding of a 'theory'. As I have indicated before, I think good terminology is important in crystallizing out ideas. Language can easily mislead, but part of the philosopher's job is to find out where it can *lead*. (Curiously enough Popper has also stated that the words you use do not matter much: what is important is what they mean in your context. Fair enough, but it can lead to Humpty-Dumpty-ism, such as Popper's interpretation of simplicity.)

(viii) WHICH IS PRIMARY, LOGICAL PROBABILITY (CREDIBILITY) OR SUBJECTIVE PROBABILITY?

It seems to me that subjective probabilities are primary because they are the ones you have to use whether you like it or not. But I think it is mentally healthy to think of your subjective probabilties as estimates of credibilities, whether these really 'exist' or not. Harold Jeffreys said that the credibilities should be laid down by an international body. He would undoubtedly be the chairman. (Much laughter.) As Henry Daniels once said (c. 1952) when I was arguing for subjectivism, 'all statisticians would like their methods to be adopted', meaning that in some sense everybody is a subjectivist.

(ix) IMAGINARY RESULTS

This matter has already been discussed but I am mentioning it again because it distinguishes between some Bayesians in practice, and so forms part of the categorization under discussion. I shall give an example of it now because this will help to shed light on the tenth facet.

It is necessary to introduce some notation. Let us suppose that we throw a sample of N things into t pigeon holes, with statistically independent physical probabilities $p_1, p_2, ..., p_t$, these being unknown, and that you obtain frequencies $n_1, n_2, ..., n_t$ in the t categories or cells. This is a situation that has much interested philosophers of induction, but for some reason (presumably lack of familiarity) they do not usually call it multinomial sampling. In common with many people in the past, I was interested (Good, 1965b, 1965/67) in estimating the physical probabilities $p_1, p_2, ..., p_t$. The philosopher William Ernest Johnson was especially interested in this problem. Now you might say that the obvious thing is to take n_i/N as your estimate, that's the maximum likelihood estimate for p_i. But if $n_i=0$, this is liable to be a bad estimate because it might lead you to give infinite odds in a bet against the next object's falling in the ith cell. So the frequencies n_i need a little 'flattening'. You can flatten them if you like by adding constants to them and many orthodox statisticians would say 'Let's just add 1 and we'll look at the operational characteristics'. That might be reasonable, but seems arbitrary and we'd like some better justification if possible. Well, William Ernest Johnson proved that under certain assumptions you must add a constant, but he wasn't able to say what the constant should be. (Also one of his assumptions was not convincing.) Now, by a generalization of a theorem of de Finetti, you can prove that adding a constant k is equivalent to assuming a unique initial distribution (see, for example, Good, 1965b, p. 22). One can then confirm that the initial density must be proportional to $(p_1 p_2 ... p_t)^{k-1}$ which is known as a symmetric Dirichlet distribution. (Johnson did not know this.) Now, I've found that this prior is not adequate for any value of k, so I made the next simplest assumption, and took a linear combination of these priors. This is equivalent to assuming a distribution for k which I called a Type III distribution. Some choices for the Type III distribution produces results consistent enough with orthodox (non-Bayesian) methods for a variety of imaginary samples. It wasn't necessary to use real data although I did happen to use real data also. Imaginary data has the disadvantage that it annoys some people, but it has the advantage that it can be generated more quickly and that it avoids much of the wishful thinking that might be associated with real data. You can just imagine some results, make some final judgments, and require that the prior distribution selected is consistent with the judgments you insist upon. That approach led me to what I think was a satisfactory prior (Good, 1965/67). No satisfactory prior had been proposed previously. That then is an example of a philosophical attitude leading to a practical solution of a statistical problem. As a matter of fact, it wasn't just the estimation of the p's that emerged from that work, but, more important, a significance test for whether the p's were all equal. The method has the pragmatic advantage that it can be used for all sample sizes, whereas the ordinary chi-squared test breaks down when the cell averages are less than 1. Once you have decided on a prior (the initial relative probabilities of the components of the non-null hypothesis), you can calculate the weight of evidence against the null hypothesis without using asymptotic theory. [This would be true for any prior that is a linear combination of Dirichlet distributions, even if they were not symmetric, because in this case the calculations involve only one-dimensional integrations.] That then was an example of the device of imaginary results, for the selection of a prior, worked out in detail.

   The successful use of the device of imaginary results for this problem *makes it obvious that it can and will also be used effectively for many other statistical problems. I believe it will revolutionize multivariate Bayesian statistics.*

(x) HIERARCHIES OF PROBABILITIES

When you make a judgment about probabilities you might sit back and say 'Is that judgment probable'. This is how the mind works – it is natural to think that way, and this leads to a hierachy of types of probabilities (Good, 1951/52) which in the example just mentioned, I found useful, as well as on other occasions. Now an objection immediately arises: There is nothing in principle to stop you integrating out the higher types of probability. But it remains a useful suggestion to help the mind in making judgments. It was used by Good (1965/67) and has now been adopted by other Bayesians, using different terminology, such as priors of the second 'order' (instead of 'type') or 'two-stage Bayesian models'. A convenient term for a parameter in a prior is 'hyperparameter'.

   New techniques arose out of the hierarchical suggestion, again apparently first in connection with the multinomial distribution (in the same paper), namely the concept of Type II maximum likelihood (maximization of the Bayes factor against the null hypothesis by allowing the hyperparameters to vary), and that of a Type II likelihood ratio for significance tests. I shall discuss these two concepts when discussing likelihood in general.

(xi) THE CHOICE OF AXIOMS

One distinction between different kinds of Bayesians is merely a mathematical one, whether the axioms should be taken as simple as possible, or whether, for example, they should include Kolmogorov's axiom, the axiom of complete additivity. I prefer the former course because I would want people to use the axioms even if they do not know what 'enumerable' means, but I am prepared to use Kolmogorov's axiom whenever it seems to be sufficiently mathematically convenient. Its interest is mathematical rather than philosophical (except perhaps for the philosophy of mathematics). This last facet by the way is related to an excellent lecture by Jimmie Savage of about 1970, called "What kind of probability do you want?".

So much for the eleven facets. Numbers (i) to (vii) and number (ix) all involve a compromise with non-Bayesian methods; and number (viii) a compromise with the 'credibilists'.

IV. EXAMPLES OF THE BAYESIAN INFLUENCE AND OF SUTC

(a) THE LOGICAL AND HISTORICAL ORIGINS OF LIKELIHOOD

One aspect of utility is communicating with other people. There are many situations where you are interested in making a decision without communicating. But there are also many situations, especially in much statistical and scientific practice where you do wish to communicate. One suggestion, 'obvious', and often overlooked as usual, is that you should make your assumptions clear and you should try to separate out the part that is disputable from the part that is less so. One immediate consequence of this suggestion is an emphasis on likelihood, because, as you all know, in Bayes' theorem you have the initial probabilities, and then you have the likelihoods which are the probabilities of the event, given the various hypotheses, and then you multiply the likelihoods by the probabilities and that gives you results proportional to the final probabilities. That is Bayes' theorem expressed neatly, the way Harold Jeffreys (1939/61) expressed it. Now the initial probability of the null hypothesis is often highly disputable. One person might judge it to be between 10^{-3} and 10^{-1} whereas another might judge it to be between 0.9 and 0.99. There is much less dispute about likelihoods. There is no dispute about the numerical values of likelihoods if your basic parametric model is accepted. Of course you usually have to use subjective judgment in laying down your parametric model. Now the *hidebound* objectivist tends to hide that fact; he will not volunteer the information that he uses judgment at all, but if pressed he will say "I do, in fact, have good judgment" [Laughter.] So there are good and bad subjectivists, the bad subjectivists are the people with bad or dishonest judgment and also the people who do not make their assumptions clear when communicating with other people. But, on the other hand, there are no good 100% (hidebound) objectivists; they are all bad (laughter) because they sweep their judgments UTC.

> *Aside:* In the spoken discussion the following beautiful interchanges took place. *Kempthorne* (who also made some complimentary comments): Now, on the likelihood business, the Bayesians discovered likelihood Goddamit! Fisher knew all this stuff. Now look Jack, you are an educated guy. Now please don't pull this stuff. This really drives me up the wall! *Lindley:* If Fisher understood the likelihood principle why did he violate it? *Kempthorne:* I'm not saying he understood it and I'm not saying you do or you – nobody understands it (laughter). But likelihood ideas, so to speak, have some relevance to the data. That's a completely non-Bayesian argument. *Good:* It dates back to the 18th century. *Kempthorne:* Oh it dates back; but there are a lot of things being (?) Doogian, you know. They started with this guy Doog. Who is this bugger? (Laughter.) Doog is the guy who spells everything backwards. (Prolonged laughter.)

In reply to this entertaining harangue, which was provoked by a misunderstanding that was perhaps my fault, although I did refer to Fisherian information, I mention the following points. Bayes' theorem (Bayes, 1763/65, 1940/58; Laplace, 1774) cannot be stated without introducing likelihoods; *therefore likelihood dates back at least to 1774.* Again, *maximum* likelihood was used by Daniel Bernoulli (1774/78/1961); see, for example, Todhunter (1865/1965, p. 236) or Eisenhart (1964, p. 29). Fisher introduced the name *likelihood* and emphasized the method of maximum likelihood. Such emphasis is important and of course merits recognition. The fact that he was anticipated in its use does not deprive him of the major part of the credit or of the blame especially as the notion of defining amount of information in terms of likelihood was his brilliant idea and it led to the Aitken-Silverstone information inequality (the minimum-variance bound).

Gauss (1798/1809/57/1963) according to Eisenhart, used inverse probability combined with a Bayes *postulate* (uniform initial distribution) and an assumption of normal error, to give one of the interpretations of the method of least squares. He could have used maximum likelihood in this context but apparently did not, so perhaps Daniel Bernoulli's use of maximum likelihood had failed to convince him or to be noticed by him. Further historical research might be required to settle this last question if it is possible to settle it at all.

So likelihood is important as all statisticians agree now-a-days, and it *takes sharper values* than initial probabilities. But some people have gone to extremes and say that initial probabilities don't mean anything. Now I think one reason for their saying so is trade unionism of a certain kind. It is very nice for a statistician to be able to give his customer absolutely clear-cut results. It is unfortunate that he can't do it so he is tempted to cover up, to pretend he has not had to use any judgment. Those Bayesians who *insist* on sharp initial probabilities are I think also guilty of 'trade unionism', unless they are careful to point out that these are intended only as crude approximations, for I do not believe that sharp initial probabilities usually correspond to their honest introspection. If, on the other hand, they agree that they are using only approximations we might need more information about the degree of the approximations, and then they would be forced to use inequality judgments, thus bringing them closer to the True Religion. (I believe Dr Kyburg's dislike of the Bayesian position, as expressed by him later in this conference, depended on his interpreting a Bayesian as one who uses sharp initial probabilities.) The use of 'vague' initial probabilities (inequality judgments) does not prevent Bayes' theorem from establishing the likelihood principle. For Dr Kempthorne's benefit, and perhaps for some others, I mention that to me the likelihood principle means that the likelihood function exhausts all the information about the parameters that can be obtained from an experiment or observation, provided of course that there is an undisputed set of exhaustive simple statistical hypotheses such as is provided, for example, by a parametric model. [In practice, such assumptions are often undisputed but are never indisputable. This is the main reason why significance tests, such as the chi-squared test, robust to changes in the model, are of value. Even here there is a Doogian interpretation that can be based on beliefs about the distribution of the test statistic when it is assumed that the null hypothesis is false. I leave this point on one side for the moment.] Given the likelihood, the inferences that can be drawn from the observations would, for example, be unaffected if the statistician arbitrarily and falsely claimed that he had a train to catch, although he really had decided to stop sampling because his favorite hypothesis was ahead of the game. (This might cause you to distrust the statistician, but if you believe his observations, this distrust would be immaterial.) On the other hand, the 'Fisherian' tail-area method for significance testing violates the likelihood principle because the statistician who is prepared to pretend he has a train to catch (optional stopping of sampling) can reach arbitrarily high significance levels, given enough time, even when the null hypothesis is true. (For example, see Good, 1955/56b, p. 13.)

(b) WEIGHT OF EVIDENCE

Closely related to the concept of likelihood is that of weight of evidence, which I mentioned before and promised to define.

Let us suppose that we have only two hypotheses under consideration (which might be because we have decided to consider hypotheses two at a time). Denote them by H and $\bar{H}$, where the bar over the second H denotes negation. (These need not be 'simple statistical hypotheses', as defined in a moment.) Suppose further that we have an event, experimental result, or observation denoted by E. The conditional probability of E is either $P(E \mid H)$ or $P(E \mid \bar{H})$, depending on whether H or $\bar{H}$ is assumed. If H and $\bar{H}$ are 'simple statistical hypotheses', then these two probabilities have sharp uncontroversial values given tautologically by the meanings of H and $\bar{H}$. Even if they are *composite* hypotheses (not 'simple' ones) the Bayesian will still be prepared to talk about these two probabilities. In either case we can see, by four applications of the product axiom, or by two applications of Bayes' theorem, that

$$\frac{P(E \mid H)}{P(E \mid \bar{H})} = \frac{O(H \mid E)}{O(H)}$$

where O denotes *odds*. (The odds corresponding to a probability p are defined as $p/(1-p)$.) Turing (1940) called the right side of this equation *the factor in favor of the hypothesis H provided by the evidence E*, for obvious reasons. Its logarithm is the *weight of evidence* in favor of H, as defined independently by Peirce (1878), Good (1947/50), and Minsky and Selfridge (1961). It was much used by Harold Jeffreys (1939/61), except that in that book he identified it with the final log-odds because his initial probabilities were taken as 1/2. He had previously (1936) used the general form of weight of evidence and had called it 'support'. The non-Bayesian uses the left side of the equation, and calls it the probability ratio, provided that H and $\bar{H}$ are simple statistical hypotheses. He SUTC the right side, because he does not talk about the probability of a hypothesis. The Bayesian, the doctor, the judge and the jury can appreciate the importance of the right side even with only the vaguest estimates of the initial odds of H. For example, the Bayesian (or at least the Doogian) can logically argue in the following manner (Good, 1974/50, p. 70): If we assume that it was sensible to start a sampling experiment in the first place, and if it has provided appreciable weight of evidence in favor of some hypothesis, and it is felt that the hypothesis is not yet convincing enough, then it is sensible to enlarge the sample since we know that the final odds of the hypothesis have increased whatever they are. Such conclusions can be reached even though judgments of the relevant initial probability and of the utilities have never been announced. Thus, even when the initial probability is extremely vague, the axioms of subjective probability (and weight of evidence) can be applied.

When one or both of H and $\bar{H}$ are composite, the Bayesian has to assume relative initial probabilities for the simple components of the composite hypothesis. Although these are subjective, they typically seem to be less subjective than the initial probability of H itself. To put the matter more quantitatively, although this is not easy in so general a context, I should say that the judgment of the factor in favor of a hypothesis H might typically differ from one person to another by up to about 5, while the initial odds of H might differ by a factor of 10 or 100 or 1000. Thus the separation of the estimation of the weight of evidence from the initial (or final) probability of H serves a useful purpose, especially for communication with other people, just as it is often advisable to separate the judgments of initial probabilities and likelihoods.

It often happens that the weight of evidence is so great that a hypothesis seems convincing almost irrespective of the initial probability. For example, in quantum mechanics, it seems convincing that the Schrödinger equation is approximately true (subject to some limitations), given the rest of some standard formulation of quantum mechanics, because of great quantities of evidence from a variety of experiments, such as the measurements of the frequencies of spectral lines to several places of decimals. The large weight of evidence makes it seem, to people who do not stop to think, that the initial probability of the equation (conditional on the rest of the theory) is irrelevant; but really there has to be an implicit judgment that the initial probability is not too low; for example, not less that 10^{-50}. (In a fuller discussion I would prefer to talk of the relative odds of *two* equations in competition.) How we judge such inequalities, whether explicitly or implicitly, is not clear: if we knew how we made judgments we would not call them judgments (Good, 1958/59). It must be something to do with the length of the equation (just as the total length of chromosomes in a cell could be used as a measure of complexity of an organism) and with its analogy with the classical wave equation and heat equation. (The latter has even suggested to some people, for example, Weizel (1953), that there is some underlying random motion that will be found to 'explain' the equation.) At any rate the large weight of evidence permits the initial probability to be SUTC and it leads to an apparent objectivism (the reliance on the likelihoods alone) that is really multisubjectivism. The same happens in many affairs of ordinary life, in perception (Good, 1947/50, p. 68), in the law, and in medical diagnosis (for example, Card and Good, 1973).

On a point of terminology, the factor in favor of a hypothesis is equal to the likelihood ratio, in the sense of Neyman, Pearson, and Wilks, only when both H and $\bar{H}$ are simple statistical hypotheses. This is another justification for using Turing's and Peirce's expressions, apart from their almost self-explanatory nature, which provides their potential for improving the reasoning powers of all people. Certainly the expression 'weight of evidence' captures one of the meanings that was intended by ordinary language. It is not surprising that it was an outstanding philosopher who first noticed this: for one of the functions of philosophy is to make such captures.

George Barnard, who is one of the Likelihood Brethren, has rightly emphasized the merits of graphing the likelihood function. A Bayesian should support this technique because the initial probability density can be combined with the likelihood afterwards. If the Bayesian is a subjectivist he will know that the initial probability density varies from person to person and so he will see the value of graphing of the likelihood function for communication. A Doogian will consider that even his own initial probability density is not unique so he should approve even more. Difficulties arise in general if the parameter space has more than two dimensions, both in picturing the likelihood hypersurface or the posterior density hypersurface. The problem is less acute when the hypersurfaces are quadratic in the neighborhood of the maximum. In any case the Bayesian can in addition reduce the data by using such quantitites as expected utilities. Thus he has all the advantages claimed by the likelihood brotherhood, but has additional flexibility.

(c) MAXIMUM LIKELIHOOD, INVARIANCE, QUASIUTILITIES, AND QUASILOSSES

Let us now consider the relationship between Bayesian methods and *maximum* likelihood.

In a 'full-dress' Bayesian estimation of parameters (allowing for utilities) you compute their final distribution and use it, combined with a loss function, to find a single recommended value, if a point estimate is wanted. When the loss function is quadratic this implies that the point estimate should be the final expectation of the parameter (even for vector parameters if the quadratic is non-singular). The final expectation is also appropriate if the parameter is a physical probability because the subjective expectation of a physical probability of an event is equal to the current subjective probability of that event.

If you do not wish to appear to assume a loss function, you can adopt the argument of Jeffreys (1939/61, Section 4.0). He points out that for a sample of size n (n observations), the final probability density is concentrated in a range of order $n^{-1/2}$, and that the difference between the maxi-

mum-likelihood value of the parameter and the mode of the final probability density is of order $1/n$. (I call this last method, the choice of this
mode, a Bayesian method 'in mufti'.) "Hence if the number of observations is large, the error committed by taking the maximum likelihood
solution as the estimate is less than the uncertainty inevitable in any case.
... the above argument shows that in the great bulk of cases its results are
indistinguishable from those given by the principle of inverse probability,
which supplies a justification for it". It also will not usually make much
difference if the parameter is assumed to have a uniform initial distribution. (Jeffreys, 1939/61, p. 145; Good, 1947/50, p. 55. L. J. Savage,
1959/62, p. 23, named estimation that depends on this last point 'stable
estimation'.)

By a slight extension of Jeffreys's argument, we can see that a point
estimate based on a loss function, whether it is the expectation of the
parameter or some other value (which will be a kind of average) induced
by the loss function, will also be approximated by using the Bayes method
in mufti, and by the maximum-likelihood estimate, when the number of
observations is large. Thus the large-sample properties of the maximum-
likelihood method cannot be used for distinguishing it from a wide class
of Bayesian methods, whether full-dress or in mufti. This is true whether
we are dealing with point estimates or interval estimates. Interval estimates and posterior distributions are generally more useful, but point
estimates are easier to talk about and we shall concentrate on them for
the sake of simplicity.

One may also regard the matter from a more geometrical point of view.
If the graph of the likelihood function is sharply peaked, then the final
density will also usually be sharply peaked at nearly the same place. This
again makes it clear that there is often not much difference between
Bayesian estimation and maximum-likelihood estimation, provided that
the sample is large. This argument applies provided that the number of
parameters is itself not large.

All this is on the assumption that the Bayesian assumptions are not
dogmatic in the sense of ascribing zero initial probability to some range
of values of the parameter; though 'provisional dogmatism' is often justifiable to save time, where you hold at the back of your mind that it might
be necessary to make an adjustment in the light of the evidence. Thus I do
not agree with the often-given dogmatic advice that significance tests *must*

be chosen before looking at the results of an experiment, although of course I appreciate the point of the advice. It is appropriate advice for people of bad judgment.

It is perhaps significant that Daniel Bernoulli introduced the method of maximum likelihood (in a special case) at almost the same time as the papers by Bayes and Laplace on inverse probability were published. But, as I said before, it is the logical rather than the historical connections that I wish to emphasize most. I merely state my belief that the influence of informal Bayesian thinking on apparently non-Bayesian methods has been considerable at both a conscious and a less conscious level, ever since 1763, and even from 1925 to 1950 when non-Bayesian methods were at their zenith relative to Bayesian ones.

Let us consider loss functions in more detail. In practice, many statisticians who do not think of themselves as Bayesians make use of 'squared-error loss', and regard it as Gauss-given, without usually being fully aware that a loss is a negative utility and smacks of Bayes. The method of least squares is not always regarded as an example of minimizing squared loss (see Eisenhart, 1964), but it *can* be thought of that way. It measures the value of a putative regression line for fitting given observations. Since statisticians might not always be happy to concur with this interpretation, perhaps a better term for 'loss' when used in this conventional way would be 'quasiloss' or 'pseudoloss'. We might use it, *faute de mieux*, when we are not sure what the utility is, although posterior distributions for the parameters of the regression line would preserve more of the information.

If the loss is an analytic function it has to be quadratic in the neighborhood of the correct answer, but it would be more realistic in most applications to assume the loss to be asymptotic to some value when the estimate is far away from the correct value. Thus a curve or surface having the shape of an upside-down normal or multinormal density would be theoretically better than 'squared loss' (a parabola or paraboloid). But when the samples are large enough the 'tails of the loss function' perhaps do not usually affect the estimates much, except when there are outliers.

Once the minimization of the sum of squared residuals is regarded as an attempt to maximize a utility, it leads us to ask what other substitutes for utility might be used, *quasiutilities* if you like. This question dates back over a quarter of a millenium in estimation problems. Like quasi-losses, which are merely quasiutilities with a change of sign, they are introduced because it is often difficult to decide what the real utilities are. This difficulty especially occurs when the applications of your work are not all known in advance, as in much pure science (the 'knowledge business' to use Kempthorne's term). A quasiutility might be somewhat *ad hoc*, used partly for its mathematical convenience in accordance with Type II rationality. It is fairly clear that this was an important reason historically for the adoption of the method of least squares on a wide scale.

Fisher once expressed scorn for economic applications of statistics, but he introduced his ingenious concept of amount of information in connection with the estimation of parameters, and it can be regarded as another quasiutility. It measures the expected value of an experiment for estimating a parameter. Then again Turing made use of expected weight of evidence for a particular application in 1940. It measures the expected value of an experiment for discriminating between two hypotheses. The idea of using the closely related Shannon information in the design of statistical experiments has been proposed a number of times (Cronbach, 1953; Lindley, 1956; Good, 1955/56a), and is especially pertinent for problems of search such as in dendroidal medical diagnosis (for example, Good, 1967/70). It measures the expected value of an experiment for distinguishing between several hypotheses. *In this medical example the doctor should switch to more 'real' utilities for his decisions when he comes close enough to the end of the search to be able to 'backtrack'.* A number of other possible quasiutilities are suggested in Good (1967/70) and in Good and Card (1970/71), some of which are invariant with respect to transformations of the parameter space.

In all these cases, it seems to me that the various concepts are introduced essentially because of the difficulty of making use of utility in its more standard economic sense. I believe the term 'quasi-utility' might be useful in helping to crystallize this fact, and thus help to clarify and unify the logic of statistical inference. The quasiutilities mentioned so far are all defined in terms of the probability model alone, but I do not regard this feature as part of the definition of a quasiutility.

Even in the law, the concept of weight of evidence (in its ordinary linguistic sense, which I think is usually the same as its technical sense though not formalized) helps to put the emphasis on the search for the truth, leaving utilities to be taken into account later. One might even conjecture that the expressions 'amount of information' and 'weight of evidence' entered the *English language* because utilities cannot always be sharply and uncontroversially estimated. Both these expressions can be given useful quantitative meanings defined in terms of probabilities alone, and so are relevant to the 'knowledge business'.

These various forms of quasiutility were not all suggested with the conscious idea of replacing utility by something else, but it is illuminating to think of them in this light, and, if the *word* 'quasi-utility' had been as old as quasiutilities themselves, the connection could not have been overlooked. It shows how influential Bayesian ideas can be in the logic if not always in the history of statistics. The history is difficult to trace because of the tendency of many writers (i) to cover up their tracks, (ii) to forget the origins of their ideas, deliberately or otherwise, and (iii) not to be much concerned with 'kudology', the fair attributions of credit, other than the credit of those they happen to like such as themselves.

The word 'quasiutility' provokes one to consider whether there are other features of utility theory that might be interesting to apply to quasiutilities, apart from the maximization of their expectations for decision purposes. One such feature is the use of minimax procedures, that is, cautious decision procedures that minimize the maximum expected loss (or quasiloss here). Although minimax procedures are controversial, they have something to be said for them. They can be used when all priors are regarded as possible, or more generally when there is a class of possible priors (Hurwicz, 1951; Good, 1951/52, where this generalized minimax procedure was independently proposed: 'Type II minimax'), so that there is no unique Bayesian decision: then the minimax procedure corresponds to the selection of the 'least favorable' prior, in accordance with a theorem of Wald (1950). When the quasiutility is invariant with respect to transformations of the parameter space, then so is the corresponding minimax procedure and it therefore has the merit of decreasing arbitrariness. When the quasiutility is Shannon (or Szilard) information, the minimax procedure involves choosing the prior of maximum entropy (Good, 1968/69b, 1968), a suggestion made for other reasons by Jaynes (1957). The maximum-entropy method was reinterpreted as a method for *selecting null hypotheses* by Good (1962/63). I especially would like to emphasize this interpretation because the formulation of hypotheses is often said to lie outside the statistician's domain of formal activity, *qua* statistician. It has been pointed out that Jeffreys's invariant prior (Jeffreys, 1946) can be regarded as a minimax choice when quasiutility is measured by weight of evidence (Good, 1968/69b, 1968). Thus other invariant priors could be obtained from other invariant quasiutilities (of which there is a one-parameter family mentioned later).

Jeffreys's invariant prior is equal to the square root of the determinant of Fisher's information matrix, although Jeffreys (1946) did not express it this way explicitly. Thus there can be a logical influence from non-Bayesian to Bayesian methods, and of course many other examples of influence in this direction could be listed.

Let us return to the discussion of Maximum Likelihood (ML) estimation. Since nearly all methods lead to ROME (Roughly Optimal Mantic Estimation) when samples are large, the real justification for choosing one method rather than another one must be based on samples that are not large.

One interesting feature of ML estimation, a partial justification for it, is its invariance property. That is, if the ML estimate of a parameter θ is denoted by $\hat{\theta}$, then the ML estimate of $f(\theta)$, for any monotonic function f, even a discontinuous one, is simply $f(\hat{\theta})$. Certainly invariant procedures have the attraction of decreasing arbitrariness to some extent, and it is a desideratum for an *ideal* procedure. But there are other invariant procedures of a more Bayesian tone to which I shall soon return: of course a completely Bayesian method would be invariant if the prior probabilities and utilities were indisputable. Invariance, like patriotism, is not enough. An example of a very bad invariant method is to choose as the estimate the least upper bound of all possible values of the parameter if it is a scalar. This method is invariant under all increasing monotonic transformations of the parameter!

Let us consider what happens to ML estimation for the physical probabilities of a multinomial distribution, which has been used as a proving ground for many philosophical ideas.

In the notation used earlier, let the frequencies in the cells be n_1, $n_2, ..., n_t$, with total sample size N. Then the ML estimates of the physical probabilities are n_i/N, $i = 1, 2, ..., t$. Now I suppose many people would say that a sample size of $N = 1,000$ is large, but even with this size it could easily happen that one of the n_i's is zero, for example, the letter Z could well be absent in a sample of 1,000 letters of English text. Thus a sample might be large in one sense but effectively small in another (Good, 1953,

1953/56, 1965b). If one of the letters is absent ($n_i = 0$), then the maximum-likelihood estimate of p_i is zero. This is an appallingly bad estimate if it is used in a gamble, because if you believed it (which you wouldn't) it would cause you to give arbitrarily large odds against that letter occurring on the next trial, or perhaps ever. Surely even the Laplace-Lidstone estimate $(n_i + 1)/(N + t)$ would be better, although it is not optimal. The estimate of Jeffreys (1946), $(n_i + 1/2)/(N + t/2)$, which is based on his 'invariant prior', is also better (in the same sense) than the ML estimate. Still better methods are available which are connected with reasonable 'Bayesian significance tests' for multinomial distributions (Good, 1965b, 1956/67).

Utility and quasi-utility functions are often invariant in some sense, although 'squared loss' is invariant only under *linear* transformations. For example, if the utility in estimating a vector parameter $\boldsymbol{\theta}$ as $\boldsymbol{\phi}$ is $u(\boldsymbol{\theta}, \boldsymbol{\phi})$, and if the parameter space undergoes some one-one transformation $\boldsymbol{\theta}^* = \boldsymbol{\psi}(\boldsymbol{\theta})$ we must have, for consistency, $\boldsymbol{\phi}^* = \boldsymbol{\psi}(\boldsymbol{\phi})$ and $\mathbf{u}^*(\boldsymbol{\theta}^*, \boldsymbol{\phi}^*) = \mathbf{u}(\boldsymbol{\theta}, \boldsymbol{\phi})$, where $\mathbf{u}^*$ denotes the utility function in the transformed parameter space.

The principle of selecting the least favorable prior when it exists (in accordance with the minimax strategy) may be called *the principle of least utility*, or, when appropriate, *the principle of least quasi-utility*. Since the minimax procedure must be invariant with respect to transformations of the problem into other equivalent languages, it follows that the principle of least utility leads to an invariant prior. This point was made by Good (1968/69b, 1968). It was also pointed out there (see also Good, 1970/71b; Good and Gaskins, 1971 and 1972 (Appendix C)) that there is a class of invariant quasiutilities *for distributions*. Namely, the quasiutility of assuming a distribution of density $g(x)$, when the true distribution of x is $F(\mathbf{x})$, was taken as

$$\int \log \{g(\mathbf{x}) [\det \Delta(\mathbf{x})]^{-1/2}\} \, dF(\mathbf{x})$$

where

$$\Delta(\boldsymbol{\theta}) = \left\{ -\frac{\partial^2 u(\boldsymbol{\theta}, \boldsymbol{\phi})}{\partial \phi_i \partial \phi_j} \bigg|_{\phi = \theta} \right\} \quad i, j = 1, 2, \ldots.$$

From this it follows further that

$$[\det \Delta(\mathbf{x})]^{1/2}$$

is an invariant prior, though it might be 'improper' (have an infinite integral). In practice improper priors can always be 'shaded off' or truncated to give them propriety (Good, 1947/50, p. 56).

If θ is the (vector) parameter in a distribution function $F(\mathbf{x} \mid \theta)$ of a random variable $\mathbf{x}$, and θ is not to be used for any other purpose, then in logic we must identify $u(\theta, \phi)$ with the utility of taking the distribution to be $F(\mathbf{x} \mid \phi)$ instead of $F(\mathbf{x} \mid \theta)$. One splendid example of an invariant utility is the expected weight of evidence per observation for discriminating between θ and ϕ, or 'dinegentropy',

$$u_0(\theta, \phi) = \int \log \frac{dF(\mathbf{x} \mid \theta)}{dF(\mathbf{x} \mid \phi)} \, dF(\mathbf{x} \mid \theta),$$

which is invariant under non-singular transformations both of the *random* variable and of the parameter space. (Its use in statistical mechanics dates back to Gibbs.) Moreover it is additive for entirely independent problems, as a utility function should be. With this quasiutility, $\Lambda(\theta)$ reduces to Fisher's information matrix, and the square root of the determinant of $\Lambda(\theta)$ reduces to Jeffreys's invariant prior. The dinegentropy was used by Jeffreys (1946) as a measure of distance between two distributions. The distance of a distribution from a correct one *can* be regarded as a kind of loss function. Another additive invariant quasiutility is (Good, 1956a; Rényi, 1961; Good and Card, 1970/71, p. 180) the 'generalized dinegentropy',

$$u_c(\theta, \phi) = \frac{1}{c} \log \int \left[\frac{dF(\mathbf{x} \mid \theta)}{dF(\mathbf{x} \mid \phi)} \right]^c dF(\mathbf{x} \mid \theta) \quad (c > 0),$$

the limit of which as $c \to 0$ is the expected weight of evidence, $u_0(\theta, \phi)$, somewhat surprising at first sight. The square root of the determinant of the absolute value of the Hessian of this utility at $\phi = \theta$ is then an invariant prior indexed by the non-negative number c. Thus there is a continuum of additive invariant priors of which Jeffreys's is an extreme case. For example, for the mean of a univariate normal distribution the invariant prior is uniform, mathematically independent of c. The invariant prior for the variance ϕ is $\sigma^{-1} \sqrt{\{2(1+c)\}}$, which is proportional to σ^{-1} and so is again mathematically independent of c.

In more general situations the invariant prior will depend on c and will therefore not be unique. In principle it might be worth while to assume a ('type III') distribution for c, to obtain an average of the various additive invariant priors. It might be best to give extra weight to the value $c=0$ since weight of evidence seems to be the best general-purpose measure of corroboration (Good, 1960/68, 1967/68).

It is interesting that Jeffreys's invariant prior, and its generalizations, and also the principles of maximum entropy and of minimum discriminability (Kullback, 1959) can all be regarded as applications of the principle of least quasiutility. This principle thus unifies more methods than has commonly been recognized. The existing criticisms of minimax procedures thus apply to these special cases.

The term 'invariance' can be misleading if the class of transformations under which invariance holds is forgotten. For the invariant priors, although this class of transformations is large, it does not include transformations to a different *application* of the parameters. For example, if θ has a physical meaning, such as height of a person, it might occur as a parameter in the distribution of her waist measurement or her bust measurement, and the invariance will not apply between these two applications. This in my opinion (and L. J. Savage's, July 1959) is a logical objection to the use of invariant priors when the parameters have clear physical meaning. To overcome this objection completely it would perhaps be necessary to consider the joint distribution of all the random variables of potential interest. In the example this would mean that the joint distribution of at least the 'vital statistics', given θ, should be used in constructing the invariant prior.

There is another argument that gives a partial justification for the use of invariant priors in spite of Savage's objection just mentioned. It is based on the notion of 'marginalism' in the sense defined by Good (1958c, pp. 808–809; 1968/69b, p. 61; 1970/71c, p. 15). I quote from the latter, "It is only in marginal cases that the choice of the prior makes much difference [when it is chosen to give the non-null hypothesis a reasonable chance of winning on the size of sample we have available]. Hence the name marginalism. It is a trick that does not give accurate final probabilities, but it protects you from missing what the data is trying to say owing to a careless choice of prior distribution". In accordance with this principle one might argue, as do Box and Tiao (1973, p. 44) that a prior should, at least on some occasions, be uninformative relative to the experiment being performed. From this idea they derive the Jeffreys invariant prior.

It is sometimes said that the aim in estimation is not necessarily to minimize loss but merely to obtain estimates close to the truth. But there is an implicit assumption here that it is better to be closer than further away, and this is equivalent to the assumption that the loss function is monotonic and has a minimum (which can be taken as zero) when the estimate is equal to the true value. This assumption of monotonicity is not enough to determine a unique estimate nor a unique interval estimate having an assigned probability of covering the true value (where the probability might be based on information before or after the observations are taken). But for large enough samples (*effectively* large, for the purpose in hand), as I said, all reasonable methods of estimation lead to Rome, if Rome is not too small.

(d) A BAYES NON-BAYES COMPROMISE FOR PROBABILITY DENSITY ESTIMATION

Up to a few years ago, the only nonparametric methods for estimating probability densities, from observations $x_1, x_2, ..., x_N$, were non-Bayesian. These, methods, on which perhaps a hundred papers have been written, are known as *window methods*. The basic idea, for estimating the density at a point x, was to see how many of the N observations lie in some interval or region around x, where the number v of such observations tends to infinity while $v/N \to 0$ when $N \to \infty$. Also less weight is given to observations far from x than to those close to x, this weighting being determined by the shape of the window.

Although the window methods have some intuitive appeal it is not clear in what way they relate to the likelihood principle. On the other hand, if the method of ML is used it leads to an unsatisfactory estimate of the density function, namely a collection of fractions $1/N$ of Dirac delta functions, one at each of the observations. [A discussant: Go all the way to infinity if they are Dirac functions. Don't be lazy! IJG: Well I drew them a little wide so they are less high to make up for it.] There is more than one objection to this estimate; partly it states that the next observation will certainly take a value that it almost certainly will not, and partly it is not smooth enough to satisfy your subjective judgment of what a density function should look like. It occurred to me that it should make sense to apply a 'muftian' Bayesian method, which in this application means finding some formula giving a posterior density in the function space of all density functions for the random variable X, and then maximizing this posterior density so as to obtain a single density function (a single 'point in function space') as the 'best' estimate of the whole density function for X. But this means that from the log-likelihood $\sum \log f(x_i)$ we should subtract a 'roughness penalty' before maximizing. (Good, 1970/71d, 1970/71b; Good and Gaskins, 1971, 1972.) There is some arbitrariness in the selection of this roughness penalty (which is a functional of the required density function f), which was reduced to the estimation of a single hyperparameter, but I omit the details. The point I would like to make here is that the method can be interpreted in a non-Bayesian manner, although it was suggested for Bayesian reasons. Moreover, in the present state of the art, only the Bayesian interpretation allows us to make a comparison between two hypothetical density functions. The weight of evidence by itself is not an adequate guide for this problem. Then again the non-Bayesian could examine the operational characteristics of the Bayesian interpretation. The Doogian should do this because it might lead him to a modification of the roughness penalty. The ball travels backwards and forwards between the Bayesian and non-Bayesian courts, the ball-game as a whole forming a plank of the Doogian platform.

It is easy to explain why the method of ML breaks down here. It was not designed for cases where there are very many parameters, and in this problem there are an infinite number of them, since the problem is nonparametric. (A nonparametric problem is one where the class of distribution functions cannot be specified in terms of a finite number of parameters, but of course any distribution can be specified in terms of an infinite number of parameters. My method of doing so is to regard the square root of the density function as a point in Hilbert space.)

To select a roughness penalty for multidimensional density functions, I find consistency appealing, in the sense that the estimate of densities that are known to factorize, such as $f(x)\,g(y)$ in two dimensions, should be the same whether f and g are estimated together or separately. This idea enabled me to propose a multidimensional roughness penalty but numerical examples of it have not yet been tried.

An interesting feature of the *subtractive roughness-penalty method* of density estimation, just described, is that it can be made invariant with respect to transformations of the **x** axes, even though such transformations could make the true density function arbitrarily rough. The method proposed for achieving invariance was to make use of the tensor calculus, by noticing that the elements of the matrix $\Delta(\theta)$ form a covariant tensor, which could be taken as the 'fundamental tensor' g_{ij} analogous to that occurring in General Relativity. For 'quadratic loss' this tensor becomes a constant, and, as in Special Relativity, it is then not necessary to use tensors. The same thing happens more generally if $u(\theta, \phi)$ is any function (with continuous second derivatives) of a quadratic.

(e) TYPE II MAXIMUM LIKELIHOOD AND THE TYPE II LIKELIHOOD RATIO

The notion of a hierarchy of probabilities, mentioned earlier, can be used to produce a compromise between Bayesian and non-Bayesian methods, by treating hyperparameters in some respects as if they were ordinary parameters. In particular, a Bayes factor can be maximized with respect to the hyperparameters, and the hyperparameters so chosen (their 'Type II ML' values) thereby fix the ordinary prior, and therefore the posterior distribution of the ordinary parameters. This *Type II ML method* could also be called the *Max Factor method*. This technique was well illustrated by Good (1965/67). It ignores only judgments you might have about the Type III distributions, but I have complete confidence that this will do far less damage than ignoring all your judgments about Type II distributions as in the ordinary method of ML. Certainly in the reference just mentioned the Type II ML estimates of the physical probabilities were far better than the Type I ML estimates.

The same reference exemplified the *Type II likelihood Ratio*. The ordinary (Neyman-Pearson-Wilks) Likelihood Ratio (LR) is defined as the ratio of two maximum likelihoods, where the maxima are taken within two spaces corresponding to two hypotheses (one space embedded in the other). The ratio is then used as a test statistic, its logarithm to base $1/\sqrt{e}$ having asymptotically (for large samples) a chi-squared distribution with a number of degrees of freedom equal to the difference of the dimensionalities of the two spaces. The Type II Likelihood Ratio is defined analogously as

$$\max_{\theta \in \omega} P\{E \mid H(\theta)\} / \max_{\theta \in \Omega} P\{E \mid H(\theta)\}$$

where θ is now a hyperparameter in a prior $H(\theta)$, Ω is the set of all values of θ and ω is a subset of Ω. In the application to multinomial distributions this led to a new statistic called G having asymptotically a chi-squared distribution with one degree of freedom (corresponding to a single hyper-parameter, namely the parameter of a symmetric Dirichlet distribution). Later calculations showed that this asymptotic distribution was accurate down to fantastically small tail-area probabilities such as 10^{-16}, see Good and Crook (1972). In this work it was found that if the Bayes factor F, based on the prior selected in Good (1965/67), were used as a non-Bayesian statistic, in accordance with the Bayes/non-Bayes compromise, it was almost equivalent to the use of G, in the sense of giving nearly the same significance levels (tail-area probabilities) to samples. It was also found that the Bayes factor based on the (less reasonable) Bayes postulate was roughly equivalent in the same sense, thus supporting my claims for the Bayes/non-Bayes compromise.

(f) THE NON-UNIQUENESS OF UTILITIES

For some decision problems the utility function can be readily measured in monetary terms; for example, in a gamble. In a moderate gamble the utility can reasonably be taken as proportional to the relevant quantities of money. Large insurance companies often take such 'linear' gambles. But in many other decision problems the utility is not readily expressible in monetary terms, and can also vary greatly from one person to another. In such cases the Doogian, and many a statistician who is not Doogian or does not know that he is, will often wish to keep the utilities separate from the rest of the statistical analysis if he can. There are exceptions because, for example, many people might assume a squared loss function, but with different matrices, yet they will all find expected values to be the optimal estimates of the parameters.

One implication of the recognition that utilities vary from one person to another is that the expected benefit of a client is not necessarily the same, *nor even of the same sign*, as that of the statistical consultant. This can produce ethical problems for the statistician, although it may be possible to reward him in a manner that alleviates the problems. (See, for example, Good, 1951/52, 1970/73.)

One example of this conflict of interests relates to the use of confidence-interval estimation. This technique enables the statistician to ensure that his interval estimates (*asserted* without reference to probability) will be correct say 95% of the time in the long run. If he is not careful he might measure his utility gain by this fact alone (especially if he learns his statistics from cookbooks) and it can easily happen that it won't bear much relation to his client's utility on a specific occasion. The client is apt to be more concerned with the final probability that the interval will contain the true value of the parameter.

Neyman has warned against dogmatism but his followers do not often give nor heed the warning. Notice further that there are *degrees* of dogmatism and that greater degrees can be justified when the principles involved are the more certain. For example, it seems more reasonable to be dogmatic that 7 times 9 is 63 than that witches exist and should be caused not to exist. Similarly it is more justifiable to be dogmatic about the axioms of subjective probability than to insist that the probabilities can be sharply judged or that confidence intervals should be used in preference to Bayesian posterior intervals. (For chrisake don't call them 'Bayesian confidence intervals', which is a contradiction in terms.)

Utilities are implicit in some circumstances even when many statisticians are unaware of it. Interval estimation provides an example of this; for it is often taken as a criterion of choice between two confidence intervals, both having the same confidence coefficient, that the shorter interval is better. Presumably this is because the shorter interval is regarded as leading to a more economical search or as being in general more informative. In either case this is equivalent to the use of an informal utility or quasiutility criterion. It will often be possible to improve the interval estimate by taking into account the customer's utility function more explicitly.

An example of this is when a confidence interval is stated for the position of a ship, in the light of direction finding. If an admiral is presented with say an elliptical confidence region, I suspect he would reinterpret it as a posterior probability density surface, with its mode in the center. (Good, 1951, 1968/69b.) The admiral would rationally give up the search when the expense per hour sank below the expected utility of locating the ship. In other words, the client would sensibly ignore the official meaning of the statistician's assertion. If the statistician knows this, it might be better (at least for his client) if he went Bayesian (in some sense) and gave the client what he wanted.

(g) TAIL-AREA PROBABILITIES

Null hypotheses are usually known in advance to be false, and the point of significance tests is usually to find out whether they are nevertheless approximately true. (Good, 1947/50, p. 90.) In other words *a null hypothesis is usually composite even if only just*. But for the sake of simplicity I shall here regard the null hypothesis as a simple statistical hypothesis, as an approximation to the usual real-life situation.

I have heard it said that the notion of tail-area probabilities, for the significance test of a null hypothesis H_0 (assumed to be a simple statistical hypothesis), can be treated as a primitive notion, not requiring further analysis. But this cannot be true irrespective of the test criterion and of the plausible alternatives to the null hypothesis, as was perhaps originally pointed out by Neyman and E. S. Pearson. A value X_1 of the test criterion X should be regarded as 'more extreme' than another one X_2 only if the observation of X_1 gives 'more evidence' against the null hypothesis. To give an interpretation of 'more evidence' it is necessary to give up the idea that tail-areas are primitive notions, as will soon be clear. One good interpretation of 'more evidence' is that the weight of evidence against H_0 provided by X_1 is greater than that provided by X_2, that is

$$\log \frac{\text{P.D.}(X_1 \mid H_1)}{\text{P.D.}(X_1 \mid H_0)} > \log \frac{\text{P.D.}(X_2 \mid H_1)}{\text{P.D.}(X_2 \mid H_0)},$$

where H_1 is the negation of H_0 and is a composite statistical hypothesis, and P.D. stands for 'probability density'. (When H_0 and H_1 are both simple statistical hypotheses there is little reason to use 'tail-area' significance tests.) This interpretation of 'more extreme' in particular provides a solution to the following logical difficulty, as also does the Neyman-Pearson technique if all the simple statistical hypotheses belonging to H_1 make the simple likelihood ratio monotonic increasing as x increases.

Suppose that the probability density of a test statistic X, given H_0, has a known shape, such as that in Figure 1a. We can transform the x axis so

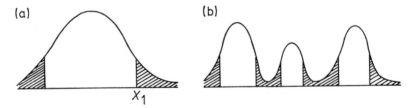

Fig. 1.

that the density function becomes any density function we like, such as that illustrated in Figure 1b. We then might not know whether the x's 'more extreme' than the observed one should be interpreted as all the shaded part of 1(b), where the ordinates are smaller than the one observed. Just as the tail-area probability wallah points out that the Bayes postulate is not invariant with respect to transformations of the x axis, the Bayesian can say *tu quoque*. (Compare, for example, Good, 1971a, p. 53; Kalbfleisch, 1971, § 7,1–8.) Of course Doogians and many other modern Bayesians are not at all committed to the Bayes postulate, though they often use it as an approximation to their honest judgment, or marginalistically.

When tail-areas are used for significance testing, we need to specify what is meant by a 'more extreme' value of the criterion. A smaller ordinate might not be appropriate, as we have just seen. I believe it is a question of ordering the values of the ordinate according to the weight of evidence against the null hypothesis, as just suggested. [Sometimes this ordering is mathematically independent of the relative initial probabilities of the simple statistical hypotheses that make up the composite non-null hypothesis H_1. In this case the interpretation of 'more extreme' is maximally robust modulo the Bayesian assumptions.] This or similar fact is often swept UTC, although a special case of it is often implicit when it is pointed out that sometimes a single tail should be used and sometimes a double tail, depending on the nature of the non-null hypotheses.

For some problems it would not be appropriate to interpret 'more extreme' to mean 'further to the right' nor 'either further to the right of one point or further to the left of another' (i.e. for 'double tails'). For example, the null hypothesis might be a bimodal distribution with mean zero, the rivals being unimodal also with mean zero. Then we might need to regard values of the random variable close to the origin as significant, in addition to large positive and negative values. We'd be using a 'triple tail' so to speak. All this comes out in the wash when 'more extreme' is interpreted in terms of weight of evidence.

It is stimulating to consider what is 'more extreme' in multivariate problems. It will be adequate to think of bivariate problems which are enough to bring out all the philosophical aspects (which are more important than the mathematical ones). We might first ask what is the analogue of being 'further to the right'. One analogue is being 'further to the north and east'. This analogue is often dubious (unless the two independent variables are like chalk and cheese, or like oil and water) even without reference to any Bayesian or Neymanian-Pearsonian ideas. For under a linear transformation of the independent variables, such as an orthogonal transformation, there are a continuous infinity of different regions that are further to the north and east. The corresponding ambiguity in one dimension refers merely to the question of whether a single tail is more or less appropriate than a double tail.

The previously mentioned elucidation of 'more extreme' in terms of weight of evidence applies just as much to multivariate problems as to univariate ones, and provides an answer to this 'north-east' difficulty.

Even when a sensible meaning is ascribed to the expression 'more extreme', my impression is that small tail-areas, such as 1/10000, are by no means as strong evidence against the null hypothesis as is often supposed, and this is one reason why I believe that Bayesian methods are important in applications where small tail areas occur, such as medical trials, and even more in ESP, radar, cryptanalysis, and ordinary life. It would be unfortunate if a radar signal were misinterpreted through overlooking this point, thus leading to the end of life on earth! *The more important a decision the more 'Bayesian' it is apt to be.*

The question has frequently been raised of how the use of tail-area significance tests can be made conformable with a Bayesian philosophy. [See, for example, Anscombe (1968/69).] An answer had already appeared in Good (1947/50, p. 94), and I say something more about it here. (See also Good, 1968/69a, p. 61.)

A reasonable informal Bayesian interpretation of tail-area probabilities can be given in some circumstances by treating the criterion X as if it were the whole of the evidence (even if it is not a sufficient statistic). Suppose that the probability density f_0 of X given H_0 is known, and that you can make a rough subjective estimate of the density f_1 given $\bar{H}_0$. (If you cannot do this at all then the tail area method is I think counterintuitive.) Then we can calculate the Bayes factor against H_0 as a ratio of ordinates $f_1(X)/f_0(X)$. It turns out that this is often of the order of magnitude of $(1/\sqrt{N}) \int_X^\infty f_1(x)\, dx / \int_X^\infty f_0(x)\, dx$, where N is the sample size, and this in its turn will be somewhat less than $1/(P\sqrt{N})$ where P is the right-hand tail-area probability on the null hypothesis. (See Good, 1957, p. 863; improved in 1965/1967, p. 416; and still further in Good and Crook, 1972.) Moreover, this argument suggests that, for a fixed sample size, there should be a roughly monotonic relationship and a *very* rough proportionality between the Bayes factor F against the null hypothesis and the reciprocal of the tail-area probability, P, provided of course that the non-null hypothesis is not at all specific. (See also Good, 1947/50, p. 94; 1965/67.)

Many elementary textbooks recommend that test criteria should be chosen before observations are made. Unfortunately this could lead to a data analyst's missing some unexpected and therefore probably important feature of the data. There is no existing substitute for examining the original observations with care. This is often more valuable than the application of formal significance tests. If it is easy and inexpensive to obtain new data then there is little objection to the usual advice, since the original data can be used to formulate hypotheses to be tested on later samples. But often a further sample is expensive or virtually impossible to obtain.

The point of the usual advice is to protect the statistician against his own poor judgment.

A person with bad judgment might produce many far-fetched hypotheses on the basis of the first sample. Thinking that they were worth testing, if he were non-Bayesian he would decide to apply standard significance tests to these hypotheses on the basis of a second sample. Sometimes these would pass the test, but some one with good judgment might be able to see that they were still improbable. It seems to me that the ordinary method of significance tests makes some sense because experimenters often have reasonable judgment in the formulation of hypotheses, so that the initial probabilities of these hypotheses are not usually entirely negligible. A statistician who believes his client is sensible might assume that the hypotheses formulated in advance by the client are plausible, without trying to produce an independent judgment of their initial probabilities.

Let us suppose that data are expensive and that a variety of different non-null hypotheses have been formulated on the basis of a sample. Then the Bayesian analyst would try, in conjunction with his client, to judge the initial probabilities $q_1, q_2, \ldots$ of these hypotheses. Each separate non-null hypothesis might be associated with a significance test if the Bayesian is Doogian. These tests might give rise to tail-area probabilities $P_1, P_2, P_3, \ldots$. How can these be combined into a single tail-area probability? (Good, 1958c.)

Let us suppose that the previous informal argument is applicable and that we can interpret these tail-area probabilities as approximate Bayes factors $C/P_1, C/P_2, C/P_3, \ldots$ against the null hypothesis, these being in turn based on the assumption of the various rival non-null hypotheses. ('Significance tests in parallel'.) By a theorem of weighted averages of Bayes factors, it follows that the resulting factor is a weighted average of these, so that the equivalent tail-area probability is about equal to a weighted harmonic mean of $P_1, P_2, P_3, \ldots$, with weights $q_1, q_2, q_3, \ldots$. This result is not much affected if C is a slowly decreasing function of P instead of being constant, which I believe is often the case. Nevertheless the harmonic-mean rule is only a rule of thumb.

But we could now apply the Bayes/non-Bayes compromise for the invention of test criteria, and use this weighted harmonic mean as a non-Bayes test criterion (Good, 1957, p. 863; 1965/67; Good and Crook, 1972).

The basic idea of the Bayes/non-Bayes compromise for the invention of test criteria is that you can take a Bayesian model, which need not be an especially good one, come up with a Bayes factor on the basis of this model, but then use it as if it were a non-Bayesian test criterion. That is, try to work out or 'Monte Carlo' its distribution based on the null hypothesis, and also its power relative to various non-null hypotheses.

An example of the Bayes/non-Bayes compromise arises in connection with discrimination between two approximately multinomial distributions. A crude Bayesian model would assume that the two distributions were precisely multinomial and this would lead to a linear discriminant function. This could then be used in a non-Bayesian manner or it might lead to the suggestion of using a linear discriminant function optimized by some other, possibly non-Bayesian, method. Similarly an approximate assumption of multinormality for two hypotheses leads to a quadratic discriminant function with a Bayesian interpretation but which can then be interpreted non-Bayesianwise. (See Good, 1965a, pp. 49–50, where there are further references.)

Let us now consider an example of an experimental design. I take this example from Finney (1953, p. 90) who adopts an orthodox (non-Bayesian) line. Finney emphasizes that, in his opinion, you should decide in advance how you are going to analyze the experimental results of a designed experiment. He considered an experimental design laid out as shown in Figure 2. The design consists of ten plots, consisting of five blocks each divided into two plots. We decide to apply treatment *A* and treatment *B* in a random order within each block, and we happen to get the design shown. Now this design could have arisen by another process: namely by selecting equiprobably the five plots for the application of treatment A from the $10!/(5!)^2 = 252$ possibilities. Finney then says, "The form of analysis depends not on the particular arrangement of plots and

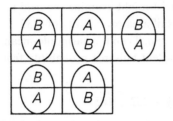

Fig. 2. An agricultural experiment.

varieties in the field ((I have been talking about *treatments* instead here but it does not affect the argument)) but on the process of randomization from which the particular one was selected". [The doubled parentheses enclose my comment.] [Perhaps one should talk of a stochastic or random design *procedure* and a *realization of the procedure*.] For one design procedure we would perhaps use the comparison within the five pairs, and for the other procedure we would compare the set of five yields from treatment *A* with the set of five yields from treatment *B*. Leaving aside the analysis of variance, we might find that every plot *A* did better than every plot *B*, thus bringing off a distribution free chance of 1/252; but we are 'permitted' to say merely that the chance in 1/32 if the design procedure was based on the five blocks. Suppose the statistician hadn't said which was his design and then he'd dropped dead after the experiment and suppose this is an important experiment organized by the government to decide whether a certain big expensive and urgent food production method was to be put into effect. Would it be reasonable to search the statistician's papers carefully to find out what his intentions had been? Or would it on the other hand be reasonable to call in agriculturalists to look at the plots in the field in order to try to decide which design would have been more reasonable? There are of course reasons for choosing one design rather than another one. So, if you entirely accept the Fisherian logic (as exemplified by Finney) you are whole-heartedly trusting the original judgment of choice of design: this is what the mystique recommends. My own feeling is that you would do better to judge the prior probabilities that each of the two designs is to be preferred, and then use these probabilities as weights in a procedure for combining significance tests. (Good, 1958c; 1971a, p. 83.)

A (living) agriculturist might examine the field and say that the design corresponding to the tail-area probability of 1/32 deserved twice as much weight as the other design. Then the harmonic-mean rule of thumb would suggest that the equivalent tail-area probability from the observations is

$$\frac{1}{\frac{2}{3} \times 32 + \frac{1}{3} \times 252} \approx \frac{1}{105}.$$

Of course we might do better by using the analysis of variance in a similar manner. I have used a distribution-free approach for the sake of simplicity. This imprecise result is better than either of the precise ones, 1/32 and 1/252. I predict that lovers of the 'precision fallacy' will ignore all this.

It is often said that non-Bayesian methods have the advantage of conveying the evidence in an experiment in a self-contained manner. But we see from the example just discussed that they depend on a previous judgment; which in the special case of the dead-dropping of the statistician, has to be a posterior judgment. So it's misleading to tell the student he must decide on his significance test in advance, although it is correct according to the Fisherian techique.

(h) RANDOMNESS, AND SUBJECTIVISM IN THE PHILOSOPHY OF PHYSICS

I would have included a detailed discussion on the use of random sampling and random numbers, but have decided not to do so because my views on the subject are explained, for example, in Good (1956b, p. 255; 1970/71a, p. 135; and 1972). The use of random sampling is a device for obtaining apparently precise objectivity but this precise objectivity is attainable, *as always*, only at the price of throwing away some information (by using a *Statistician's Stooge* who knows the random numbers but does not disclose them). But the use of sampling without randomization involves the pure Bayesian in such difficult judgments that, at least if he is at all Doogian, he might decide, by Type II rationality, to use random sampling to save time. As Cornfield (1968/70, p. 108) points out, this can be covered within the Bayesian framework.

Since this conference is concerned with physics as well as with statistics I should like to mention a connection between something I have been saying and a point that is of interest in the philosophy of physics. (This point is also discussed in Good, 1972.)

When discussing the probability that the millionth digit of π is a 7, I could have pointed out that similar statements can be made about pseudorandom numbers. These are deterministic sequences that are complicated enough so that they appear random at least superficially. It would be easy to make them so complicated that it would be practically impossible to find the law of generation when you do not know it. Pseudorandom numbers are of value in computer applications of the so-called Monte Carlo method. They are better than random numbers for this purpose because it is possible to check a program exactly and conveniently when pseudorandom numbers are used. One could well say that the probabilities of the various sequences are the same as for a random sequence until the law of generation is known. This is an example of the shifting or evolving probabilities that I mentioned before. The probabilities would all become 1 or 0 if the law of generation were discovered.

Now it seems to me that this is precisely the situation in classical statistical mechanics. Since the exact positions and velocities of the molecules are not known there is no practical difference between the assumptions of determinism and indeterminism in classical statistical mechanics.

Let us imagine that the world really is physically deterministic. The subjective probability (of the physicist) is almost equal to 1 that the entropy of a closed system will not decrease. He tends to believe that this subjective probability is also a physical probability but it is difficult to see how it can be on the assumption of determinism, any more than it can be a physical probability that the millionth digit of π is a 7. The world is so complicated that pseudoindeterminism is indistinguishable in practice from strict physical indeterminism. Thus the physicist is 'safe' when he says that certain of his subjective probabilities are physical, because he cannot be refuted. Yet he is really sweeping his subjective probabilities UTC. I claim that von Mises (1957, pp. 184–186) made a logical error when he argued that the effectiveness of classical statistical mechanics proves that the world is really indeterministic. (Compare Good, 1958b, and 1959, p. 447.) Of course quantum mechanics is something else.

V. DISCUSSION

I have allowed for some of the discussion already in this write-up. Here I mention a point raised in the discussion by Peter Finch. He asked in what way initial probabilities come into statistical practice. Apart from the fact that Bayesian methods are gradually coming to the fore, I mentioned the following example.

Fisher (1959, p. 47) argued that, at the time he wrote, there was little evidence that smoking was a cause of lung cancer. The research by Doll *et al.* had shown a tail-area probability of about 1/80 suggesting such a relationship, but had also shown a similar level of significance suggesting that inhalers less often get lung cancer than non-inhalers. Fisher's ironical punch line then was that therefore the investigators should have recommended smokers to inhale! I believe that the investigators (unconsiously?) assumed that the initial probability here was too low to justify giving this advice. (Good, 1961; but see also Good, 1962b.)

Peter Finch also said that the purpose of avoiding far-fetched hypotheses was to get one's work published. My reply was that I had frequently used initial probabilities in my own classified statistical work to help decide what hypotheses I should follow up, with no question of publication.

Virginia Polytechnic Institute and State University
Blacksburg, Virginia 24061

REFERENCES

Anscombe, F. J., (1968/69), 'Discussion of Paper by I. J. Good, April 9, 1968', *J. Amer. Statist. Assoc.* **64** (1969), 50–51.

Bayes, T., (1763/65, 1940/58), 'An Essay Toward Solving a Problem in the Doctrine of Chances', *Phil. Trans. Roy Soc.* **53**, 370–418, **54**, 296–325. Reprints (i) The Graduate School, Department of Agriculture, Washington, D.C. (1940), (ii) *Biometrika* **45** (1958), 293–315.

Bernoulli, Daniel, (1774/78/1961), 'The Most Probable Choice Between Several

Discrepant Observations and the Formation Therefrom of the Most Likely Induction' (in Latin), *Acta Academiae Scientiorum Petropolitanae* **1** (1777), 3–23. English tr. by C. G. Allen in *Biometrika* **48** (1961), 3-13.

Bernstein, S., (1921/22), 'Versuch einer axiomatischen Begründung der Wahrscheinlichkeitsrechnung', *Mitt. Charkow* **15**, 209–274. Abstract in *Jahrbuch der Math.* **48.**

Box, G. E. P. and G. C. Tiao (1973), *Bayesian Inference in Statistical Analysis* (Addison-Wesley).

Card, W. I. and I. J. Good, (1973), 'A Logical Analysis of Medicine', in *A Companion to Medical Studies* (eds. R. Passmore and J. S. Robson; Oxford, Blackwell's), Vol. 3, Chapter 60.

Cornfield, J., (1968/70), 'The Frequency Theory of Probability, Bayes' Theorem, and Sequential Clinical Trials', in *Bayesian Statistics* (ed. D. L. Meyer and R. O. Collier Jr.; Itasca, Illinois: Peacock), 1–28.

Cox, R. T., (1946), 'Probability, Frequency and Reasonable Expectation', *Amer. J. Physics* **14**, 1–13.

Cox, R. T., (1961), *The Algebra of Probable Inference* (Baltimore: John Hopkins Press).

Cronbach, L. J., (1953), 'A Consideration of Information Theory and Utility Theory as Tools for Psychometric Problems' (Technical Report, College of Education, University of Illinois, Urbana).

Eisenhart, C., (1964), 'The Meaning of 'least' in Least Squares', *J. Washington Acad. Sci.* **54**, 24–33.

de Finetti, B., (1937/64), 'Foresight: Its Logical Laws, Its Subjective Sources', tr. from the French of 1937 by H. Kyburg, in *Studies in Subjective Probability* (ed. H. E. Kyburg and H. E. Smokler; Wiley), 95–158.

de Finetti, B., (1968/70), 'Initial Probabilities: a Prerequisite for any Valid Induction', in *Induction, Physics, and Ethics* (ed. P. Weingartner and G. Zecha; D. Reidel, Dordrecht), 3–17. Also in *Synthese* **20** (1969).

Finney, D. J., (1953), *Statistical Science in Agriculture* (Wiley, and Oliver and Boyd).

Fisher, R. A., (1959), *Smoking: the Cancer Controversy* (Oliver and Boyd, Edinburgh and London).

Gauss, K. F., (1789/1809/57/1963), 'Theory of the Motion of Heavenly Bodies Moving About the Sun in Sections' (Latin) in Carl Friedrich Gauss Werke; Eng. tr. by Charles Henry Davis (1857), Dover reprint, 1963. (Reference from Eisenhart, 1964).

Good, I. J., (1947/50), *Probability and the Weighing of Evidence* (London, Charles Griffin; New York, Hafners: 1950, pp. 119).

Good, I. J., (1951), Contribution to the discussion on H. E. Daniels', 'Theory of Position Finding', *J. Roy. Statist. Soc.* **B13**, 203.

Good, I. J., (1951/52), 'Rational Decisions', *J. Roy. Statist. Soc.* **B14**, 107–114.

Good, I. J., (1953), 'On the Population Frequencies of Species and the Estimation of Population Parameters', *Biometrika* **40**, 237-264.

Good, I. J., (1953/56), 'On the Estimation of Small Frequencies in Contingency Tables', *J. Roy. Statist. Soc.* **B18**, 113–124.

Good, I. J., (1955/56a), 'Some Terminology and Notation in Information Theory', *Proc. Institution Elec. Engrs.*, Part C (3), **103**, 200–204; or Monograph 155R (1955).

Good, I. J., (1955/56b), Contributions to the discussion in the Third London Symposium on Information Theory, *Information Theory, Third London Symposium 1955* (Butterworth's 1956).

Good, I. J., (1956a), 'The Surprise Index for the Multivariate Normal Distribution', *Annals Math. Statist.* **27**, 1130–1135; **28** (1957), 1055

Good, I. J., (1956b), 'Which Comes First, Probability or Statistics?', *J. Inst. Actuaries* **82**, 249–255.

Good, I. J., (1957), 'Saddle-point Methods for the Multinomial Distribution', *Annals Math. Statist.* **28**, 861–881.

Good, I. J., (1958a), 'The Interaction Algorithm and Practical Fourier Analysis', *J. Roy. Statist. Soc.* **B20**, 361–372; **22** (1960), 372–375.

Good, I. J., (1958b), Review of R. von Mises *Probability, Statistics and Truth* (2nd. edn.), *J. Roy. Statist. Soc.* **A121**, 238–240.

Good, I. J., (1958c), 'Significance Tests in Parallel and in Series', *J. Amer. Statist. Assoc.* **53**, 799–813.

Good, I. J., (1958/59), 'Could a Machine Make Probability Judgements?', *Computers and Automation* **8**, 14–16 and 24–26.

Good, I. J., (1959), 'Kinds of Probability', *Science* **129**, 443–447; *L'Industria* (1959, Ital. tr. by Fulvia de Finetti).

Good, I. J., (1960/62), 'Subjective Probability as the Measure of a Non-measurable Set', International Conference, Stanford. In *Logic, Methodology, and Philosophy of Science* (ed. E. Nagel, P. Suppes, and A. Tarski; Stanford Univ. Press), 319–329.

Good, I. J., (1960/68), 'Weight of Evidence, Corroboration, Explanatory Power, Information, and the Utility of Experiments', *J. Roy. Statist. Soc.* **B22**, 319–331; *Corrigenda* **30**, 203.

Good, I. J., (1961), Contribution to the discussion of a paper by C. A. B. Smith, 'Consistency in Statistical Inference and Decision', *J. Roy Statist. Soc.* **B23**, 28–29.

Good, I. J., (1962a), Review of Richard T. Cox, 'The Algebra of Probable Inference' (John Hopkins Press, Baltimore, 1961); *Math. Gaz.* (1962), 246–247; *MR* **24** (1962), 107, Rev. No. A563.

Good, I. J., (1962b), 'The Bitter End', *The Scientist Speculates: an Anthology of Partly-Baked Ideas* (London: Heinemann; New York: Basic Books; paperback: Putnam's), 365–367. (Also in various newspapers, c. March 1963.)

Good, I. J., (1962/63), 'Maximum Entropy for Hypothesis Formulation, Especially for Multidimensional Contingency Tables', *Ann. Math. Statist.* **34**, 911–934.

Good, I. J., (1965a), 'Speculations Concerning the First Ultra-intelligent Machine', *Advances in Computers* **6**, 31–88.

Good, I. J., (1965b), *The Estimation of Probabilities: An Essay on Modern Bayesian Methods*, MIT Press (pp. xii+109).

Good, I. J., (1965/67), 'A Bayesian Significance Test for Multinomial Distributions', *J. Roy. Statist. Soc.* **B29**, 399–431.

Good, I. J., (1967/68), 'Corroboration, Explanation, Evolving Probability, Simplicity, and a Sharpened Razor', *Brit. J. Phil. Sc.* **19**, 123–143.

Good, I. J., (1967/70), 'Some Statistical Methods in Machine-intelligence Research', in the Thirteenth Conference on the Design of Experiments in Army Research, and Testing (an unclassified conference); *Virginia J. Sci.* **19**, 101–110; *Math. Biosc.* **6**, 185–208.

Good, I. J., (1968), 'The Utility of a Distribution', *Nature* **219**, 1392.

Good, I. J., (1968/69a), 'A Subjective Analysis of Bode's Law and an 'Objective' Test for Approximate Numerical Rationality', *J. Amer. Statist. Assoc.* **64**, 23–66 (with discussion).

Good, I. J., (1968/69b), 'What is the Use of a Distribution?', Invited paper in the Second International Symposium on Multivariate Statistics, Dayton, Ohio; *Multivariate Analysis II* (ed. P. R. Krishnaiah; New York: Academic Press), 183–203.

Good, I. J., (1968/70), 'Discussion of Bruno de Finetti's paper "Initial Probabilities: a Prerequisite for Any Valid Induction",' 1968 Salzburg Colloquium in Philosophy of Science (International Union of History and Philosophy of Science; Division of Logic, Methodology and Philosophy of Science), *Synthese* **20** (1969), 17–24; and in *Induction, Physics and Ethics* (ed. P. Weingartner and G. Zecha, Synthese-Library, D. Reidel Pub. Comp., Dordrecht, Holland), 18–25.

Good, I. J., (1970/71a), 'The Probabilistic Explication of Information, Evidence, Surprise, Causality, Explanation, and Utility', an invited lecture at the Waterloo conference, April 1970, with appendix 'Twenty-seven Principles of Rationality', in *Foundations of Statistical Inference: Proc. Symp. on the Foundations of Statistical Inference prepared under the auspices of the Réné Descartes Foundation and held at the Dept. of Statistics. Univ. of Waterloo, Ontario, Canada, from March 31 to April 9, 1970* (ed. V. P. Godambe & D. A. Sprott; Holt, Rinehart and Winston of Canada, Ltd., Toronto and Montreal, 1971), 108–141 (with discussion).

Good, I. J., (1970/71b), 'Nonparametric Roughness Penalty for Probability Densities', *Nature, Physical Science* **229**, 29–30.

Good, I. J., (1970/71c), Contribution to the discussion of a paper by J. Neyman, in the Waterloo conference, 14–15.

Good, I. J., (1970/71d), Contribution to the discussion of a paper by J. Orear and D. Cassel, in the Waterloo conference, 284–286.

Gooα, l. J., (1970/73), 'Information, Rewards, and Quasi-utilities', lecture invited by Jakob Marschak, Second World Congress of the Econometric Society, Cambridge, England; revision in *Science, Decision, and Value* (ed. J. J. Leach, R. Butts, and J. Pearch; Dordrecht: D. Reidel, 1973), 115–127; Abstract, *Econometrica* **46**, 195–196.

Good, l. J., (1971a), 'The Bayesian Influence', mimeographed notes of lectures at VPI & SU; pp. 122.

Good, I. J., (1971b), '46656 Varieties of Bayesians', letter in *The Amer. Statist.* **25**, No. 5, 62–63.

Good, I. J., (1972), 'Random Thoughts About Randomness', invited lecture in the Symp. on the Concept of Randomness dedicated to the memory of L. J. Savage in the Biennial Meeting of the Phil. Science Assoc., Olds Plaza Hotel, Michigan. To appear in *Boston Studies in the Philosophy of Science* (Dordrecht: D. Reidel).

Good, I. J. and W. I. Card (1970/71), 'The Diagnostic Process with Special Reference to Errors', *Methods of Information in Medicine* **10**, No. 3, 176–188.

Good, I. J. and J. F. Crook, (1972), 'The Bayes/Non-Bayes Compromise and the Multinomial Distribution', submitted to *J. Amer. Statist. Assoc.*, September 6, 1972. No referee's report as of September 25, 1973!

Good, I. J. and R. A. Gaskins, (1971), 'Non-parametric Roughness Penalties for Probability Densities', *Biometrika* **58**, No. 2, 255–277.

Good, I. J. and R. A. Gaskins, (1972), 'Global Non-parametric Estimation of Probability Densities', *Virginia J. Sci.* **23**, 171–193.

Good, I. J. and G. H. Toulmin, (1956), 'The Number of New Species, and the Increase in Population Coverage, when a Sample is Increased', *Biometrika* **43**, 45–63.

Hurwicz, L., (1951), 'Some Specification Problems and Applications to Econometric Models', *Econometrica* **19**, 343–344.

Jaynes, E. T., (1957), 'Information Theory and Statistical Mechanics', *Phys. Rev.* **106**, 620–630.

Jeffreys, H., (1936), 'Further Significance Tests', *Proc. Cambridge Philos. Soc.* **32**, 416–445.

Jeffreys, H., (1939/61), *Theory of Probability* (1st and 3rd edns. Oxford: University Press).

Jeffreys, H., (1946), 'An Invariant Form for the Prior Probability in Estimation Problems', *Proc. Roy. Soc.* **A186**, 453–461.

Kalbfleish, J., (1971), *Probability and Statistical Inference* (Lecture Notes for Mathematics, 233; Dept. of Statistics, Univ. of Western Ontario).

Keynes, John Maynard, (1921), *A Treatise on Probability* (London: Macmillan).

Keynes, John Maynard, (1933), *Essays in Biography* (London: Rupert Hart Davis).

Koopman, Bernard Osgood, (1940a), 'The Axioms and Algebra of Intuitive Probability', *Ann. Math.* **41**, 269–292.

Koopman, Bernard Osgood, (1940b), 'The Bases of Probability', *Bull. Amer. Math. Soc.* **46**, 763–774.

Kullback, S. (1959), *Information Theory and Statistics* (Wiley).

Laplace, P. S. de, (1774), 'Mémoire sur la probabilité des causes par les événements', *Mémoires ... par divers Sarans* **6**, 621–656. (Reference obtained from Todhunter, 1865/1965).

Lindley, D. V., (1956), 'On the Measure of Information Provided by an Experiment', *Ann. Math. Statist.* **27**, 986–1005.

Minsky, Marvin and Oliver G. Selfridge, (1961), 'Learning in Random Nets', in *Information Theory* (ed. Colin Cherry; London: Butterworths), 335–347.

Mises, R. von, (1957), *Probability, Statistics and Truth* (2nd revised edn., prepared by Hilda Geiringer; London: Allen and Unwin; New York: MacMillan).

Peirce, Charles Sanders, (1878), 'The Probability of Induction', *Popular Science Monthly*, reprinted in *The World of Mathematics* **2** (ed. James R. Newman; New York: Simon and Schuster, 1956), 1341–1354.

Ramsey, F. P., (1926/31/50/64), 'Truth and Probability', a 1926 lecture published in *The Foundations of Mathematics and Other Logical Essays* (Routledge and Kegan Paul, London, 1931; The Humanities Press, New York, 1950). Reprinted in *Studies in Subjective Probability* (ed. H. E. Kyburg and H. E. Smokler, Wiley, 1963), 63–92.

Rényi, A., (1961), 'On Measures of Entropy and Information', *Proc. Fourth Berkeley Symp. Math. Statist. and Prob.* **1**, 547–561 (Berkeley, California; Univ. of California Press).

Robbins, Herbert E., (1968), 'Estimating the Total Probability of the Unobserved Outcomes of an Experiment', *Ann. Math. Statistics* **39**, 256–257.

Savage, L. J., (July, 1959), private communication.

Savage, L. J., (1954), *The Foundations of Statistics* (Wiley).

Savage, L. J., (1959/62), 'Subjective Probability and Statistical Practice', in *The Foundations of Statistical Inference* (ed. G. A. Barnard and D. R. Cox; London and New York).

Smith, C. A. B., (1961), 'Consistency in Statistical Inference and Decision', *J. Roy. Statist. Soc.* **B23**, 1–25.

Todhunter, I., (1865/1965), *A History of the Mathematical Theory of Probability* (Chelsea, New York, 1965).

Turing, A. M., (1940), private communication.

Wald, A., (1950), *Statistical Decision Functions* (Wiley).

Weizel, W., (1953), 'Ableitung der Quantentheorie aus klassischem, kausal determiniertem Modell', *Zeit. Phys.* **134**, 264–285.

E. T. JAYNES

CONFIDENCE INTERVALS VS
BAYESIAN INTERVALS

ABSTRACT. For many years, statistics textbooks have followed this 'canonical' procedure: (1) the reader is warned not to use the discredited methods of Bayes and Laplace, (2) an orthodox method is extolled as superior and applied to a few simple problems, (3) the corresponding Bayesian solutions are *not* worked out or described in any way. The net result is that no evidence whatsoever is offered to substantiate the claim of superiority of the orthodox method.

To correct this situation we exhibit the Bayesian and orthodox solutions to six common statistical problems involving confidence intervals (including significance tests based on the same reasoning). In every case, we find that the situation is exactly the opposite; i.e., the Bayesian method is easier to apply and yields the same or better results. Indeed, the orthodox results are satisfactory only when they agree closely (or exactly) with the Bayesian results. No contrary example has yet been produced.

By a refinement of the orthodox statistician's own criterion of performance, the best confidence interval for any location or scale parameter is proved to be the Bayesian posterior probability interval. In the cases of point estimation and hypothesis testing, similar proofs have long been known. We conclude that orthodox claims of superiority are totally unjustified; today, the original statistical methods of Bayes and Laplace stand in a position of proven superiority in actual performance, that places them beyond the reach of mere ideological or philosophical attacks. It is the continued teaching and use of orthodox methods that is in need of justification and defense.

I. INTRODUCTION[1]

The theme of our meeting has been stated in rather innocuous terms: how should probability theory be (1) formulated, (2) applied to statistical inference; and (3) to statistical physics? Lurking behind these bland generalities, many of us will see more specific controversial issues: (1) frequency vs. nonfrequency definitions of probability, (2) 'orthodox' vs. Bayesian methods of inference, and (3) ergodic theorems vs. the principle of maximum entropy as the basis for statistical mechanics.

When invited to participate here, I reflected that I have already held forth on issue (3) at many places, for many years, and at great length. At the moment, the maximum entropy cause seems to be in good hands and advancing well, with no need for any more benedictions from me; in any event, I have little more to say beyond what is already in print.[2] So it seemed time to widen the front, and enter the arena on issue (2).

Harper and Hooker (eds.), Foundations of Probability Theory, Statistical Inference, and Statistical Theories of Science, Vol. II, 175–257. All Rights Reserved. Copyright © 1976 by D. Reidel Publishing Company Dordrecht-Holland.

Why a physicist should have the temerity to do this, when no statistician has been guilty of invading physics to tell us how we ought to do our jobs, will become clear only gradually; but the main points are: (A) we were here first, and (B) because of our past experiences, physicists may be in a position to help statistics in its present troubles, well described by Kempthorne (1971). More specifically:

(A) Historically, the development of probability theory in the 18'th and early 19'th centuries from a gambler's amusement to a powerful research tool in science and many other areas, was the work of people – Daniel Bernoulli, Laplace, Poisson, Legendre, Gauss, and several others – whom we would describe today as mathematical physicists. In the 19'th century, a knowledge of their work was considered an essential part of the training of any scientist, and it was taught largely as a part of physics.

A radical change took place early in this century when a new group of workers, not physicists, entered the field. They proceeded to reject virtually everything done by Laplace and sought to develop statistics anew, based on entirely different principles. Simultaneously with this development, the physicists – with Sir Harold Jeffreys as almost the sole exception – quietly retired from the field, and statistics disappeared from the physics curriculum.

This departure of physicists from the field they had created was not, of course, due to the new competition; rather, it was just at this time that relativity theory burst upon us, X-rays and radioactivity were discovered, and quantum theory started to develop. The result was that for fifty years physicists had more than enough to do unravelling a host of new experimental facts, digesting these new revolutions of thought, and putting our house back into some kind of order. But the result of our departure was that this extremely aggressive new school in statistics soon dominated the field so completely that its methods are now known as 'orthodox statistics'. For these historical reasons, I ask you to think with me, that for a physicist to turn his attention now to statistics, is more of a home-coming than an invasion.

(B) Today, a physicist revisiting statistics to see how it has fared in our absence, sees quickly that something has gone wrong. For over fifteen years now, statistics has been in a state of growing ideological crisis – literally a crisis of conflicting ideas – that shows no signs of resolving itself, but yearly grows more acute; but it is one that physicists can recognize as basically the same thing that physics has been through several times (Jaynes, 1967). Having seen how these crises work themselves out, I think physicists may be in a position to prescribe a physic that will speed up the process in statistics.

The point we have to recognize is that issues of the kind facing us are never resolved by mere philosophical or ideological debate. At that level of discussion, people will persist in disagreeing, and nobody will be able to prove his case. In physics, we have our own ideological disputes, just as deeply felt by the protagonists as any in statistics; and at the moment I happen to be involved in one that strikes at the foundations of quantum theory (Jaynes, 1973). But in physics we have been perhaps more fortunate in that we have a universally recognized Supreme Court, to which all disputes are taken eventually, and from whose verdict there is no appeal. I refer, of course, to direct experimental observation of the facts.

This is an exciting time in physics, because recent advances in technology (lasers, fast computers, etc.) have brought us to the point where issues which have been debated fruitlessly on the philosophical level for 45 years, are at last reduced to issues of fact, and experiments are now underway testing controversial aspects of quantum theory that have never before been accessible to direct check. We have the feeling that, very soon now, we are going to know the real truth, the long debate can end at last, one way or the other; and we will be able to turn a great deal of energy to more constructive things. Is there any hope that the same can be done for statistics?

I think there is, and history points the way. It is to Galileo that we owe the first demonstration that ideological conflicts are ·resolved, not by debate, but by observation of fact. But we also recall that he ran into some difficulties in selling this idea to his contemporaries. Perhaps the most striking thing about his troubles was not his eventual physical persecution, which was hardly uncommon in those days; but rather the quality of logic that was used by his adversaries. For example, having turned his new telescope to the skies, Galileo announced discovery of the moons of Jupiter. A contemporary scholar ridiculed the idea, asserted that his theology had proved there could be *no* moons about Jupiter; and steadfastly refused to look through Galileo's telescope. But to everyone who did take a look, the evidence of his own eyes somehow carried more convincing power than did any amount of theology.

Galileo's telescope was able to reveal the truth, in a way that transcended all theology, because it could *magnify* what was too small to be perceived by our unaided senses, up into the range where it could be seen directly by all. And that, I suggest, is exactly what we need in statistics if this conflict is ever to be resolved. Statistics cannot take its dispute to the Supreme Court of the physicist; but there is another. It was recognized by Laplace in that famous remark, "Probability theory is nothing but common sense reduced to calculation".

Let me make what, I fear, will seem to some a radical, shocking suggestion: *the merits of any statistical method are not determined by the ideology which led to it*. For, many different, violently opposed ideologies may all lead to the same final 'working equations' for dealing with real problems. Apparently, this phenomenon is something new in statistics; but it is so commonplace in physics that we have long since learned how to live with it. Today, when a physicist says, "Theory A is better than theory B", he does not have in mind any ideological considerations; he means simply, "There is at least one specific application where theory A leads to a better result than theory B".

I suggest that we apply the same criterion in statistics: *the merits of any statistical method are determined by the results it gives when applied to specific problems*. The Court of Last Resort in statistics is simply our commonsense judgment of those results. But our common sense, like our unaided vision, has a limited resolving power. Given two different statistical methods (e.g., an orthodox and a Bayesian one), in many cases they lead to final numerical results which are so nearly alike that our common sense is unable to make a clear decision between them. What we need, then, is a kind of Galileo telescope for statistics; let us try to invent an extreme case where a small difference is magnified to a large one, or if possible to a qualitative difference in the conclusions. Our common sense will then tell us which method is preferable, in a way that transcends all ideological quibbling over 'subjectivity', 'objectivity', the 'true meaning of probability', etc.

I have been carrying out just this program, as a hobby, for many years, and have quite a mass of results covering most areas of statistical practice. They all lead to the same conclusion, and I have yet to find one exception to it. So let me give you just a few samples from my collection.

(a) INTERVAL ESTIMATION

Time not permitting even a hurried glimpse at the entire field of statistical inference, it is better to pick out a small piece of it for close examination. Now we have already a considerable Underground Literature on the relation of orthodox and Bayesian methods in the areas of point estimation and hypothesis testing, the topics most readily subsumed under the general heading of Decision Theory. [I say underground, because the orthodox literature makes almost no mention of it. Not only in textbooks, but even in such a comprehensive treatise as that of Kendall and Stuart (1961), the reader can find no hint of the existence of the books of Good (1950), Savage (1954), Jeffreys (1957), or Schlaifer (1959), all of which are landmarks in the modern development of Bayesian statistics].

It appears that much less has been written about this comparison in the case of interval estimation; so I would like to examine here the orthodox principle of confidence intervals (including significance tests based on the same kind of reasoning), as well as the orthodox criteria of performance and method of reporting results; and to compare these with the corresponding Bayesian reasoning and results, with magnification.

The basic ideas of interval estimation must be ancient, since they occur inevitably to anyone involved in making measurements, as soon as he ponders how he can most honestly communicate what he has learned to others, short of giving the entire mass of raw data. For, if you merely give your final best number, some troublesome fellow will demand to know how accurate the number is. And you will not appease him merely by answering his question; for if you reply, "It is within a tenth of a percent", he will only ask, "How sure are you of that? Will you make a 10:1 bet on it?"

It is not enough, then, to give a number or even an interval of possible error; at the very minimum, one must give both an interval and some indication of the reliability with which one can assert that the true value lies within it. But even this is not really enough; ideally (although this goes beyond current practice) one ought to give many different intervals – or even a continuum of all possible intervals – with some kind of statement about the reliability of each, before he has fully described his state of knowledge. This was noted by D. R. Cox (1958), in producing a nested sequence of confidence intervals; evidently, a Bayesian posterior probability accomplishes the same thing in a simpler way.

Perhaps the earliest formal quantitative treatment of interval estimation was Laplace's analysis of the accuracy with which the mass of
Saturn was known at the end of the 18'th century. His method was to
apply Bayes' theorem with uniform prior density; relevant data consist
of the mutual perturbations of Jupiter and Saturn, and the motion of
their moons, but the data are imperfect because of the finite accuracy
with which angles and time intervals can be measured. From the posterior
distribution $P(M)\,dM$ conditional on the available data, one can determine the shortest interval which contains a specified amount of posterior
probability, or equally well the amount of posterior probability contained in a specified interval. Laplace chose the latter course, and announced his result as follows: "... it is a bet of 11 000 against 1 that the
error of this result is not 1/100 of its value". In the light of present knowledge, Laplace would have won his bet; another 150 years' accumulation
of data has increased the estimate by 0.63 percent.

Today, orthodox teaching holds that Laplace's method was, in Fisher's
words, "founded upon an error". While there are some differences of
opinion within the orthodox school, most would hold that the proper
method for this problem is the confidence interval. It would seem to me
that, in order to substantiate this claim, the orthodox writers would have
to (1) produce the confidence interval for Laplace's problem, (2) show
that it leads us to numerically different conclusions, and (3) demonstrate
that the confidence interval conclusions are more statisfactory than
Laplace's. But, in some twenty years of searching the orthodox literature,
I have yet to find one case where such a program is carried out, on any
statistical problem.

Invariably, the superiority of the orthodox method is asserted, not by
presenting evidence of superior performance, but by a kind of ideological
invective about 'objectivity' which perhaps reached its purple climax in an
astonishing article of Bross (1963), whose logic recalls that of Galileo's
colleague. In his denunciation of everything Bayesian, Bross specifically
brings up the matter of confidence intervals and orthodox significance
tests (which are based on essentially the same reasoning, and often amount
to one-sided confidence intervals). So we will do likewise; in the following,
we will examine these same methods and try to supply what Bross omitted; the demonstrable facts concerning them.

We first consider three significance tests appearing in the recent literature of reliability theory. The first two, which turn out to be so clear that no magnification is needed, will also bring out an important point concerning orthodox methods of reporting results.

II. SIGNIFICANCE TESTS

> Significance tests, in their usual form, are not compatible with a Bayesian attitude.
>
> C. A. B. Smith (1962)

> At any rate, what I feel quite sure at the moment to be needed is simple illustration of the new [i.e., Bayesian] notions on real, everyday statistical problems.
>
> E. S. Pearson (1962)

(a) EXAMPLE 1. DIFFERENCE OF MEANS

One of the most common of the 'everyday statistical problems' concerns the difference of the means of two normal distributions. A good example, with a detailed account of how current orthodox practice deals with such problems, appears in a recent book on reliability engineering (Roberts, 1964).

Two manufacturers, A and B, are suppliers for a certain component, and we want to choose the one which affords the longer mean life. Manufacturer A supplies 9 units for test, which turn out to have a (mean $\pm$ standard deviation) lifetime of (42 ± 7.48) hours. B supplies 4 units, which yield (50 ± 6.48) hours.

I think our common sense tells us immediately, without any calculation, that this constitutes fairly substantial (but not overwhelming) evidence in favor of B. While we should certainly prefer a larger sample, B's units did give a longer mean life, the difference being appreciably greater than the sample standard deviation; and so if a decision between them must be made on this basis, we should have no hesitation in choosing B. However, the author warns against drawing any such conclusion, and says that, if you are tempted to reason this way, then "perhaps statistics is not for you!" In any event, when we have so little evidence, it is imperative that we analyze the data in a way that does not throw any of it away.

The author then offers us the following analysis of the problem. He first asks whether the two variances are the same. Applying the F-test,

the hypothesis that they are equal is not rejected at the 95 percent significance level, so without further ado he assumes that they *are* equal, and
pools the data for an estimate of the variance. Applying the *t*-test, he
then finds that, at the 90 percent level, the sample affords no significant
evidence in favor of either manufacturer over the other.

Now, any statistical procedure which fails to extract evidence that is
already clear to our unaided common sense, is certainly *not* for me! So, I
carried out a Bayesian analysis. Let the unknown mean lifetimes of A's
and B's components be a, b respectively. If the question at issue is whether
$b > a$, the way to answer it is to calculate the *probability* that $b > a$, conditional on all the available data. This is

$$(1) \qquad \text{Prob}\,(b > a) = \int_{-\infty}^{\infty} da \int_{a}^{\infty} db\, P_n(a)\, P_m(b)$$

where $P_n(a)$ is the posterior distribution of a, based on the sample of
$n = 9$ items supplied by A, etc. When the variance is unknown, we find
that these are of the form of the 'Student' *t*-distribution:

$$(2) \qquad P_n(a) \sim [s_A^2 + (a - \bar{t}_A)^2]^{-n/2}$$

where $\bar{t}_A$, $s_A^2 = \overline{t_A^2} - \bar{t}_A^2$ are the mean and variance of sample A. Carrying out
the integration (1), I find that the given data yield a probability of 0.920,
or odds of 11.5 to 1, that B's components *do* have a greater mean life – a
conclusion which, I submit, conforms nicely to the indications of common
sense.[3]

But this is far from the end of the story; for one feels intuitively that if
the variances are assumed equal, this ought to result in a more selective
test than one in which this is not assumed; yet we find the Bayesian test
without assumption of equal variance yielding an apparently sharper
result than the orthodox one with that assumption. This suggests that we
repeat the Bayesian calculation, using the author's assumption of equal
variances. We have again an integral like (1), but a and b are no longer
independent, their joint posterior distribution being proportional to

$$(3) \qquad P\,(a, b) \sim \{n\,[s_A^2 + (a - \bar{t}_A)^2] + m\,[s_B^2 + (b - \bar{t}_B)^2]\}^{-1/2\,(n+m)}$$

Integrating this over the same range as in (1) – which can be done simply
by consulting the *t*-tables after carrying out one integration analytically –

I find that the Bayesian analysis now yields a probability of 0.948, or odds of 18:1, in favor of *B*.

How, then, could the author have failed to find significance at the 90 percent level? Checking the tables used we discover that, without having stated so, he has applied the *equal tails* *t*-test at the 90 percent level. But this is surely absurd; it was clear from the start that there is no question of the data supporting *A*; the only purpose which can be served by a statistical analysis is to tell us *how strongly* it supports *B*.

The way to answer this is to test the null hypothesis $b=a$ against the one-sided alternative $b>a$ already indicated by inspection of the data; using the 90 percent equal-tails test throws away half the 'resolution' and gives what amounts to a one-sided test at the 95 percent level, where it just barely fails to achieve significance.

In summary, the data yield clear significance at the 90 percent level; but the above orthodox procedure (which is presumably now being taught to many students) is a compounding of two errors. Assuming the variances equal makes the difference $(\bar{t}_B - \bar{t}_A)$ appear, unjustifiedly, even more significant; but then use of the equal tails criterion throws away more than was thus gained, and we still fail to find significance at the 90 percent level.

Of course, the fact that orthodox methods are capable of being misused in this way does not invalidate them; and Bayesian methods can also be misused, as we know only too well. However, there must be something in orthodox teaching which predisposes one toward this particular kind of misuse, since it is very common in the literature and in everyday practice. It would be interesting to know why most orthodox writers will not use – or even mention – the Behrens-Fisher distribution, which is clearly the correct solution to the problem, has been available for over forty years (Fisher, 1956; p. 95), and follows immediately from Bayes' theorem with the Jeffreys prior (Jeffreys, 1939; p. 115).

(b) EXAMPLE 2. SCALE PARAMETERS

A recent Statistics Manual (Crow *et al.*, 1960) proposes the following problem: 31 rockets of type 1 yield a dispersion in angle of 2237 mils^2, and 61 of type 2 give instead 1347 mils^2. Does this constitute significant evidence for a difference in standard deviation of the two types?

I think our common sense now tells us even more forcefully that, in view of the large samples and the large observed difference in dispersion, this constitutes absolutely unmistakable evidence for the superiority of type 2 rockets. Yet the authors, applying the equal-tails F-test at the 95 percent level, find it not significant, and conclude: "We need not, as far as this experiment indicates, differentiate between the two rockets with respect to their dispersion".

Suppose you were a military commander faced with the problem of deciding which type of rocket to adopt. You provide your statistician with the above data, obtained at great trouble and expense, and receive the quoted report. What would be your reaction? I think that you would fire the statistician on the spot; and henceforth make decisions on the basis of your own common sense, which is evidently a more powerful tool than the equal-tails F-test.

However, if your statistician happened to be a Bayesian, he would report[4] instead: "These data yield a probability of 0.9574, or odds of 22.47:1, in favor of type 2 rockets". I think you would decide to keep this fellow on your staff, because his report not only agrees with common sense; it is stated in a far more useful form. For, you have little interest in being told merely whether the data constitute 'significant evidence for a difference'. It is already obvious without any calculation that they *do* constitute highly significant evidence in favor of type 2; the only purpose that can be served by a statistical analysis is, again, to tell us quantitatively *how significant* that evidence is. Traditional orthodox practice fails utterly to do this, although the point has been noted recently by some.

What we have found in these two examples is true more generally. The orthodox statistician conveys little useful information when he merely reports that the null hypothesis is or is not rejected at some arbitrary preassigned significance level. If he reports that it is rejected at the 90 percent level, we cannot tell from this whether it would have been rejected at the 92 percent, or 95 percent level. If he reports that it is not rejected at the 95 percent level, we cannot tell whether it would have been rejected at the 50 percent, or 90 percent level. If he uses an equal-tails test, he in effect throws away half the 'resolving power' of the test, and we are faced with still more uncertainty as to the real import of the data.

Evidently, the orthodox statistician would tell us far more about what the sample really indicates if he would report instead *the critical significance level at which the null hypothesis is just rejected in favor of the one-*

sided alternative indicated by the data; for we then know what the verdict would be at all levels, and no resolution has been lost to a superfluous tail. Now two possible cases can arise: (I) the number thus reported is identical with the Bayesian posterior probability that the alternative is true; (II) these numbers are different.

If case (I) arises (and it does more often than is generally realized), the Bayesian and orthodox tests are going to lead us to exactly the same numerical results and the same conclusions, with only a verbal disagreement as to whether we should use the word 'probability' or 'significance' to describe them. In particular, the orthodox t-test and F-test against one-sided alternatives would, if their results were reported in the manner just advocated, be precisely equivalent to the Bayesian tests based on the Jeffreys prior $d\mu d\sigma/\sigma$. Thus, if we assume the variances equal in the above problem of two means, the observed difference is just significant by the one-sided t-test at the 94.8 percent level; and in the rocket problem a one-sided F-test just achieves significance at the 95.74 percent level.

It is only when case (II) is found that one could possibly justify any 'objective' claim for superiority of either approach. Now it is just these cases where we have the opportunity to carry out our 'magnification' process; and if we can find a problem for which this difference is magnified sufficiently, the issue cannot really be in doubt. We find this situation, and a number of other interesting points of comparison, in one of the most common examples of acceptance tests.

(c) EXAMPLE 3. AN ACCEPTANCE TEST

The probability that a certain machine will operate without failure for a time t is, by hypothesis, $\exp(-\lambda t)$, $0 < t < \infty$. We test n units for a time t, and observe r failures; what assurance do we then have that the mean life $\theta = \lambda^{-1}$ exceeds a preassigned value θ_0?

Sobel and Tischendorf (1959) (hereafter denoted ST) give an orthodox solution with tables that are reproduced in Roberts (1964). The test is to have a critical number C (i.e., we accept only if $r \leqslant C$). On the hypothesis that we have the maximum tolerable failure rate, $\lambda_0 = \theta_0^{-1}$, the probability that we shall see r or fewer failures is the binomial sum

$$(4) \qquad W(n, r) = \sum_{k=0}^{r} \binom{n}{k} e^{-(n-k)\lambda_0 t} (1 - e^{-\lambda_0 t})^k$$

and so, setting $W(n, C) \leqslant 1 - P$ gives us the sample size n required in order that this test will assure $\theta \geqslant \theta_0$ at the $100 P$ percent significance level. From the ST tables we find, for example, that if we wish to test only for a time $t = 0.01 \, \theta_0$ with $C = 3$, then at the 90 percent significance level we shall require a test sample of $n = 668$ units; while if we are willing to test for a time $t = \theta_0$ with $C = 1$, we need test only 5 units.

The amount of testing called for is appalling if $t \ll \theta_0$; and out of the question if the units are complete systems. For example, if we want to have 95 percent confidence (synonymous with significance) that a space vehicle has $\theta_0 \geqslant 10$ years, but the test must be made in six months, then with $C = 1$, the ST tables say that we must build and test 97 vehicles! Suppose that, nevertheless, it had been decreed on the highest policy level that this degree of confidence *must* be attained, and you were in charge of the testing program. If a more careful analysis of the statistical problem, requiring a few man-years of statisticians' time, could reduce the test sample by only one or two units, it would be well justified economically. Scrutinizing the test more closely, we note four points:

(1) We know from the experiment not only the total number r of failures, but also the particular times $\{t_1 \ldots t_r\}$ at which failure occurred. This informaion is clearly relevant to the question being asked; but the ST test makes no use of it.

(2) The test has a 'quasi-sequential' feature; if we adopt an acceptance number $C = 3$, then as soon as the fourth failure occurs, we know that the units are going to be rejected. If no failures occur, the required degree of confidence will be built up long before the time t specified in the ST tables. In fact, t is the *maximum possible* testing time, which is actually required only in the marginal case where we observe exactly C failures. A test which is 'quasi-sequential' in the sense that it terminates when a clear rejection or the required confidence is attained, will have an expected length less than t; conversely, such a test with the expected length set at t will require fewer units tested.

(3) We have relevant prior information; after all, the engineers who designed the space vehicle knew in advance what degree of reliability was needed. They have chosen the quality of materials and components, and the construction methods, with this in mind. Each sub-unit has had its own tests. The vehicles would never have reached the final testing stage unless the engineers knew that they were operating satisfactorily. In

other words, we are not testing a completely unknown entity. The ST test (like most orthodox procedures) ignores all prior information, except perhaps when deciding which hypotheses to consider, or which significance level to adopt.

(4) In practice, we are usually concerned with a different question than the one the ST test answers. An astronaut starting a five-year flight to Mars would not be particularly comforted to be told, "We are 95 percent confident that the average life of an imaginary population of space vehicles like yours, is at least ten years". He would much rather hear, "There is 95 percent probability that *this* vehicle will operate without breakdown for ten years". Such a statement might appear meaningless to an orthodox statistician who holds that (probability)≡(frequency). But such a statement would be very meaningful indeed to the astronaut.

This is hardly a trivial point; for if it were *known* that $\lambda^{-1} = 10$ yr, the probability that a particular vehicle will actually run for 10 yrs would be only $1/e = 0.368$; and the period for which we are 95 percent sure of success would be only $-10 \ln(0.95)$ years, or 6.2 months. Reports which concern only the 'mean life' can be rather misleading.

Let us first compare the ST test with a Bayesian test which makes use of exactly the same information; i.e., we are allowed to use only the total number of failures, not the actual failure times. On the hypothesis that the failure rate is λ, the probability that exactly r units fail in time t is

$$(5) \qquad p(r \mid n, \lambda, t) = \binom{n}{r} e^{-(n-r)\lambda t} (1 - e^{-\lambda t})^r.$$

I want to defer discussion of nonuniform priors; for the time being suppose we assign a uniform prior density to λ. This amounts to saying that, before the test, we consider it extremely unlikely that our space vehicles have a mean life as long as a microsecond; nevertheless it will be of interest to see the result of using this prior. The posterior distribution of λ is then

$$(6) \qquad p(d\lambda \mid n, r, t) = \frac{n!}{(n-r-1)! \, r!} e^{-(n-r)\lambda t} (1 - e^{-\lambda t})^r \, d(\lambda t).$$

The Bayesian acceptance criterion, which ensures $\theta \geqslant \lambda_0^{-1}$ with 100 P

percent probability, is then

$$(7) \qquad \int_{\lambda_0}^{\infty} p(d\lambda \mid n, r, t) \leqslant 1 - P.$$

But the left-hand side of (7) is identical with $W(n, r)$ given by (4); this is just the well-known identity of the incomplete Beta function and the incomplete binomial sum, given already in the original memoir of Bayes (1763).

In this first comparison we therefore find that the ST test is mathematically identical with a Bayesian test in which (1) we are denied use of the actual failure times; (2) because of this it is not possible to take advantage of the quasi-sequential feature; (3) we assign a ridiculously pessimistic prior to λ; (4) we still are not answering the question of real interest for most applications.

Of these shortcomings, (2) is readily corrected, and (1) undoubtedly could be corrected, without departing from orthodox principles. On the hypothesis that the failure rate is λ, the probability that r specified units fail in the time intervals $\{dt_1 \ldots dt_r\}$ respectively, and the remaining $(n-r)$ units do not fail in time t, is

$$(8) \qquad p(dt_1 \ldots dt_r \mid n, \lambda, t) = [\lambda^r e^{-\lambda r \bar{t}} dt_1 \ldots dt_r] [e^{-(n-r)\lambda t}]$$

where $\bar{t} \equiv r^{-1} \sum t_i$ is the mean life of the units which failed. There is no single 'statistic' which conveys all the relevant information; but r and $\bar{t}$ are jointly sufficient, and so an optimal orthodox test must somehow make use of both. When we seek their joint sampling distribution $p(r, d\bar{t} \mid n, \lambda, t)$ we find, to our dismay, that for given r the interval $0 < \bar{t} < t$ is broken up into r equal intervals, with a different analytical expression for each. Evidently a decrease in r, or an increase in $\bar{t}$, should incline us in the direction of acceptance; but at what rate should we trade off one against the other? To specify a definite critical region in both variables would seem to imply some postulate as to their relative importance. The problem does not appear simple, either mathematically or conceptually; and I would not presume to guess how an orthodox statistician would solve it.

The relative simplicity of the Bayesian analysis is particularly striking in this problem; for all four of the above shortcomings are corrected effortlessly. For the time being, we again assign the pessimistic uniform prior to λ; from (8), the posterior distribution of λ is then

$$(9) \qquad p(d\lambda \mid n, t, t_1 \ldots t_r) = \frac{(\lambda T)^r}{r!} e^{-\lambda T} d(\lambda T)$$

where

$$(10) \qquad T \equiv r\bar{t} + (n - r)t$$

is the total unit-hours of failure-free operation observed. The posterior probability that $\lambda \geqslant \theta_0$ is now

$$(11) \qquad B(n, r) = \frac{1}{r!} \int_{\lambda_0 T}^{\infty} x^r e^{-x} dx = e^{-\lambda_0 T} \sum_{k=0}^{r} \frac{(\lambda_0 T)^k}{k!}$$

and so, $B(n, r) \leqslant 1 - P$ is the new Bayesian acceptance criterion at the 100 P percent level; the test can terminate with acceptance as soon as this inequality is satisfied.

Numerical analysis shows little difference between this test and the ST test in the usual range of practical interest where we test for a time short compared to θ_0 and observe only a very few failures. For, if $\lambda_0 t \ll 1$, and $r \ll n$, then the Poisson approximation to (4) will be valid; but this is just the expression (11) except for the replacement of T by nt, which is itself a good approximation. In this region the Bayesian test (11) with maximum possible duration t generally calls for a test sample one or two units smaller than the ST test. Our common sense readily assents to this; for if we see only a few failures, then information about the actual failure time adds little to our state of knowledge.

Now let us magnify. The big differences between (4) and (11) will occur when we find many failures; if all n units fail, the ST test tells us to reject at all confidence levels, even though the observed mean life may have been thousands of times our preassigned θ_0. The Bayesian test (11) does not break down in this way; thus if we test 9 units and all fail, it tells us to accept at the 90 percent level if the observed mean life $\bar{t} \geqslant 1.58 \, \theta_0$. If we test 10 units and 9 fail, the ST test says we can assert with 90 percent confidence that $\theta \geqslant 0.22 \, t$; the Bayesian test (11) says there is 90 percent probability that $\theta \geqslant 0.63 \, \bar{t} + 0.07 \, t$. Our common sense has no difficulty in deciding which result we should prefer; thus taking the actual failure times into account leads to a clear, although usually not spectacular, improvement in the test. The person who rejects the use of Bayes' theorem in the manner of Equation (9) will be able to obtain a comparable improvement only with far greater difficulty.

But the Bayesian test (11) can be further improved in two respects. To correct shortcoming (4), and give a test which refers to the reliability of the individual unit instead of the mean life of an imaginary 'population' of them, we note that if λ were known, then by our original hypothesis the probability that the lifetime θ of a given unit is at least θ_0, is

(12) $p(\theta \geqslant \theta_0 \mid \lambda) = e^{-\lambda\theta_0}$.

The probability that $\theta \geqslant \theta_0$, conditional on the evidence of the test, is therefore

(13) $p(\theta \geqslant \theta_0 \mid n, t_1 \dots t_r) =$

$$= \int_0^\infty e^{-\lambda\theta_0}\, p(d\lambda \mid n, t_1 \dots t_r) = \left(\frac{T}{T + \theta_0}\right)^{r+1}.$$

Thus, the Bayesian test which ensures, with 100 P percent probability, that the life of an *individual unit* is at least θ_0, has an acceptance criterion that the expression (13) is $\geqslant P$; a result which is simple, sensible, and as far as I can see, utterly beyond the reach of orthodox statistics.

The Bayesian tests (11) and (13) are, however, still based on a ridiculous prior for λ; another improvement, even further beyond the reach of orthodox statistics, is found as a result of using a reasonable prior. In 'real life' we usually have excellent grounds based on previous experience and theoretical analyses, for predicting the general order of magnitude of the lifetime in advance of the test. It would be inconsistent from the standpoint of inductive logic, and wasteful economically, for us to fail to take this prior knowledge into account.

Suppose that initially, we have grounds for expecting a mean life of the order of t_i; or a failure rate of about $\lambda_i \cong t_i^{-1}$. However, the prior information does not justify our being too dogmatic about it; to assign a prior centered sharply about λ_i would be to assert so much prior knowledge that we scarcely need any test. Thus, we should assign a prior that, while incorporating the number t_i, is still as 'spread out' as possible, in some sense.

Using the criterion of maximum entropy, we choose that prior density $p_i(\lambda)$ which, while yielding an expectation equal to λ_i, maximizes the 'measure of ignorance' $H = -\int p_i(\lambda) \log p_i(\lambda) \, d\lambda$. The solution is: $p_i(\lambda) = t_i \exp(-\lambda t_i)$. Repeating the above derivation with this prior, we find that the posterior distribution (9) and its consequences (11)–(13) still hold, but that Equation (11) is now to be replaced by

(14) $T = r\bar{t} + (n - r) t + t_i.$

Subjecting the resulting solution to various extreme conditions now shows an excellent correspondence with the indications of common sense. For example, if the total unit-hours of the test is small compared to t_i, then our state of knowledge about λ can hardly be changed by the test, unless an unexpectedly large number of failures occurs. But if the total unit-hours of the test is large compared to t_i, then for all practical purposes our final conclusions depend only on what we observed in the test, and are almost independent of what we thought previously. In intermediate cases, our prior knowledge has a weight comparable to that of the test; and if $t_i \gtrsim \theta_0$, the amount of testing required is appreciably reduced. For, if we were already quite sure the units *are* satisfactory, then we require less additional evidence before accepting them. On the other hand, if $t_i \ll \theta_0$, the test approaches the one based on a uniform prior; if we are initially very doubtful about the units, then we demand that the test itself provide compelling evidence in favor of them.

These common-sense conclusions have, of course, been recognized qualitatively by orthodox statisticians; but only the Bayesian approach leads automatically to a means of expressing all of them explicitly and quantitatively in our equations. As noted by Lehmann (1959), the orthodox statistician can and does take his prior information into account, in some degree, by moving his significance level up and down in a way suggested by the prior information. But, having no formal principle like maximum entropy that tells him how much to move it, the resulting procedure is far more 'subjective' (in the sense of varying with the taste of the individual) than anything in the Bayesian approach which recognizes the role of maximum entropy and transformation groups in determining priors.

No doubt, the completely indoctrinated orthodoxian will continue to reject priors based even on the completely impersonal (and parameter-independent) principles of maximum entropy and transformation groups, on the grounds that they are still 'subjective' because they are not frequencies [although I believe I have shown (Jaynes, 1968, 1971) that if a random experiment is involved, the probabilities calculated from maximum entropy and transformation groups have just as definite a connection with frequencies as probabilities calculated from any other principle of probability theory]. In particular, he would claim that the prior just introduced into the ST test represents a dangerous loss of 'objectivity' of that test.

To this I would reply that the judgment of a competent engineer, based on data of past experience in the field, represents information fully as 'objective' and reliable as anything we can possibly learn from a random experiment. Indeed, most engineers would make a stronger statement; since a random experiment is, by definition, one in which the outcome – and therefore the conclusion we draw from it – is subject to uncontrollable variations, it follows that the only fully 'objective' means of judging the reliability of a system is through analysis of stresses, rate of wear, etc., which avoids random experiments altogether.

In practice, the real function of a reliability test is to check against the possibility of completely unexpected modes of failure; once a given failure mode is recognized and its mechanism understood, no sane engineer would dream of judging its chances of occurring merely from a random experiment.

(d) SUMMARY

In the article of Bross (1963) – and in other places throughout the orthodox literature – one finds the claim that orthodox significance tests are 'objective' and 'scientific', while the Bayesian approach to these problems is erroneous and/or incapable of being applied in practice. The above comparisons have covered some important types of tests arising in everyday practice, and in no case have we found any evidence for the alleged superiority, or greater applicability, of orthodox tests. In every case, we have found clear evidence of the opposite.

The mathematical situation, as found in these comparisons and in many others, is just this: some orthodox tests are equivalent to the Bayesian ones based on non-informative priors, and some others, when sufficiently improved both in procedure and in manner of reporting the results, can be made Bayes-equivalent. We have found this situation when the orthodox test was (A) based on a sufficient statistic, and (B) free of nuisance parameters. In this case, we always have asymptotic equivalence for tests of a simple hypothesis against a one-sided alternative. But we often find exact equivalence for all sample sizes, for simple mathematical reasons; and this is true of almost all tests which the orthodox statistician himself considers fully satisfactory.

The orthodox t-test of the hypothesis $\mu = \mu_0$ against the alternative $\mu > \mu_0$ is exactly equivalent to the Bayesian test for reasons of symmetry; and there are several cases of exact equivalence even when the distribution is not symmetrical in parameter and estimator. Thus, for the Poisson distribution the orthodox test for $\lambda = \lambda_0$ against $\lambda > \lambda_0$ is exactly equivalent to the Bayesian test because of the identity

$$\frac{1}{n!} \int_{\lambda}^{\infty} x^n e^{-x} \, dx = \sum_{k=0}^{n} \frac{e^{-\lambda} \lambda^k}{k!}$$

and the orthodox F-test for $\sigma_1 = \sigma_2$ against $\sigma_1 > \sigma_2$ is exactly Bayes-equivalent because of the identity

$$\frac{(n + m + 1)!}{n! \, m!} \int_{0}^{P} x^n (1 - x)^m \, dx = \sum_{k=0}^{m} \frac{(n + k)!}{n! \, k!} P^{n+1} (1 - P)^k.$$

In these cases, two opposed ideologies lead to just the same final working equations.

If there is no single sufficient statistic (as in the ST test) the orthodox approach can become extremely complicated. If there are nuisance parameters (as in the problem of two means), the orthodox approach is faced with serious difficulties of principle; it has not yet produced any unambiguous and fully satisfactory way of dealing with such problems.

In the Bayesian approach, neither of these circumstances caused any difficulty; we proceeded in a few lines to a definite and useful solution. Furthermore, Bayesian significance tests are readily extended to permit us to draw inferences about the specific case at hand, rather than about some purely imaginary 'population' of cases. In most real applications, it is just the specific case at hand that is of concern to us; and it is hard to see how frequency statements about a mythical population or an imaginary experiment can be considered any more 'objective' than the Bayesian statements. Finally, no statistical method which fails to provide any way of taking prior information into account can be considered a full treatment of the problem; it will be evident from our previous work (Jaynes, 1968) and the above example, that Bayesian significance tests are extended just as readily to incorporate any testable prior information.

III. TWO-SIDED CONFIDENCE INTERVALS

> The merit of the estimator is judged by the distribution of estimates to which it gives rise, i.e., by the properties of its sampling distribution.

> We must content ourselves with formulating a rule which will give good results 'in the long run' or 'on the average' ....
>
> Kendall and Stuart (1961)

The above examples involved some one-sided confidence intervals, and they revealed some cogent evidence concerning the role of sufficiency and nuisance parameters; but they were not well adapted to studying the principle of reasoning behind them. When we turn to the general principle of two-sided confidence intervals some interesting new features appear.

(a) EXAMPLE 4. BINOMIAL DISTRIBUTION

Consider Bernoulli trials B_2 (i.e., two possible outcomes at each trial, independence of different trials). We observe r successes in n trials, and asked to estimate the limiting frequency of success f, and give a statement about the accuracy of the estimate. In the Bayesian approach, this is a very elementary problem; in the case of a uniform prior density for f [the basis of which we have indicated elsewhere (Jaynes, 1968) in terms of transformation groups; it corresponds to prior knowledge that it is *possible* for the experiment to yield either success or failure], the posterior distribution is proportional to $f^r(1-f)^{n-r}$ as found in Bayes' original memoir, with mean value $\bar{f}=(r+1)/(n+2)$ as given by Laplace (1774), and variance $\sigma^2=\bar{f}(1-\bar{f})/(N+3)$.

The $(\bar{f}\pm\sigma)$ thus found provide a good statement of the 'best' estimate of f, and if $\bar{f}$ is not too close to 0 or 1, an interval within which the true value is reasonably likely to be. The full posterior distribution of f yields more detailed statements; if $r \gg 1$ and $(n-r) \gg 1$, it goes into a normal distribution $(\hat{f}, \sigma)$. The $100\,P$ percent interval (i.e., the interval which contains $100\ P$ percent of the posterior probability) is then simply $(\hat{f} \pm q\sigma)$, where q is the $(1+P)/2$ percentile of the normal distribution; for the 90, 95, and 99% levels, $q = 1.645, 1.960, 2.576$ respectively.

When we treat this same problem by confidence intervals, we find that it is no longer an undergraduate-level homework problem, but a research project. The final results are so complicated that they can hardly be expressed analytically at all, and we require a new series of tables and charts.

In all of probability theory there is no calculation which has been subjected to more sneering abuse from orthodox writers than the Bayesian one just described, which contains Laplace's rule of succession. But suppose we take a glimpse at the final numerical results, comparing, say, the 90% confidence belts with the Bayesian 90% posterior probability belts.

This must be done with caution, because published confidence intervals all appear to have been calculated from approximate formulas which yield wider intervals than is needed for the stated confidence level. We use a recently published (Crow *et al.*, 1960) recalculated table which, for the case $n = 10$, gives intervals about 0.06 units smaller than the older Pearson-Clopper values.

If we have observed 10 successes in 20 trials, the upper 90% confidence limit is given as 0.675; the above Bayesian formula gives 0.671. For 13 successes in 26 trials, the tabulated upper confidence limit is 0.658; the Bayesian result is 0.652.

Continued spot-checking of this kind leads one to conclude that, quite generally, the Bayesian belts lie just inside the confidence belts; the difference is visible graphically only for wide belts for which, in any event, no accurate statement about f was possible. The inaccuracy of published tables and charts is often greater than the difference between the Bayesian interval and the correct confidence interval. Evidently, then, claims for the superiority of the confidence interval must be based on something other than actual performance. The differences are so small that I could not magnify them into the region where common sense is able to judge the issue.

Once aware of these things the orthodox statistician might well decide to throw away his tables and charts, and obtain his confidence intervals from the Bayesian solution. Of course, if one demands very accurate intervals for very small samples, it would be necessary to go to the incomplete Beta-function tables; but it is hard to imagine any real problem where one would care about the exact width of a very wide belt. When $r \gg 1$ and $(n-r) \gg 1$, then to all the accuracy one can ordinarily use, the required interval is simply the above ($\bar{f} \pm q\sigma$). Since, as noted, published confidence intervals are 'conservative' – a common euphemism – he can even improve his results by this procedure.

Let us now seek another problem, where differences can be magnified to the point where the equations speak very clearly to our common sense.

(b) EXAMPLE 5. TRUNCATED EXPONENTIAL DISTRIBUTION

The following problem has occurred in several industrial quality control situations. A device will operate without failure for a time θ because of a protective chemical inhibitor injected into it; but at time θ the supply of this chemical is exhausted, and failures then commence, following the exponential failure law. It is not feasible to observe the depletion of this inhibitor directly; one can observe only the resulting failures. From data on actual failure times, estimate the time θ of guaranteed safe operation by a confidence interval. Here we have a continuous sample space, and we are to estimate a location parameter θ, from the sample values $\{x_1 \ldots x_N\}$, distributed according to the law

$$(15) \qquad p(dx \mid \theta) = \begin{cases} \exp(\theta - x)\, dx, & x > \theta \\ 0, & x < \theta \end{cases}.$$

Let us compare the confidence intervals obtained from two different estimators with the Bayesian intervals. The population mean is $E(x) = = \theta + 1$, and so

$$(16) \qquad \theta^*(x_1 \ldots x_N) \equiv \frac{1}{N} \sum_{i=1}^{N} (x_i - 1)$$

is an unbiased estimator of θ. By a well-known theorem, it has variance $\sigma^2 = N^{-1}$, as we are accustomed to find. We must first find the sampling distribution of θ^*; by the method of characteristic functions we find that it is proportional to $y^{N-1} \exp(-Ny)$ for $y>0$, where $y \equiv (\theta^* - \theta + 1)$. Evidently, it will not be feasible to find the shortest confidence interval in closed analytical form, so in order to prevent this example from growing into another research project, we specialize to the case $N=3$, suppose that the observed sample values were $\{x_1, x_2, x_3\} = \{12, 14, 16\}$; and ask only for the shortest 90% confidence interval.

A further integration then yields the cumulative distribution function $F(y) = [1 - (1 + 3y + 9y^2/2) \exp(-3y)]$, $y > 0$. Any numbers y_1, y_2 satisfying $F(y_2) - F(y_1) = 0.9$ determine a 90% confidence interval. To find the shortest one, we impose in addition the constraint $F'(y_1) = F'(y_2)$. By computer, this yields the interval

(17) $\qquad \theta^* - 0.8529 < \theta < \theta^* + 0.8264$

or, with the above sample values, the shortest 90% confidence interval is

(18) $\qquad 12.1471 < \theta < 13.8264.$

The Bayesian solution is obtained from inspection of (15); with a constant prior density [which, as we have argued elsewhere (Jaynes, 1968) is the proper way to express complete ignorance of location parameter], the posterior density of θ will be

(19) $\qquad p(\theta \mid x_1 \ldots x_N) = \begin{cases} N \exp N(\theta - x_1), & \theta < x_1 \\ 0 & , \quad \theta > x_1 \end{cases}$

where we have ordered the sample values so that x_1 denotes the least one observed. The shortest posterior probability belt that contains 100 P percent of the posterior probability is thus $(x_1 - q) < \theta < x_1$, where $q = -N^{-1} \log(1-P)$. For the above sample values we conclude (by slide-rule) that, with 90% probability, the true value of θ is contained in the interval

(20) $\qquad 11.23 < \theta < 12.0.$

Now what is the verdict of our common sense? The Bayesian interval corresponds quite nicely to our common sense; the confidence interval (18) is over twice as wide, and *it lies entirely in the region $\theta > x_1$ where it is obviously impossible for θ to be*!.

I first presented this result to a recent convention of reliability and quality control statisticians working in the computer and aerospace industries; and at this point the meeting was thrown into an uproar, about a dozen people trying to shout me down at once. They told me, "This is complete nonsense. A method as firmly established and thoroughly worked over as confidence intervals couldn't possibly do such a thing. You are maligning a very great man; Neyman would never have advocated a method that breaks down on such a simple problem. If you can't do your arithmetic right, you have no business running around giving talks like this".

After partial calm was restored, I went a second time, very slowly and carefully, through the numerical work leading to (18), with all of them leering at me, eager to see who would be the first to catch my mistake [it is easy to show the correctness of (18), at least to two figures, merely by applying parallel rulers to a graph of $F(y)$]. In the end they had to concede that my result was correct after all.

To make a long story short, my talk was extended to four hours (all afternoon), and their reaction finally changed to: "My God – why didn't somebody tell me about these things before? My professors and textbooks never said anything about this. Now I have to go back home and recheck everything I've done for years".

This incident makes an interesting commentary on the kind of indoctrination that teachers of orthodox statistics have been giving their students for two generations now.

(c) WHAT WENT WRONG?

Let us try to understand what is happening here. It is perfectly true that, *if* the distribution (15) is indeed identical with the limiting frequencies of various sample values, and *if* we could repeat all this an indefinitely large number of times, then use of the confidence interval (17) *would* lead us, in the long run, to a correct statement 90% of the time. But it would lead us to a wrong answer 100% of the time in the subclass of cases where $\theta^* > x_1 + 0.85$; and *we know from the sample whether we are in that subclass*.

That there must be a very basic fallacy in the reasoning underlying the principle of confidence intervals, is obvious from this example. The difficulty just exhibited is generally present in a weaker form, where it escapes detection. The trouble can be traced to two different causes.

Firstly, it has never been a part of 'official' doctrine that confidence intervals must be based on sufficient statistics; indeed, it is usually held to be a particular advantage of the confidence interval method that it leads to exact frequency-interpretable intervals without the need for this. Kendall and Stuart (1961), however, noting some of the difficulties that may arise, adopt a more cautious attitude and conclude (loc. cit., p. 153): "... confidence interval theory is possibly not so free from the need for sufficiency as might appear".

We suggest that the general situation, illustrated by the above example, is the following: whenever the confidence interval is not based on a sufficient statistic, it is possible to find a 'bad' subclass of samples, *recognizable from the sample,* in which use of the confidence interval would lead us to an incorrect statement more frequently than is indicated by the confidence level; and also a recognizable 'good' subclass in which the confidence interval is wider than it needs to be for the stated confidence level. The point is not that confidence intervals fail to do what is claimed for them; the point is that, if the confidence interval is not based on a sufficient statistic, it is possible to do better in the individual case by taking into account evidence from the sample that the confidence interval method throws away.

The Bayesian literature contains a multitude of arguments showing that it is precisely the original method of Bayes and Laplace which does take into account all the relevant information in the sample; and which will therefore always yield a superior result to any orthodox method not based on sufficient statistics. That the Bayesian method does have this property (i.e., the 'likelihood principle') is, in my opinion, now as firmly established as any proposition in statistics. Unfortunately, many orthodox textbook writers and teachers continue to ignore these arguments; for over a decade hardly a month has gone by without the appearance of some new textbook which carries on the indoctrination by failing to present both sides of the story.

If the confidence interval *is* based on a sufficient statistic, then as we saw in Example 4, it turns out to be so nearly equal to the Bayesian interval that it is difficult to produce any appreciable difference in the numerical results; in an astonishing number of cases, they are identical. That is the case in the example just given, where x_1 is a sufficient statistic, and it yields a confidence interval identical with the Bayesian one (20).

Similarly, the shortest confidence interval for the mean of a normal distribution, whether the variance is known or unknown; and for the variance of a normal distribution, whether the mean is known or un-known; and for the width of a rectangular distribution, all turn out to be identical with the shortest Bayesian intervals at the same level (based on a uniform prior density for location parameters and the Jeffreys prior $d\sigma/\sigma$ for scale parameters). Curiously, these are just the cases cited most often by textbook writers, after warning us not to use those erroneous Bayesian methods, as an illustration of their more 'objective' orthodox methods.

The second difficulty in the reasoning underlying confidence intervals concerns their criteria of performance. In both point and interval estimation, orthodox teaching holds that the reliability of an estimator is measured by its performance 'in the long run', i.e., by its sampling distribution. Now there are some cases (e.g., fixing insurance rates) in which long-run performance *is* the sole, all-important consideration; and in such cases one can have no real quarrel with the orthodox reasoning (although the same conclusions are found just as readily by Bayesian methods). However, in the great majority of real applications, long-run performance is of no concern to us, because it will never be realized.

Our job is not to follow blindly a rule which would prove correct 90% of the time in the long run; there are an infinite number of radically different rules, all with this property. Our job is to draw the conclusions that are most likely to be right in the specific case at hand; indeed, the problems in which it is most important that we get this theory right are just the ones (such as arise in geophysics, econometrics, or antimissile defense) where we know from the start that the experiment can *never* be repeated.

To put it differently, the sampling distribution of an estimator is not a measure of its reliability in the individual case, because considerations about samples that have *not* been observed, are simply not relevant to the problem of how we should reason from the one that *has* been observed. A doctor trying to diagnose the cause of Mr. Smith's stomachache would not be helped by statistics about the number of patients who complain instead of a sore arm or stiff neck.

This does not mean that there are no connections at all between individual case and long-run performance; for if we have found the procedure which is 'best' in each individual case, it is hard to see how it could fail to be 'best' also in the long run.

The point is that the converse does not hold; having found a rule whose long-run performance is proved to be as good as can be obtained, it does not follow that this rule is necessarily the best in any particular individual case. One can trade off increased reliability for one class of samples against decreased reliability for another, in a way that has no effect on long-run performance; but has a very large effect on performance in the individual case.

Now, if I closed the discussion of confidence intervals at this point, I know what would happen; because I have seen it happen several times. Many persons, victims of the aforementioned indoctrination, would deny and ridicule what was stated in the last five paragraphs, claim that I am making wild, irresponsible statements; and make some reference like that of Bross (1963) to the 'first-rate mathematicians' who have already looked into these matters.

So, let us turn to another example, in which the above assertions are demonstrated explicitly, and so simple that all calculations can be carried through analytically.

(d) EXAMPLE 6. THE CAUCHY DISTRIBUTION

We sample two members $\{x_1, x_2\}$ from the Cauchy population

$$(21) \qquad p(\mathrm{d}x \mid \theta) = \frac{1}{\pi} \frac{\mathrm{d}x}{1 + (x - \theta)^2}$$

and from them we are to estimate the location parameter θ. The translational and permutation symmetry of this problem suggests that we use the estimator

$$(22) \qquad \theta^*(x_1, x_2) = \tfrac{1}{2}(x_1 + x_2)$$

which has a sampling distribution $p(\mathrm{d}\theta^* \mid \theta)$ identical with the original distribution (21); an interesting feature of the Cauchy law.

It is just this feature which betrays a slight difficulty with orthodox criteria of performance. For x_1, x_2, and θ^* have identical sampling distributions; and so according to orthodox teaching it cannot make any difference which we choose as our estimator, for either point or interval estimation. They will all give confidence intervals of the same length, and in the long run they will all yield correct statements equally often.

But now, suppose you are confronted with a *specific* problem; the first measurement gave $x_1 = 3$, the second $x_2 = 5$. You are not concerned in the slightest with the 'long run', because you know that, if your estimate of θ *in this specific case* is in error by more than one unit, the missile will be upon you, and you will not live to repeat the measurement. Are you now going to choose $x_1 = 3$ as your estimate when the evidence of that $x_2 = 5$ stares you in the face? I hardly think so! Our common sense thus forces us to recognize that, contrary to orthodox teaching, the reliability of an estimator is not determined merely by its sampling distribution.

The Bayesian analysis tells, us, in agreement with common sense, that for this sample, by the criterion of any loss function which is a monotonic increasing function of $|\theta^* - \theta|$ (and, of course, for which the expected loss converges), the estimator (22) is uniquely determined as the optimal one. By the quadratic loss criterion, $L(\theta^*, \theta) = (\theta^* - \theta)^2$, it is the unique optimal estimator whatever the sample values.

The confidence interval for this problem is easily found. The cumulative distribution of the estimator (22) is

$$(23) \qquad p(\theta^* < \theta' \mid \theta) = \tfrac{1}{2} + \frac{1}{\pi} \tan^{-1}(\theta' - \theta)$$

and so the shortest $100\,P$ percent confidence interval is

$$(24) \qquad (\theta^* - q) < \theta < (\theta^* + q)$$

where

$$(25) \qquad q = \tan(\pi P/2).$$

At the 90% level, $P = 0.9$, we find $q = \tan(81°) = 6.31$. Let us call this the 90% CI.

Now, does the CI make use of all the information in the sample that is relevant to the question being asked? Well, we have made use of $(x_1 + x_2)$; but we also know $(x_1 - x_2)$. Let us see whether this extra information from the individual sample can help us. Denote the sample half-range by

$$(26) \qquad y = \tfrac{1}{2}(x_1 - x_2).$$

The sampling distribution $p(dy \mid \theta)$ is again a Cauchy distribution with the same width as (21) but with zero median.

Next, we transform the distribution of samples, $p(dx_1, dx_2 \mid \theta) = = p(dx_1 \mid \theta) p(dx_2 \mid \theta)$ to the new variables (θ^*, y). The jacobian of the transformation is just 2, and so the joint distribution is

$$(27) \qquad p(d\theta^*, dy \mid \theta) = \frac{2}{\pi^2} \frac{d\theta^* \, dy}{[1 + (\theta^* - \theta + y)^2][1 + (\theta^* - \theta - y)^2]}.$$

While (x_1, x_2) are independent, (θ^*, y) are not. The conditional cumulative distribution of θ^*, when y is known, is therefore not (23), but

$$(28) \qquad p(\theta^* < \theta' \mid \theta, y) = \tfrac{1}{2} + \frac{1}{2\pi}\left[\tan^{-1}(\theta' - \theta + y) + \tan^{-1} \times \right.$$
$$\left. \times (\theta' - \theta - y)\right] + \frac{1}{4\pi y} \log\left[\frac{1 + (\theta' - \theta + y)^2}{1 + (\theta' - \theta - y)^2}\right]$$

and so, in the subclass of samples with given $(x_1 - x_2)$, the probability that the confidence interval (24) will yield a correct statement is not $P = (2/\pi) \tan^{-1} q$, but

$$(29) \qquad \begin{aligned} w(y, q) &= \frac{1}{\pi}\left[\tan^{-1}(q + y) + \tan^{-1}(q - y)\right] + \\ &+ \frac{1}{2\pi y} \log\left[\frac{1 + (q + y)^2}{1 + (q - y)^2}\right]. \end{aligned}$$

Numerical values computed from this equation are given in Table I,

TABLE I

Performance of the 90% confidence
interval for various sample
half-ranges y

y	$w(y, 6.31)$	$F(y)$
0	0.998	1.000
2	0.991	0.296
4	0.952	0.156
6	0.702	0.105
8	0.227	0.079
10	0.111	0.064
12	0.069	0.053
14	0.047	0.046
>14	$\dfrac{4q}{\pi(1 + y^2)}$	$\dfrac{2}{\pi y}$

in which we give the actual frequency $w(y, 6.31)$ of correct statements obtained by use of the 90% confidence interval, for various half-ranges y. In the third column we give the fraction of all samples, $F(y) = (2/\pi)$ $\tan^{-1}(1/y)$ which have half-range greater than y.

It appears that information about $(x_1 - x_2)$ was indeed relevant to the question being asked. In the long run, the 90% CI will deliver a right answer 90% of the time; however, its merits appear very different in the individual case. In the subclass of samples with reasonably small range, the 90% CI is too conservative; we can choose a considerably smaller interval and still make a correct statement 90% of the time. If we are so unfortunate as to get a sample with very wide range, then it is just too bad; but the above confidence interval would have given us a totally false idea of the reliability of our result. In the 6% of samples of widest range, the supposedly '90%' confidence interval actually yields a correct statement less than 10% of the time – a situation that ought to alarm us if confidence intervals are being used to help make important decisions.

The orthodox statistician can avoid this dangerous shortcoming of the confidence interval (24), without departing from his principles, by using instead a confidence interval based on the conditional distribution (28). For every sample he would choose a different interval located from (29) so as to be the shortest one which *in that subclass* will yield a correct statement 90% of the time. For small-range samples this will give a narrower interval, and for wide-range samples a correct statement more often, than will the confidence interval (24). Let us call this the 90% 'uniformly reliable' (UR) estimation rule.

Now let us see some numerical analysis of (29), showing how much improvement has been found. The 90% UR rule will also yield a correct statement 90% of the time; but for 87% of all samples (those with range less than 9.7) the UR interval is shorter than the confidence interval (24). For samples of very small range, it is 4.5 times shorter, and for half of all samples, the UR interval is less than a third of the confidence interval (24). In the 13% of samples of widest range, the confidence interval (24) yields correct statements less than 90% of the time, and so in order actually to achieve the claimed reliability, the UR interval must be wider, if we demand that it be simply connected. But we can find a UR region of two disconnected parts, whose total length remains less than a third of the CI (24) as $y \to \infty$.

The situation, therefore, is the following. For the few 'bad' samples of very wide range, no accurate estimate of θ is possible, and the confidence interval (24), being of fixed width, cannot deliver the presumed 90% reliability. In order to make up for this and hold the average success for all samples at 90%, it is then forced to cheat us for the great majority of 'good' samples by giving us an interval far wider than is needed. The UR rule never misleads us as to its reliability, neither underestimating it nor overestimating it for any sample; and for most samples it gives us a much shorter interval.

Finally, we note the Bayesian solution to this problem. The posterior distribution of θ is, from (21) in the case of a uniform prior density,

$$(30) \qquad p\left(d\theta \mid x_1, x_2\right) = \frac{2}{\pi} \frac{(1 + y^2)\, d\theta}{[1 + (\theta - x_1)^2]\,[1 + (\theta - x_2)^2]}$$

and, to find the shortest 90% posterior probability interval, we compute the cumulative distribution:

$$(31) \qquad p\left(\theta < \theta' \mid x_1, x_2\right) = \tfrac{1}{2} + \frac{1}{2\pi}\left[\tan^{-1}(\theta' - x_1) + \tan^{-1} \times\right.$$

$$\left. \times\, (\theta' - x_2)\right] + \frac{1}{4\pi y}\log\left[\frac{1 + (\theta' - x_2)^2}{1 + (\theta' - x_1)^2}\right]$$

and so, – but there is no need to go further. At this point, simply by comparing (31) with (28), the horrible truth appears: the uniformly reliable rule is precisely the Bayesian one! And yet, if I had simply introduced the Bayesian solution *ab initio,* the orthodox statistician would have rejected it instantly on grounds that have nothing to do with its performance.

(e) GENERAL PROOF

The phenomenon just illustrated is not peculiar to the Cauchy distribution or to small samples; it holds for any distribution with a location parameter. For, let the sampling distribution be

$$(32) \qquad p\left(dx_1 \ldots dx_n \mid \theta\right) = f(x_1 \ldots x_n; \theta)\, dx_1 \ldots dx_n.$$

The statement that θ is a location parameter means that

$$(33) \qquad f(x_1 + a, x_2 + a, \ldots x_n + a; \theta + a) = f(x_1 \ldots x_n; \theta),$$
$$-\infty < a < \infty.$$

Now transform the sample variables $\{x_1 \ldots x_n\}$ to a new set $\{y_1 \ldots y_n\}$:

$$(34) \qquad y_1 \equiv \bar{x} = n^{-1} \sum x_i$$

$$(35) \qquad y_i = x_i - x_1, \quad i = 2, 3, \ldots n.$$

From (33), (34), (35), the sampling distribution of the $\{y_1 \ldots y_n\}$ has the form

$$(36) \qquad p(dy_1 \ldots dy_n \,|\, \theta) = g(y_1 - \theta; y_2 \ldots y_n)\, dy_1 \ldots dy_n.$$

If y_1 is not a sufficient statistic, a confidence interval based on the sampling distribution $p(dy_1 \,|\, \theta)$ will be subject to the same objection as was (24); i.e., knowledge of $\{y_2 \ldots y_n\}$ will enable us to define 'good' and 'bad' subclasses of samples, in which the reliability of the confidence interval is better or worse than indicated by the stated confidence level. To obtain the Uniformly Reliable interval, we must use instead the distribution conditional on all the 'ancillary statistics' $\{y_2 \ldots y_n\}$. This is

$$(37) \qquad p(dy_1 \,|\, y_2 \ldots y_n; \theta) = Kg(y_1 - \theta; y_2 \ldots y_n)\, dy_1$$

where K is a normalizing constant. But the Bayesian posterior distribution of θ based on uniform prior is:

$$p(d\theta \,|\, x_1 \ldots x_n) = p(d\theta \,|\, y_1 \ldots y_n) =$$
$$(38) \qquad = Kg(y_1 - \theta; y_2 \ldots y_n)\, d\theta$$

which has exactly the same density function as (37). Therefore, by a refined orthodox criterion of performance, the 'best', (i.e., Uniformly Reliable) confidence interval for any location parameter is identical with the Bayesian posterior probability interval (based on a uniform prior) at the same level.

With a scale parameter σ, data $\{q_1 \ldots q_n\}$, set $\theta = \log \sigma$, $x_i = \log q_i$, and the above argument still holds; the UR confidence interval for any scale parameter is identical with the Bayesian interval based on the Jeffreys prior $d\sigma/\sigma$.

IV. POLEMICS

Seeing the above comparisons, one naturally asks: on what grounds was it ever supposed that confidence intervals represent an advance over the original treatment of Laplace? On this point the record is clear and abundant; orthodox arguments against Laplace's use of Bayes' theorem, and in favor of confidence intervals, have never considered such mundane things as demonstrable facts concerning performance. They consist of ideological slogans, such as "Probability statements can be made only about random variables. It is meaningless to speak of the probability that θ lies in a certain interval, because θ is not a random variable, but only an unknown constant".

On such grounds we are to be denied the derivation via Equations (1), (6), (9), (19), (30), (38) which in each case leads us in a few lines to a result that is either the same as the best orthodox result or demonstrably superior to it. On such grounds it is held to be very important that we use the words, "the probability that the interval covers the true value of θ" and we must *never, never* say, "the probability that the true value of θ lies in the interval". Whenever I hear someone belabor this distinction, I feel like the little boy in the fable of the Emperor's New Clothes.

Suppose someone proposes to you a new method for carrying out the operations of elementary arithmetic. He offers scathing denunciations of previous methods, in which he never examines the results they give, but attacks their underlying philosophy. But you discover that application of the new method leads to the conclusion that $2+2=5$. I think all protestations to the effect that, "Well, the case of $2+2$ is a peculiar pathological one, and I didn't intend the method to be used there", will fall on deaf ears. A method of reasoning which leads to an absurd result in *one* problem is thereby proved to contain a fallacy. At least, that is a rule of evidence universally accepted by scientists and mathematicians.

Orthodox statisticians appear to use different rules of evidence. It is clear from the foregoing that one can produce any number of examples, at first sight quite innocent-looking, in which use of confidence intervals or orthodox significance tests leads to absurd or dangerously misleading results. And, note that the above examples are not pathological freaks; every one of them is an important case that arises repeatedly in current practice. To the best of my knowledge, nobody has ever produced an example where the Bayesian method fails to yield a reasonable result; indeed, in the above examples, and in those noted by Kendall and Stuart (1961), the only cases where confidence intervals appear satisfactory at all are just the ones where they agree closely (or often exactly) with the Bayesian intervals. From our general proof, we understand why. And, year after year, the printing presses continue to pour out textbooks whose authors extoll the virtues of confidence intervals and warn the student against the thoroughly discredited method of Bayes and Laplace.

A physicist viewing this situation finds it quite beyond human understanding. I don't think the history of science can offer any other example in which a method which has always succeeded was rejected on doctrinaire grounds in favor of one which often fails.

Proponents of the orthodox view often describe themselves, as did Bross (1963), as 'objective', and 'fact-oriented', thereby implying that Bayesians are not. But the foundation-stone of the orthodox school of thought is this dogmatic insistence that the word 'probability' *must* be interpreted as 'frequency in some random experiment'; and that any other meaning is metaphysical nonsense. Now, assertions about the 'true meaning of probability', whether made by the orthodox or the Bayesian, are not statements of demonstrable fact. They are statements of ideological belief about a matter that cannot be settled by logical demonstration, or by taking votes. The only fully objective, fact-oriented criterion we have for deciding issues of this type, is just the one scientists use to test any theory: sweeping aside all philosophical clutter, which approach leads us to the more reasonable and useful results? I propose that we make some use of this criterion in future discussions.

Mathematically, or conceptually, there is absolutely nothing to prevent us from using probability theory in the broader Laplace interpretation, as the 'calculus of inductive reasoning'. Evidence of the type given above indicates that to do so greatly increases both the power and the simplicity of statistical methods; in almost every case, the Bayesian result required far less calculation. The main reason for this is that both the *ad hoc* step of 'choosing a statistic' and the ensuing mathematical problem of finding its sampling distribution, are eliminated. In particular, the F-test and the t-test, which require considerable mathematical demonstration in the orthodox theory, can each be derived from Bayesian principles in a few lines of the most elementary mathematics; the evidence of the sample is already fully displayed in the likelihood function, which can be written down immediately.

Now, I understand that there are some who are not only frightened to death by a prior probability, they do not even believe this last statement, the so-called 'likelihood principle', although a proof has been given (Birnbaum, 1962). However, I don't think we need a separate formal proof if we look at it this way. Nobody questions the validity of applying Bayes' theorem in the case where the parameter θ is itself a 'random variable'. But in this case the entire evidence provided by the sample *is* contained in the likelihood function; independently of the prior distribution, different intervals $d\theta$ are indicated by the sample to an extent precisely proportional to $L(\theta)\, d\theta$. It is already conceded by all that the likelihood function has this property when θ is a random variable with an arbitrary frequency distribution; is it then going to lose this property in the special case where θ is a constant? Indeed, isn't it a matter of the most elementary common sense to recognize that, in the specific problem at hand, θ is always just an unknown constant? Whether it would or would not be different in some other case that we are not reasoning about, is just not relevant to our problem; to adopt different methods on such grounds is to commit the most obvious inconsistency.

I am unable to see why 'objectivity' requires us to interpret every probability as a frequency in some random experiment; particularly when we note that in virtually every problem of real life, the direct probabilities are not determined by any real random experiment; they are calculated from a theoretical model whose choice involves 'subjective' judgment. The most 'objective' probabilities appearing in most problems are, therefore, frequencies only in an *ad hoc,* imaginary universe invented just for the purpose of allowing a frequency interpretation. The Bayesian could also, with equal ease and equal justification, conjure up an imaginary universe in which all his probabilities are frequencies; but it is idle to pretend that a mere act of the imagination can confer any greater objectivity on our methods.

According to Bayes' theorem, the posterior probability is found by multiplying the prior probability by a numerical factor, which is determined by the data and the model. The posterior probabilities therefore partake of whatever 'qualities' the priors have:

(A) If the prior probabilities are real frequencies, then the posterior probabilities are also real frequencies.

(B) If the prior probabilities are frequencies in an imaginary universe, then the posterior probabilities are frequencies in that same universe.

(C) If the prior probabilities represent what it is reasonable to believe before the experiment, by any criterion of 'reasonable', then the posterior probabilities will represent what it is equally reasonable to believe after the experiment, by the same criterion.

In no case are there any grounds for questioning the use of Bayes' theorem, which after all is just the condition for consistency of the product rule of probability theory; i.e., $p(AB \mid C)$ is symmetric in the propositions A and B, and so it can be expanded two different ways: $p(AB \mid C)=$ $=p(A \mid BC)p(B \mid C)=p(B \mid AC)p(A \mid C)$. If $p(B \mid C)\neq0$, the last equality is just Bayes' theorem:

$$P(A \mid BC) = p(A \mid C)\frac{P(B \mid AC)}{P(B \mid C)}.$$

To recognize these things in no way forces us to accept the 'personalistic' view of probability (Savage, 1954, 1962). 'Objectivity' clearly does demand at least this much: the results of a statistical analysis ought to be independent of the personality of the user. In particular, our prior probabilities should describe the prior information; and not anybody's vague personal feelings.

At present, this is an ideal that is fully achieved only in particularly simple cases where all the prior information is testable in the sense defined previously (Jaynes, 1968). In the case of the aforementioned 'competent engineer' the determination of the exact prior is, of course, not yet completely formalized. But, as stressed before, the measure of our success in achieving 'objectivity' is just the extent to which we are able to eliminate all personalistic elements, and approach a completely 'impersonalistic' theory of inference or decision; on this point I must agree whole-heartedly with orthodox statisticians.

The real issue facing us is not an absolute value judgment but a relative one; it is not whether Bayesian methods are 100% perfect, or whether their underlying philosophy is opprobrious; but simply whether, at the present time, they are better or worse than orthodox methods in the results they give in practice. Comparisons of the type given here and in the aforementioned Underground Literature – and the failure of orthodoxy to produce any counter-examples – show that the original statistical methods of Laplace stand today in a position of proven superiority, that places them beyond the reach of attacks on the philosophical level, and *a fortiori* beyond any need for defense on that level.

Presumably, the future will bring us still better statistical methods; I predict that these will be found through further refinement and generalization of our present Bayesian principles. After all, the unsolved problems of Bayesian statistics are ones (such as treatment of nontestable prior information) that, for the most part, go so far beyond the domain of orthodox methods that they cannot even be formulated in orthodox terms.

It would seem to me, therefore, that instead of attacking Bayesian methods because we still have unsolved problems, a rational person would want to be constructive and recognize the unsolved problems as the areas where it is important that further research be done. My work on maximum entropy and transformation groups is an attempt to contribute to, and not to tear down, the beautiful and powerful intellectual achievement that the world owes to Bayes and Laplace.

Dept. of Physics, Washington University,
St. Louis, Missouri 63130

REFERENCES

Note: Two recent objections to the principle of maximum entropy (Rowlinson, 1970; Friedman and Shimony, 1971) appear to be based on misunderstandings of work done seventeen years ago (Jaynes, 1957). In the meantime, these objections had been anticipated and answered in other articles (particularly Jaynes, 1965, 1967, 1968), of which these authors take no note. To help avoid further misunderstandings of this kind, the following references include a complete list of my publications in which maximum entropy is discussed, although not all are relevant to the present topic of Bayesian interval estimation.

Bayes, Rev. Thomas, 'An Essay Toward Solving a Problem in the Doctrine of Chances', *Phil. Trans. Roy. Soc.* 330–418 (1763). Reprint, with biographical note by G. A. Barnard, in *Biometrika* **45**, 293–315 (1958) and in *Studies in the History of Statistics and Probability*, E. S. Pearson and M. G. Kendall, (eds), C. Griffin and Co. Ltd., London, (1970). Also reprinted in *Two Papers by Bayes with Commentaries*, (W. E. Deming, ed.), Hafner Publishing Co., New York, (1963).

Birnbaum, Allen, 'On the Foundations of Statistical Inference', *J. Am. Stat. Ass'n* **57** 269 (1962).

Bross, Irwin D. J., 'Linguistic Analysis of a Statistical Controversy', *The Am. Statist.* **17**, 18 (1963).

Cox, D. R., 'Some Problems Connected with Statistical Inference', *Ann. Math. Stat.* **29**, 357 (1958).

Crow, E. L., Davis, F. A., and Maxfield, M. W., *Statistics Manual*, Dover Publications, Inc., New York (1960).

Fisher, R. A., *Statistical Methods and Scientific Inference*, Hafner Publishing Co., New York (1956).

Friedman, K. and Shimony, A., 'Jaynes' Maximum Entropy Prescription and Probability Theory', *J. Stat. Phys.* **3**, 381–384 (1971).

Good, I. J., *Probability and The Weighing of Evidence*, C. Griffin and Co. Ltd., London (1950).

Good, I. J., *The Estimation of Probabilities*, Research Monograph #30, The MIT Press, Cambridge, Mass. (1965); paperback edition, 1968.

Jaynes, E. T., 'Information Theory and Statistical Mechanics, I, II', *Phys. Rev.* **106**, 620–630; **108**, 171–190 (1957).

Jaynes, E. T., *Probability Theory in Science and Engineering*, No. 4 of *Colloquium Lectures on Pure and Applied Science*, Socony-Mobil Oil Co., Dallas, Texas (1958).

Jaynes, E. T., 'Note on Unique Decipherability', IRE Trans. on Information Theory, p. 98 (September 1959).

Jaynes, E. T., 'New Engineering Applications of Information Theory', in *Engineering Uses of Random Function Theory and Probability*, J. L. Bogdanoff and F. Kozin, (eds.), J. Wiley & Sons, Inc., N.Y. (1963); pp. 163-203.

Jaynes, E. T., 'Information Theory and Statistical Mechanics', in *Statistical Physics*, K. W. Ford, (ed.), W. A. Benjamin, Inc., (1963); pp. 181–218.

Jaynes, E. T., 'Gibbs vs. Boltzmann Entropies', *Am. J. Phys.* **33**, 391 (1965).

Jaynes, E. T., 'Foundations of Probability Theory and Statistical Mechanics', Chap. 6 in *Delaware Seminar in Foundations of Physics*, M. Bunge, (ed.), Springer-Verlag, Berlin (1967); Spanish translation in *Modern Physics*, David Webber, (ed.), Alianza Editorial s/a, Madrid 33 (1973).

Jaynes, E. T., 'Prior Probabilities', IEEE Trans. on System Science and Cybernetics, SSC-4, (September 1968), pp. 227–241.

Jaynes, E. T., 'The Well-Posed Problem', in *Foundations of Statistical Inference*, V. P. Godambe and D. A. Sprott, (eds.), Holt, Rinehart and Winston of Canada, Toronto (1971).

Jaynes, E. T., 'Survey of the Present Status of Neoclassical Radiation Theory', in *Coherence and Quantum Optics*, L. Mandel and E. Wolf, (eds.), Plenum Publishing Corp., New York (1973), pp. 35–81.

Jeffreys, H., *Theory of Probability*, Oxford University Press (1939).

Jeffreys, H., *Scientific Inference*, Cambridge University Press (1957).

Kempthorne, O., 'Probability, Statistics, and the Knowledge Business', in *Foundations of Statistical Inference*, V. P. Godambe and D. A. Sprott, (eds.), Holt, Rinehart and Winston of Canada, Toronto (1971).

Kendall, M. G. and Stuart, A., *The Advanced Theory of Statistics*, Volume 2, C. Griffin and Co., Ltd., London (1961).

Lehmann, E. L., *Testing Statistical Hypotheses*, J. Wiley & Sons, Inc., New York (1959), p. 62.

Pearson, E. S., Discussion in Savage (1962); p. 57.

Roberts, Norman A., *Mathematical Methods in Reliability Engineering*, McGraw-Hill Book Co., Inc., New York (1964) pp. 86–88.

Rowlinson, J. S., 'Probability, Information and Entropy', *Nature* **225**, 1196–1198 (1970).

Savage, L. J., *The Foundations of Statistics*, John Wiley, & Sons, Inc., New York (1954).

Savage, L. J., *The Foundations of Statistical Inference*, John Wiley & Sons, Inc., New York (1962).

Schlaifer, R., *Probability and Statistics for Business Decisions*, McGraw-Hill Book Co., Inc., New York (1959).

Sobel, M. and Tischendorf, J. A., Proc. Fifth Nat'l Symposium on Reliability and Quality Control, I.R.E., pp. 108–118 (1959).

Smith, C. A. B., Discussion in Savage (1962); p. 60.

NOTES

[1] Supported by the Air Force Office of Scientific Research, Contract No. F44620-60-0121.

[2] For those who had hoped, or at least expected, to hear instead a summary of the present status of maximum entropy, see the Note at the beginning of the References.

[3] This analysis is mathematically equivalent to use of the Behrens-Fisher distribution; however, the numerical work was done directly from Equation (1) rather than relying on tables which have been so little used and which would require a risky kind of interpolation. The first integration can be done analytically, and the second is easily done numerically to all the accuracy needed. Tail areas for $a < 0$ need not be truncated, since they contribute to (1) only in the sixth decimal place.

[4] IBM 7092 calculation by Mr. Robert Schainker. Using the Jeffreys prior, $d\sigma/\sigma$, the posterior distributions have the form $p(d\sigma \mid s) = x^r e^{-x} dx/r!$, where $x \equiv ns^2/2\sigma^2$, $2r = n-3$, and $s_1^2 = 2,237$, etc. The required probability is then an integral like (1), which can be expressed as a finite sum for numerical work. Alternatively, it can be expressed in terms of the incomplete Beta function, so that in principle the F-tables could be used; however, these tables use too widely separated values of the significance level for accurate interpolation.

DISCUSSION

INTERFACES BETWEEN STATISTICS AND CONTENT

(Remarks on the paper 'Confidence Intervals vs Bayesian Intervals' by E. T. Jaynes by Margaret W. Maxfield)

Professor Jaynes recommends common sense as a 'Court of Last Resort' for statistics. He calls for applying competing statistical methods and choosing the one whose result conforms best with our common sense.

However, one of the main reasons we apply statistical methods at all is to inform our sense in an area where we find it hazy. We want to know 'how big is big'.

Under Part II, Significance Tests, Example 1, Professor Jaynes explains an example from Roberts (1964; references are to the bibliography of Jaynes' paper):

Two manufacturers, A and B, are suppliers for a certain component, and we want to choose the one which affords the longer mean life. Manufacturer A supplies 9 units for test, which turn out to have a (mean $\pm$ standard deviation) lifetime of (42 ± 7.48) hours. B supplies 4 units, which yield (50 ± 6.48) hours.

Roberts concludes from an F test that the two variances are not significantly different, whereupon he pools the estimates of variance, an error in Jaynes' judgment. In any case, Roberts pools the estimates incorrectly, using a formula

$$s^2_{\text{pooled}} = \frac{n_A s^2_A + n_B s^2_B}{n_A + n_B - 2},$$

instead of the correct formula

$$s^2_{\text{pooled}} = \frac{(n_A - 1) s^2_A + (n_B - 1) s^2_B}{n_A + n_B - 2},$$

(See Crow, page 68, for instance), thus sufficiently overestimating the pooled variance to yield a conclusion of no significant difference in means at the 90% level, with an equal-tails *t*-test. With the correct estimate, the difference in sample means proves significant.

(a) POOLING VARIANCES AS AN INTERFACE

Neither Roberts nor Jaynes mentions the use of the problem context in the decision whether to pool variance estimates. The reason for pooling estimates is that we consider that there is a common variance to be estimated. Ideally, we base this model, not on inspection of the data, but on the content of the application. For instance, there may be a 'state of the art' limitation in production of components, and experience may suggest that both suppliers are near that limit. Or, the tolerances on components that fulfill other specifications may be quite tight, so that unless the variance is 'in control', the components will fail grosser tests.

In Roberts' problem we might suppose that the buyer expects the variance to be the same, but does not trust his common sense as to whether the ratio of the observed estimates is improbably large. This use of statistics and their distributions to guide common sense is a good example of an interface between statistics and content.

A nonstatistical data analysis is quite appropriate, also. If the buyer must make his decision on the basis of the submitted samples alone, he may observe that the components from manufacturer *B* have both the better (longer) mean component life and the smaller estimate of variance, which would fit a rationale of better quality control by manufacturer *B* – better and less variable product.

(b) ALTERNATIVE HYPOTHESES AS AN INTERFACE

Jaynes' second objection to Roberts' solution is to his use of an equal-tails test, of which Jaynes says: "But this is surely absurd; it was clear from the start that there is no question of the data supporting *A*; the only purpose which can be served by a statistical analysis is to tell us *how strongly* it supports *B*". Of course, there is nothing immoral about abandoning all statistical procedures and awarding the contract to the winner, whether he won by a nose or a mile. Undoubtedly, most decisions are made in this commonsense way, a comparison of means with no examination of dispersions. Dispersions, and their effect on reliability of estimation, are demonstrably beyond sense that is common, as any introductory statistics class will reveal.

If a buyer does realize, however, that fluctuation affects the reliability of his estimate of the difference in means, he will want to consult t tables, and in entering those tables, use an equal-tails model.

Full decision-theoretic analyses are rare, partly because of buyers' inability or reluctance to quantify their loss estimates. Either the losses are very one-sided (the buyer awards the contract to his brother-in-law), or they are as hard to estimate as the difference in means. In the absence of any utility or loss functions, the Bayesian analysis is perfectly suitable. However, if there are any lower-order criteria than difference in 'mean component life available for the choice between manufacturers – transportation costs, tie-in sales, etc., – the buyer needs to know not only who won, but by how much.

Jaynes next attacks a problem from Crow *et al.* (1960) about comparison of variances. The problem in this 'Manual' is introduced explicitly as a mere exercise in calculating an F statistic and entering the F table. There is no surrounding information offered. The F statistic is calculated as 1.66, for a ratio of standard deviations of 1.3 to 1. The finding that the F statistic in this case is not significant at the 5% level violates Jaynes' common sense.

(c) SIGNIFICANCE AS AN INTERFACE

Upon learning that an F statistic is not especially improbable, we are surprised at what different variance estimates can arise from the same variance, perhaps. We may recheck our calculations. Then we revise our common sense.

Since in both the problems quoted, the experimenter originally does not know which competitor to choose, and, in fact, that is the point of the problem, a one-tailed model is inappropriate.

The policy Jaynes recommends, of reporting the significance level at which the result would be just significant, may seem to take the binary curse from significance testing for us. However, it must be emphasized that results in a critical region are improbable *in the aggregate,* not as individuals, every single one of which has zero probability, wherever it lies.

(d) PRIMITIVE NOTIONS

People use common nontechnical connotative understanding to draw commonsense conclusions about points, lines, sets, and so on – the primitive undefined terms in mathematics. Until 'random' and 'probable', or their successors as basic terms in statistics, are understood and used in nontechnical language, we must slowly develop common sense about statistics from experience.

(Reply of E. T. Jaynes to comments of Margaret Maxfield)

It has been recognized by all, beginning with Laplace, that the purpose of a statistical analysis is to aid our common sense by giving a quantitative measure to what we feel intuitively. If it is thought that there is an inconsistency between this and my program, please note the distinction between (1) using a statistical method to help our common sense; and (2) judging the relative merits of two statistical methods by magnifying their differences up to the range where common sense needs no help.

Whether we use the coefficient n or $(n-1)$ in the pooled variance estimate depends on whether we define the symbol s^2 as the sample variance or the unbiased estimate of the population variance. Since Roberts uses n as the weighting coefficient and (p. 86) explicitly calls s the sample standard deviation, I assumed that he was using the former convention. If we reinterpret s^2 as does Ms Maxfield, then all the numerical results – both Roberts' and mine – will be changed slightly, but in the same direction and by nearly the same amount. This will not affect the comparison of our methods.

The remarks about one-sided and equal-tails tests, and about reporting critical significance levels, ignore some elementary facts that I tried to point out. That in an equal-tails test "the F statistic in this case is not significant at the 5% [my 95%] level," is just a mathematical fact, and in no way violates my common sense. On the contrary, it confirms my common sense by demonstrating the folly of using an equal-tails test. Ponder the proper fate of a Public Health official who obtains evidence for a difference in side effects of two polio vaccines that is significant at the 95.7% level by a one-sided test, and concludes: "We need not differentiate between the vaccines."

The purpose of these tests is to give an indication whether our data are consistent with some nominal value θ_0 for a parameter, or whether there is statistically significant evidence for a departure from θ_0. In deciding which test will best serve this purpose, then, we need to ask, "How much information bearing on this question do you convey when you report the result of the test?" This establishes an ordering much like the notion of admissibility.

Thus, for the t-test, denote the cumulative distribution by $\text{Prob}[t < t\,(P)] = P$, and consider what we learn from the results of one-sided and equal-tails tests. If Mr A tells us that the null hypothesis $[H_0 : \theta = \theta_0]$ was not rejected by the equal-tails test at the 90% level, then we know t was somewhere in $|t| < t\,(0.95)$. We can't tell from this whether H_0 would be rejected at the 80% level. If he tells us that H_0 was rejected at the 90% level, then we know t was in one of the tails, $|t| \geqslant t\,(0.95)$; but we don't know whether H_0 would have been rejected at the 92% level, or which alternative $[H_1 : \theta > \theta_0]$ or $[H_2 : \theta < \theta_0]$ is favored by the data.

If Mr A would report instead the critical level P for the equal-tails test, we would know far more. This determines that $t = \pm t\,(P')$, where $2P' = 1 + P$, and we know what the verdict would be at any level, for the equal-tails test. The critical level P_1 for the one-sided test $(H_0 \text{ vs. } H_1)$ is either P' or $(1 - P')$, but we can't tell which.

Evidently, we would know still more if Mr A would report the critical level P_1 for the one-sided test, instead of the equal-tails test. From P_1 we know what the verdict would be, at any level, for the equal-tails test *and* for both of the one-sided tests. But it is straightforward mathematics to show that P_1 is identical with the Bayesian posterior probability that H_1 is true.

All these considerations apply equally well to the F-test, and many others. A one-sided test tells us everything an equal-tails test does; and more. Where, then, is the justification for ever using an equal-tails test, or for claiming that "a one-tailed model is inappropriate?"

More importantly, it is not a matter of personal opinion, but a mathematically demonstrable fact, that the Bayesian method of significance testing, originated by Laplace, leads us at once to the maximum information given by the optimal orthodox test. Obviously, then, orthodox rejection of Bayesian tests cannot be justified on grounds of their actual performance.

Finally, I call the readers' attention to the devastating criticisms of orthodox hypothesis testing theory by Pratt (1961) and L. J. Savage (1962) which, to the best of my knowledge, remain unanswered to this day.

COMMENTS ON PAPER BY DR E. T. JAYNES

Oscar Kempthorne

(1) The paper by Jaynes is clearly a very seriously developed discussion of some of the problems and obscurities of statistical inference. My own views are given briefly in my presentation at this confernce. I shall not give these here but shall attempt a critique, not necessarily critical, of Jaynes' paper.

(2) I am very concerned that the picture of 'orthodox' statistics presented by Jaynes will lead philosophers of science and physicists to the view that statistics *as it is often practised* is stupid. It is a hard, bare fact that workers in noisy sciences use statistical methods, as presented, for example, by Snedecor and Cochran, very widely. Part of the thesis of my own presentation is that there has been very little attention to unavoidable noise in philosophy and physics. I refuse to discuss the matter with anyone who does not admit the problem of noise (or error, or wandering, or variability, whatever term appeals).

(3) At the beginning we are presented with a polarity, *the* orthodox solution versus *the* Bayesian solution. This needs clarification.

(a) What is the orthodox solution? There are in fact at least two streams of thought and statistical practice arising from a common origin. Fisher in his first paper used a Bayesian argument, but then by 1922 ('The Mathematical Foundations of Theoretical Statistics') had rejected Bayesian ideas. Instead, he produced (i) a large number of significance tests and (ii) a theory of statistical estimation. A natural step was then to construct some sort of statistical interval of uncertainty by inverting a significance test. This led later to Fisher's fiducial inference, which is a mystery to all but a very few, and has, I think, been rejected by almost all statisticians (including, perhaps, myself). In 1928, Neyman and Pearson tried to give more exact and more mathematical structure to the idea of tests of singificance. By 1933 they had replaced, in my opinion, significance tests by accept-reject rules and had cast the whole matter into a simple decision theoretic-structure. This led to some of the procedures which Jaynes justly castigates. The decision-theoretic approach with emphasis on frequency of errors of the two types was seen by Fisher to be in conflict with a need for quantifying in a reasonable way what may be termed the evidential content of data. There has been much ferment on the basic issue. The literature is now huge. The upshot has been a rejection by very many of the idea that Neyman-Pearson accept-reject rule theory necessarily leads to a valid quantification of evidence from data. I reviewed (*Biometrics* **25**, 647–654, 1969, discussion of paper by Cornfield) in an elementary way some of the problems and gave an example very much like Jaynes' Example 5. A serious attempt was made in the book *Probability, Statistics and Data Analysis,* by O. Kempthorne and J. L. Folks, Iowa State University Press, 1971 to present the problems. I think it is described there in what ways the Neyman-Pearson processes break down, for example, how the idea of unbiased *tests of significance* breaks down for the simplest case of independent Bernouilli trials. The story is a very long one. If, however, one wishes to enter deep discussion of the controversies a huge amount of literature, due to Fisher, Barnard, Cox, Birnbaum and others (perhaps including myself), must be read with a critical mind. At the same time one must read with a critical mind the writing of L. J. Savage, Lindley, I. R. Savage, Box and others. And also one must read with a critical mind the theory of games (especial Section 4.8.2 of the von Neumann-Morgenstern classic) and the theory of decision which is closely related to the theory of utility (whence the theory of preferences). I have found *Games and Decisions* by Luce and Raiffa very informative and I would like readers to pay especial attention to pages 33 to 37. The overwhelmingly strong message for me from the London conference was that a large portion of the ideas of outstanding workers can reasonably be subjected to severe criticism. *No one's work is sacrosanct.* We heard criticisms of the basic work in modern physics, and we know of the extreme doubts of Einstein about quantum physics and also of the extreme doubts of the validity of Einstein's criticisms. So I put forward a guiding principle: *It is complete naiveté to assume that a presentation by worker X (very good though he may be) of a theory of physics or a theory of statistics, or a theory of decision, or whatever can be taken as definitive and totally forcing.*

Any writer presents, of course, as convincing a case as possible, and almost every writer has contributed to understanding.

The basic point about 'orthodox' statistics is that there is an orthodoxy in the books on mathematical statistics, but this orthodoxy is present only mildly, and almost tangentially, in the orthodoxy (if there is one) of practising statisticians, as represented by a number of texts on statistical

methods. To attack Neyman-Pearson orthodoxy is one thing (which may
be accomplished with some success, I think) but to assume that this
orthodoxy has had any deep influence on *working* statistical practice is,
I think, largely fallacious.

It is a bad thing, I believe, to suggest, for example, that working
statisticians will estimate a probability by a formula that can give an
answer of (2), as a leading text suggests.

Or to suggest that working statisticians will estimate a probability to be
a negative number, as a Nobel laureate in physics, P. A. M. Dirac,
suggested in his Baker lecture to the Royal Society of London in 1942.

Dr. Jaynes and the neo-Bayesians have a very difficult problem arising
from the fact that there are few (perhaps no) books which describe
orthodox statistical practice from a theoretical viewpoint (though I suggest, perhaps naturally, that the book by Kempthorne and Folks attempts
this).

I repeat that the whole field is very, very difficult, and give my view that
the difficulties are not mitigated by the existence of many books on
mathematical statistics that give a picture, which is ludicrous, as our
neo-Bayesian friends are asserting, and correctly so, I think. I now turn
to a detailed reaction to the Jaynes paper.

(b) Jaynes talks about *the* Bayesian solution. I state emphatically
with no fear of being proved wrong, that there is *not* a definite Bayes
solution. There is a Bayes 'solution' associated with each choice for a
prior distribution. This ambiguity would be resolved if there were a
completely compelling way to produce a particular prior distribution
for each problem by purely logical analysis. The attempts to do this have
failed, I believe. Professor Lindley himself has so admitted, I believe.

If this is accepted, then it is incorrect to talk about *the* Bayesian result.
If the attitude to the whole problem is changed, so that the prior
distribution is a summarization of the *beliefs* of the individual investigator
(a viewpoint which L. J. Savage proposed, I believe), then the result must
be labelled as the result of the individual's prior. So if Jaynes is using his
personal prior, the result of the Bayes algorithm should be called 'Jaynes'
probability' or 'Jaynes' interval'. I do not find such statements offensive
or misleading. In a real sense, all probabilities of future contingencies are
personal. I have no objection at all to stating 'Kempthorne's probability
that it will rain tomorrow is 0.3'. Whether anyone else should accept this

as a reasonable probability *for him* is another problem. I do accept probabilities of authorities, e.g. in the genetic counselling area. If Professor Lindley would state 'My probability of hypothesis *H* is 0.7', I cannot possibly quarrel with him. He is making an assertion about himself and I regard him as the best available authority on himself and his beliefs.

(4) When Jaynes talks about the 'orthodox' solution I am of the opinion that he is talking about the Neyman-Pearson decision-theory based solution. That this became orthodoxy was a frequently and strongly voiced objection by R. A. Fisher. References are very well known and are manifold. Hence to use a phrase such as the Fisher-Neyman-Pearson approach is an obvious calumny of Fisher's views and situation. In his name I protest most strongly. For the benefit of readers who were not present at the conference, it may be stated that *precisely* this phrase was used.

(5) That the obscurities of modern physics will be resolved shortly may well be hoped, but a message I obtained from the London conference was that the foundations of modern physics are in a shambles. We have the appeal to the two-hole experiment, which we are then told is a 'Gedanken' or 'thought' experiment which has never been done. [Though again, there is a suggestion that this basic experiment was done about a year ago by someone, this 'year ago' being some fifty years after the theory was promulgated as *the final answer*. (See the quotations in my own essay.)] So I regret that I do not share Jaynes' optimism, though I do share a wonderment at the advances in technology that have grown out of modern physics, and I support strongly research in physics.

(6) Jaynes appeals to 'common sense'. I suggest most strongly that our problem is to decide what 'common sense' is. We have seen 'common sense' to be totally fallacious a huge number of times in the history of human thought. And to attempt to push the point home, I suggest that Jaynes would have very great, and perhaps insuperable, difficulties in reconciling quantum physics with common sense. It is true that we all have a vague idea of something which we call 'common sense', but our problem is to make this vague feeling sufficiently precise to use in scientific discourse. One cannot help recalling that appeals to common sense, to motherhood, to land and order and so on have led to some of the worst atrocities that humanity has perpetrated.

So I assert my opinion that Jaynes is being strongly misleading in talking about what common sense tells us. I am sure that Jaynes is highly proficient in branches of physics and that he would reject my 'common sense' and justifiably so, perhaps.

A 'publicly agreed verdict' may well be terribly wrong, and history is replete with examples.

(7) Jaynes and I are agreed that a statistical method should be judged 'by the results which it gives in practice'. It was precisely on this basis that the Bayesian idea was rejected by Boole, Venn, Fisher, Neyman and so on. I hope that this list of names gives Jaynes pause. The simple fact is that a Bayesian interval has *no* predictive verifiability. The neo-Bayesian cannot successfully back his probability assertions by accepting money challenges. On the other hand, the Neyman-Pearson interval assertion is straightforward: he asserts: I will bet $ 19 to $ 1 that my interval contains the true unknown parameter value. It is a matter of mathematics that this claim can be sustained. Neyman can issue this assertion and will apart from random fluctuations remain solvent. The neo-Bayesian will, on the other hand, 'lose his shirt'. He will be coherent in his whole battery of probability assertions, but he will be *coherently wrong,* in a situation in which an individual *A* chooses a probability structure and valid data, individual *B*, a neo-Bayesian, makes his probability assertions, and individual *C* (say Neyman) challenges *B*'s assertions, with individual *A* then verifying correctness of assertion.

(8) The reader may check whether the book by Kempthorne and Folks mentions the works of Good, Savage and Jeffreys.

(9) Jaynes says "The basic ideas of interval estimation must be ancient". We will agree with him, but the problem is surely to give some logically satisfying structure for this basic idea. He gives a view, the need for an interval of uncertainty or a final best number, with which I agree but a view with which L. J. Savage who is surely one of the originators of neo-Bayesianism, disagrees in his foundations. Savage's book should be read, and the assertion of this view will be found.

(10) While Neyman-Pearson orthodoxy states that the "proper method for this problem is the confidence interval", Fisher objected strongly and consistently that the proper method was *not* the confidence interval. A view is expressed in the Kempthorne-Folks book which is along the same lines. It seems to be agreed by a sizeable group of practising statisticians that one cannot *necessarily* have confidence in confidence

intervals. It was for this reason that Folks and I introduced the term 'consonance interval' and this suggestion has met with approval by *some* workers.

(11) The words subjectivity and objectivity should be banned from the literature. But the situation is not simple. We cannot take Jaynes' common sense to be a substitute for a requirement of interpersonal validation of subjectivety formed assertions which is what 'objectivity' usually means. The comments of Bross cannot, I believe, be dismissed by a few words and a wave of the hand. I do not thereby imply that I agree *totally* with him.

(12) A question that recurs again and again is "What is a significance test?" There has been in the opinion of some statisticians, a vast confusion on the matter. A distinction between a significance test and an accept-reject rule has been made by myself, by Kempthorne and Folks in their book, by Kalbfleisch in his book, and by Kalbfleisch and Sprott in a paper presented at this conference. The confusion has been generated by a mixing of phrases which should be kept separate. One meets an infelicity of phrase of an accept-reject rule at 5% significance level. This mixes up significance tests and accept-reject rules to the confusion of most readers. A significance test is a quantification of 'strength of evidence against' expressed on the probability (ie. the $[0, 1]$) scale. The confusion is not confined to Jaynes' paper but permeates the mathematical statistics area.

(13) On Example 1, Jaynes' common sense tells him what he says. I suggest, however, that the standard error of the difference of means is $\sqrt{7.48^2 + 6.48^2} \doteq \sqrt{98} \doteq 10$. Hence the difference of means is 8 with a standard error of 10. There is evidence that B is better than A, but not 'fairly substantial' in my opinion. No one would quarrel, I believe, with a *decision* to choose B, but the situation evidentially is by no means as clear as Jaynes with his 'common sense' approach suggests.

(14) The statement "any statistical procedure which fails to extract evidence that is already clear to our unaided common sense, is certainly *not* for me", is excellent polemics, but, I suggest, no more than that. If Jaynes' circulated document is his 'common sense', then I believe we must reject a sizeable portion of this polemic.

(15) In Example 1 and elsewhere, Jaynes uses improper priors. Hacking in 1962, or thereabouts, raised the question of how a quantity

which is not a probability could be convoluted with a probability to yield a probability. Many of us have been utterly queasy about improper priors. In discussion of Fraser's presentation, Lindley mentioned recent work by Dawid, Stone and Zidek which appears to show that Fraser's structural inference has consistency problems. I am very sorry that the fact that *the same paper* appears to show irremovable defects in the neo-Bayesian process using improper priors was not mentioned. If the force of the Dawid-Stone-Zidek paper is accepted, as Lindley does, it seems, because he used it as an argument against Fraser's structural inference, then it would seem that the 1967 book of Lindley and the bulk of the Jaynes paper should be rejected, as well as other recent Bayesian books. At the conference, I attempted to obtain Lindley's view of the effect of the Dawid-Stone-Zidek paper on his own earlier work, and hence on Jaynes' work reported at this conference, but was *unsuccessful.* I hope that Lindley will contribute a definitive statement to the proceedings of this conference.

(16) We see the sentence 'How then, could the author have failed to find significance at the 90% level?'. This raises the question of what a significance test is. Some obliquely directed views of Jaynes seem to be aimed in the direction put forward in the Kempthorne-Folks book. But most of the mathematical-statistics literature is unclear on the matter. I recall Wolfowitz saying twenty years ago that Fisher had never defined a test of significance. He was correct, I believe. I would also add that I am not entirely clear in my own mind on what a significance test is, though what is given in the Kempthorne-Folks book and in my Snedecor essay represents an initial stab.

(17) In Example 2 we are told there is "absolutely unmistakable evidence for the superiority of type 2 rockets". I believe the critical reader should jib at this. I certainly do. I would give a statement of "quantitatively *how significant* that evidence is". I would report "*the* critical significance level at which etc". I am on record with Folks as advocating precisely this mode of statement. But, then, perhaps I and Folks are not 'orthodox' statisticians, or perhaps we are not statisticians at all!

(18) Throughout the examples we are given in Jaynes' words *the* Bayesian solution. But Dawid-Stone-Zidek tell us that *the* Bayesian solution (as given by Jaynes) has defects, and we can only surmise, in the absence of clear statement, that Lindley agrees on the presence of defects.

(19) We see that certain results are based on a 'ridiculous prior for λ'. This raises the question of what is a poor prior. Clearly Jaynes admits that possibility. On the other hand, Lindley states that the quesion of accuracy of a prior does not arise. And, finally, we heard a lecture by Patrick Suppes which cast extreme (for me) doubts on the whole Savage prescription, which Lindley has stated *at several times* to be his intellectual basis. To place the matter in focus, a prior that Jaynes uses may be 'ridiculous' to me and vice versa, and that is *precisely* why the Bayesian process was rejected by Boole, Venn, etc., etc.

(20) It is surely the case that a Bayesian interval makes predictions about an *imaginary* population of repetitions. The neo-Bayesians do not like such a phrase, but their opponents state something which is as close to this as the looseness of language permits.

(21) On Example 5, the points that Jaynes makes are extremely close to those that Fisher made in 1934 (39 *years ago*) in rejecting the Neyman-Pearson prescription. The notion of *recognizable* subsets was put forward very early, if not first, by Fisher. I *infer* that the literature has not been read.

(22) I have already indicated implicitly that there are difficulties with the Savage axiomatic system. I call Patrick Suppes to witness.

(23) I am of the opinion that the consideration of axiomatic structures is a very important part of logical thinking. But I believe it is exceedingly dangerous to accept any set of axioms, no matter how 'true' and 'correct' they appear to be. I have seen writings recently that indicate that certain workers are very doubtful of 'the sure-thing principle'. I suggest that we listen a bit to these workers.

(24) The neo-Bayesians quote F. P. Ramsey as the originator. They fail, however, to record that Ramsey recommended to refer to Fisher on applicational matters. The last sentence of Ramsey goes like "For all this see Fisher".

(25) We await with eagerness the Fisher lecture of L. J. Savage, which was not as assertive as his followers might well expect. (I was chairman of the session.)

(26) Jaynes says that the neo-Bayesian prescription 'works'. Just what does he mean and what is the evidence, apart from *his* common sense.

(27) That multiple comparison procedures are thought to be defective by many practitioners (and non-Bayesians) is well-known.

(28) L. J. Savage was crystal clear in his presentation that he was giving a *theory of decision for one person,* who, of course, has to be L. J. Savage. I do not, thereby, denigrate Savage. He was a fine individual with a very fine mind.

(29) Ultimately, the basic antithesis is between evidence and decision. To some everything is decision. To others and indeed everyone, decision is important. Every human and *animal* makes decisions. But there is another aspect, the accumulation and weighing of *evidence.* Just what this is, is not clear. But the lesson of all human thought is not to dismiss a vague but pervasive idea because one cannot formulate it tightly.

Statistical Laboratory, Iowa State University

REFERENCES

Cornfield, J., 1969, 'The Bayesian Outlook and its Applications', *Biometrics* **25**, 617–642, for a good listing of background material.

Dawid, A. P., M. Stone and J. V. Sidek, 1973, 'Marginalization Paradoxes in Bayesian and Structural Inference', *J. Roy. Stat. Soc. B.*, in press.

Dawid, A. P. and M. Stone, 1972, 'Expectation Consistency of Inverse Probability Distributions', *Biometrika* **59**, 486–489.

Easterling, R. G., 1972, 'A Personal View of the Bayesian Controversy in Reliability and Statistics', *IEEE Trans.* R-21, 186–194.

Fisher, R. A., 1959, *Statistical Methods and Scientific Inference*, 2nd ed., Hafner, New York.

Kalbfleisch, J. G., 1971, *Probability and Statistical Methods*, Dept. of Statistics, Waterloo.

Kalbfleisch, J. G., and D. A. Sprott, 1973, 'On Tests of Significance', this volume, p. 259.

Kempthorne, O., 1969, *Biometrics* **25**, 647–654. Discussion of paper by J. Cornfield.

Kempthorne, O., 1971, 'Probability, Statistics, and the Knowledge Business', in *Foundations of Statistical Inference*, V. P. Godambe, and D. A. Sprott, (eds.), Holt, Rinehart and Winston, New York; 1971.

Kempthorne, O., 1972, 'Theories of Inference and Data Analysis', in *Statistical Papers in Honor of George W. Snedecor*, Iowa State Univ. Press, Ames.

Kempthorne, O. and J. L. Folks, 1971, *Probability, Statistics, and Data Analysis*, Iowa State Univ. Press, Ames.

Luce, R. D. and H. Raiffa, 1957, *Games and Decisions*, Wiley, New York.

Ramsey, F. P., 1926, 'Truth and Probability', in *The Foundations of Mathematics and Other Logical Essays*, Kegan, London.

Savage, L. J., 1954, *The Foundations of Statistics*, Wiley, New York.

Snedecor, G. W. and W. G. Cochran, 1967, *Statistical Methods*, 6th ed., Iowa State Univ. Press, Ames.

Stone, M. and A. P. Dawid, 1972, 'Un-Bayesian Implications of Improper Bayes Inference in Routine Statistical Problems', *Biometrika* **59**, 369–375.

JAYNES' REPLY TO KEMPTHORNE'S COMMENTS

I am most grateful to Professor Kempthorne for this lengthy commentary. Such a magnificent confirmation of my main thesis could hardly have been hoped for; he has surely silenced those critics who thought that my account of the orthodox position was exaggerated.

Before venturing into areas where we presently differ I want to say that, during our five days acquaintance at this Conference, I have developed a warm personal affection for Oscar Kempthorne, and came to seek him out for many between-sessions and after-dinner discussions, all pleasant and valuable to me for reasons ranging from his interesting comments to the aroma of his cigars. Although it may not be apparent to the casual reader, there is a very wide area of agreement between us; on most of the issues discussed at this Conference, we would stand together.

For example, we both see at a glance the sterility of efforts to refine the mathematics without refining the concepts; or to axiomatize old ideas without any creative development of new ones. We are, I think, equally appalled at the prospect of changing the principles of logic to accommodate an illogical theory of physics.

We both tend to place more emphasis on the practical working rules and less on highflown mathematical and philosophical aspects of statistics than some of our younger colleagues, because we have seen enough ambitious but short-lived efforts with the generic title: 'A New Foundation for Statistics' to become a bit weary of them. And we have seen enough putative 'foundations' develop a fluid character unlike real foundations and adapt themselves to the unyielding practical realities, to become a bit wary of them.

It is clear to me that, on a much deeper level than the superficial differences being aired here, Oscar Kempthorne and I are kindred souls, with the same basic outlook and value judgments. On studying his comments, I am convinced that our differences arise almost entirely from misapprehensions concerning the nature of Bayesian methods *as they exist today,* which could have been cleared up if only we had more time to thresh matters out. Surely, there is no difference in our real aims to improve the power and scope of statistical methods at the practical, working level.

But granting all this, the differences between us do involve issues of crucial importance to statistics, and it would be a disservice to minimize them. This 120-year-old hangup over prior probabilities, started by Boole, must come to an end, because it is the direct cause of the troubles that today prevent orthodox statistics from giving any useful solutions to many important, real statistical problems.

Thus, linear regression with both variables subject to error is one of the most common statistical problems faced by experimenters; yet orthodox theory is helpless to deal with it because with n data points we have $(n+2)$ nuisance parameters. In irreversible statistical mechanics, and in some mathematically similar problems of communication theory and business decisions, the only probabilities involved are prior probabilities. The possibility of any useful solutions at all depends on principles such as maximum entropy, for translating prior information into prior distributions.

This debate has gone on for over 100 years, with the same old arguments and counter-arguments repeated back and forth for generations, without ever getting anywhere. Philosophical disputation may be great fun; but through recorded history its score for actually solving problems is, I believe, precisely zero. Anybody who genuinely wants to see these issues resolved must recognize the need for a better method.

Now the present condition of statistics is just the condition physics was in until the late 16th century, when Galileo showed us a better method – the direct cause of the advances that physics has made since. Instead of arguing about how objects 'ought' to move according to some philosophical or theological preconceptions, or by quoting ancient authorities such as Aristotle, why don't we just use the evidence of our own eyes? We are surrounded daily by moving objects; so any proposed theory about how they move can be tested by direct observation of the facts.

But, as this Conference showed very dramatically, 400 years of 'enlightenment' have not changed basic human nature. Today, statisticians regard themselves as the guardians of 'scientific objectivity' in drawing conclusions from data. Yet when I suggested that their own methods be judged, not by the philosophical preconceptions underlying them, but by examination of the facts of their actual performance, this appeared to many – as I knew it would – just as radical and shocking at as it did to Galileo's contemporaries. After my talk, a half-dozen people remonstrated with me, trying to inform me about the terrible defects of Bayesian methods by repeating the same tired old Boole-Venn clichés that we all learned as children. Not one of these individuals took the slightest note of the contrary facts (the mathematically demonstrable relations between actual performance of Bayesian and orthodox methods) that I had just pointed out. So we had an exact 20'th century repetition of Galileo's experience with the colleague who refused to look through his telescope.

To answer fully every point raised by Kempthorne would require a document much longer than my original presentation. Therefore, this reply must be confined to a brief summary of the situation, followed by specific comments only on those points of fact which are of general interest, and which would propagate confusion if they were allowed to go unanswered.

SUMMARY

My presentation was concerned with examining the relative merits of orthodox and Bayesian statistical methods by considering specific real problems, giving for each *an* orthodox solution which has been advocated in the recent literature, and adding what cannot be found in that literature, namely *the* Bayesian solution *which makes use of the same information* (i.e., is based on a noninformative prior). In Example 3, we also examined the further improvement obtainable when definite prior information is put in by maximum entropy. From these comparisons, several substantive conclusions emerge, which can be summed up as follows: Orthodox methods, when improved to the maximum possible extent (by using one-sided tests, reporting critical significance levels, using sufficient statistics or conditioning on all ancillary information, etc.) become mathematically equivalent to the Bayesian methods based on noninformative priors, provided that no nuisance parameters are present, and a sufficient statistic or complete set of ancillary statistics exists. Otherwise, mathematical equivalence cannot be achieved, and magnification then shows the Bayesian result to be superior.

This conclusion is supported in part by general theorems, in part by examination of specific cases. By now, we have a multitude of specific worked-out examples supporting it; and anyone who has understood my analysis can see that we are prepared to mass-produce any number of additional examples. Orthodox statistics has yet to produce *one* counter-example. The reason for this is clear to one who has studied the theorems of R. T. Cox (1946, 1961). He shows that any method of plausible reasoning in which we represent degrees of plausibility by real numbers, is necessarily either equivalent to Laplace's, or inconsistent.

Even though an orthodox statistician may, in the words between his equations, vociferously denounce the use of Bayes' theorem, it is nevertheless a matter of straightforward mathematics to see if his actual conclusions can be derived from Bayes' theorem. Either they can or they cannot. If they can, then it is obvious that his rejection of the Bayesian method is not based on its actual performance. If the conclusions are different, then we have the opportunity to judge that difference by Galileo's method. If we can magnify the difference sufficiently, it will become quite obvious which method is giving sensible results, and which is not.

Let me stress this point. Doubtless, some readers will jump to the conclusion that I deliberately chose examples to support my prejudices; and that one can just as easily produce examples on the other side. In fact, I hope that every reader of the orthodox persuasion will come to exactly that conclusion, and set about immediately to produce six examples where an orthodox method yields a result that simple common sense can see is preferable to the Bayesian result. For it is not in the passive reading of my words, but in the active attempt to produce these counter-examples, that one's eyes will be opened.

Professor Kempthorne's appraisal of my efforts falls somewhat short of the warm approbation that I had naturally hoped for. As he notes (Item 3), any writer presents 'as convincing a case as possible'. Presumably, therefore, if he was in a position to refute any of my claims – whether by exhibiting an error in mathematics, a counter-example, or a documentable contrary fact – he would have done so. Yet with a single exception, discussed below (Item 13), he does not even mention any of the substantive issues raised. Instead he favors us with pleasantries about my choice of words and phrases.

Kempthorne complains, with some justice, that I did not criticize orthodox methods as they exist today, but rather as they existed before publication of his recent book (Kempthorne and Folks, 1971; hereafter referred to as KF). But then what shall one say about reaching back to Boole (1854) and Venn (1866) for criticisms of Bayesian methods as they existed over 100 years ago; thereby ignoring not only my recent work on these problems (1968, 1971), but also that of Jeffreys (1939)? I can well imagine the howls of outrage and cries of 'unfair' that would issue forth if I went back only 37 years to quote Karl Pearson (1936) as my authority in criticizing maximum likelihood. Now let us take up some of Kempthorne's more specific comments.

(2) My topic was the relative merits of orthodox and Bayesian methods, and not how they correlate with intelligence. Not having studied the latter topic, I have nothing more to add to the conclusions already reported by professional statisticians, viz:

I believe, for example, that it would be very difficult to persuade an intelligent physicist that current statistical practice was sensible, but that there would be much less difficulty with an approach via likelihood and Bayes' theorem.

G. E. P. Box (1962)

A student of statistical methods tends to be one of two types; either he accepts the technique in its entirety and applies it to every conceivable situation, or he is more intelligent and questions the applicability at all.

O. Kempthorne (1952)

With regard to the other remark, I think an historical study would show that the reasons for the interest of both Laplace and Jeffreys in probability theory arose from the problem of extracting 'signals' (i.e., new systematic effects) from the 'noise' of imperfect observations, in astronomy and geophysics respectively. The procedures would today be called 'significance tests', and I wish every one who has not already done so, would read Jeffreys' (1939) beautiful and comprehensive chapters on significance tests, then compare them from the standpoint of solid content and usefulness in real problems, with any work ever written on the subject from the orthodox point of view.

Likewise, my own interest in statistics arose from problems of extracting signals from noise in several applications ranging from optimum design of radar receivers and magnetic resonance probes, to land mine detectors. I am on record (Jaynes, 1963) as claiming that there is no area of physics, from elementary particle theory to cyclotron design, in which the phenomenon of noise does not present itself.

In view of all this, one can imagine my consternation at the suggestion that "there has been little attention to unavoidable noise" in physics. Physicists were actively studying noise and, thanks to Laplace, knew the proper way to deal with it, long before there was any such thing as a Statistician.

(3) (a) Of course, by 'the orthodox solution' I mean the particular one *which I am describing*; and likewise for 'the Bayesian solution'. Of course, there are many different orthodox solutions to a given problem – but I think that is the last thing a defender of orthodoxy would wish to bring to our attention.

Dirac did not in any way suggest that "working statisticians would estimate a probability to be a negative number", as a reading of his lecture will show. On the other hand, it *is* a matter of documentable fact that some orthodox statisticians suggest estimating a parameter known to be positive by an estimator which can become negative for some samples [KF, p. 203, Equation (7.42)].

(b) It is really discouraging to find – 25 years after the birth of information theory (Shannon, 1948), 17 years after its bearing on the prior probability problem was shown (Jaynes, 1957), ten years after the generalization to continuous distributions (Jaynes, 1963), six years after the resulting functional analysis generalization of Gibb's work to irreversible statistical mechanics was given (Jaynes, 1967), five years after it was shown that the theory becomes parameter-independent if one uses the entropy relative to the invariant measure on the parameter space (Jaynes, 1968), and two years after the frequency interpretation of that invariant measure was demonstrated (Jaynes, 1971) – that an eminent worker in statistics is still writing that attempts to produce prior distributions by logical analysis have 'failed'.

It is true that the principles of maximum entropy and transformation groups have not yet led to the solutions of every conceivable statistical problem; and I know that there are some who reject the entire program just for that reason. Presumably these same critics do not condemn the use of insulin on the grounds that it will not cure all diseases. The point is that we have solved *some* problems, in a way which I believe will be recognized by history as the final answer; and in fact we have succeeded in a wide enough class of problems to cover perhaps 90% of current applications. Criticisms of Bayesian methods on the grounds that we still have unsolved problems, come with particularly ill grace from those who have in the past, by their discouraging negative attitude, done everything in their power to prevent these problems from being solved.

I would think that anyone might recognize that a meaningful comparison of Bayesian and orthodox solutions must use the Bayesian solution which makes use of the *same* information as does the orthodox solution. A Bayesian solution which makes use of extra prior information that the orthodox method cannot use at all, will of course be superior for that reason alone; it is more instructive – and in a sense fairer – to make comparisons using a Bayesian solution based on a noninformative prior. Now, a noninformative prior is one which is uniform, not necessarily with respect to Lebesgue measure for any particular choice of the parameter, but with respect to the invariant measure defined by the transformation group on the parameter space. As explained in my work referred to, this is just the mathematical statement of the basic desideratum of consistency: in two problems where we have the same prior information, we should assign the same prior probabilities.

My previous work (1968) shows how to construct priors for location and scale parameters, the rate constant of a Poisson process, and the parameter of a binomial distribution, by logical analysis. Evidently, the point needs to be made repeatedly and with more examples; so let me show briefly how to find the prior in the parameter space (α, β) of the standard regression problem $y = \alpha + \beta x$, by logical analysis, for the case that x, y are variables of the same kind (for example, the departure from average barometric pressure at New York and Boston), so that it is as natural to consider regression of $(x$ on $y)$ as $(y$ on $x)$. Given any proposed element of prior probability $f(\alpha, \beta) \, d\alpha \, d\beta$, interchange x and y. The estimated line becomes $x = \alpha' + \beta' y$, with a prior probability element $g(\alpha', \beta') \, d\alpha' \, d\beta'$. From the Jacobian of the transformation $\alpha' = -\beta^{-1}\alpha$, $\beta' = \beta^{-1}$, we find $g(\alpha'\beta') = \beta^3 f(\alpha, \beta)$. This transformation equation holds whatever the function f.

Now if we are 'completely ignorant' of (α, β), the interchange of (x, y) shouldn't matter; we are also 'completely ignorant' of (α', β'). But consistency demands that in two problems where we have the same state of knowledge, we must assign the same probabilities. Therefore f and g must be the same function; i.e., the prior density representing 'complete ignorance' must satisfy the functional equation $\beta^3 f(\alpha, \beta) = f(-\beta^{-1}\alpha, \beta^{-1})$, which has the solution $f(\alpha, \beta) = (1 + \beta^2)^{-3/2}$. Thus, setting $\beta = \tan\theta$, the invariant measure of the parameter space is

$$d\mu = d\alpha \, d\sin\theta.$$

Why is this not uniformly distributed in θ rather than in $\sin\theta$? Answer: it is uniform in $\sin\theta$ only for fixed α; but under rotations of the (x, y) plane α also varies [indeed, under any Euclidean transformation $(x, y) \to \to (x', y')$, where $x = x' \cos\phi - y' \sin\phi + x_0$, $y = y' \cos\phi + x' \sin\phi + y_0$, the estimated line $y = \alpha + \beta x$ goes into $y' = \alpha' + \beta' x'$, where $\alpha' = (\alpha - y_0 + \beta x_0)/(\cos\phi + \beta \sin\phi)$, $\beta' = (\beta \cos\phi - \sin\phi)/(\cos\phi + \beta \sin\phi) = \tan\theta'$; and we readily verify the invariance: $d\alpha'\, d\sin\theta' = d\alpha\, d\sin\theta$, while $d\alpha\, d\theta$ is not invariant].

This invariance of the measure $d\mu$ means that, however we draw the x and y axes, the prior $d\mu = d\alpha\, d\sin\theta$ expresses exactly the same state of prior knowledge about the position of the regression line. It thus leaves the entire decision to the subsequent evidence of the sample – which, of course, is exactly what Fisher insisted that a method of inference ought to do. But as we see, if this is the property we want to have, the goal is not achieved by closing our eyes to the very existence of a prior. It can be achieved only by logical analysis showing us *which* prior has the desired property. If we do have relevant prior information, it can now be incorporated into the problem by finding the probability measure dp that maximizes the entropy relative to $d\mu$: $H = -\int dp \, \log(dp/d\mu)$, subject to whatever constraint the prior information imposes on dp; if the constraints take the form of mean values, this reduces to the canonical ensemble formalism of statistical mechanics of J. Willard Gibbs.

Now the simple facts, made understandable by Cox's theorems, illustrated in my presentation and in many other examples throughout the Bayesian literature, explain what we have observed throughout the history of orthodox statistics; every advance in orthodox practice has brought the actual procedures back closer and closer to the original methods of Laplace. The rise of decision theory was, in fact, the main spark that touched off the present 'Bayesian Revolution'. Other examples are Fisher's introduction of conditioning, discussed below, and his introduction of notion of sufficiency.

The discovery of sufficiency was, of course, a great advance *in orthodox statistics*; because in an important class of problems it removed the ambiguity in deciding which statistic should be used; if a sufficient statistic for θ exists, it is rather hard to justify using any other for inference about θ, for reasons illustrated in my Example 5 and explained under 'What Went Wrong?' But in Bayesian statistics there never was any ambiguity of this type to resolve. Fisher's definition of sufficiency can be stated more succinctly (and in my view, more meaningfully) as: If the posterior distribution of θ depends on the sample $(x_1 \ldots x_n)$ only through the value of a certain function $\theta^*(x_1 \ldots x_n)$, then θ^* is a sufficient statistic for θ. Evidently, if a sufficient statistic exists, application of Bayes' theorem will lead us to it automatically without our having to take any special note of the idea. But Bayes' theorem will lead us to the optimum inference whether or not any sufficient statistic exists; i.e., sufficiency is a convenience affecting the amount of calculation but not the quality of the inference.

I am afraid that to castigate Bayesian methods, but not orthodox ones, on grounds of lack of uniqueness, is to get it exactly backwards. It is orthodox statistics that offers us many different solutions to a single problem, (i.e., given prior information, sampling distribution, and sample), depending on whose school of thought, whose textbook within that school, and even which chapter of that textbook, you read. An estimator ought to be unbiased, efficient, consistent, etc.; but in general orthodoxy gives us no criterion as to the relative importance of these, nor any method by which a 'best' estimator can be constructed. The use of an unbiased estimator or a shortest confidence interval will lead us to different conclusions with different choices of parameters. KF (p. 316) cannot make up their minds about whether to accept the principle of conditioning, and advocate significance tests in which the conclusions depend on the arbitrary ordering you or I might assign to data sets *which were not observed!* Indeed, there is scarcely any problem of inference for which KF offer any definite preferred solution; in most cases there is an inconclusive discussion that terminates abruptly with the remark that 'it is all very difficult', leaving the reader in utter confusion as to which method should be used. But with all this ambiguity, orthodox methods provide no means for taking prior information into account.

In sharp contrast to this, for a given sampling distribution and sample, different Bayesian results correspond, as rational inferences should, to and *only* to, differences in the prior information. When priors are determined by the principles of maximum entropy and transformation groups, Bayesian methods achieve complete invariance under parameter changes (Jaynes, 1968).

(4) We are now told that even to utter the words 'Fisher-Neyman-Pearson theory' is a calumny on Fisher's views (but apparently not on

Neyman's or Pearson's); and again for the 'benefit' (precious little) of readers not present at the Conference, may I state that I first heard this phrase from the lips of Professor Oscar Kempthorne, shortly before my talk was given. I repeated it only to say that I would follow common practice by using the word 'orthodox' as an approximate synonym.

However, since the issue has been raised, I would like to state that the term 'Fisher-Neyman-Pearson approach' appears to me as an entirely accurate and appropriate term for a certain area of statistical thought. To use it is in no way to ignore, much less deny, the fact that there were differences between Fisher on the one hand, and Neyman-Pearson on the other. However, this should not blind us to the fact that there is a very much larger area of agreement; i.e., a corpus of ideas which are not in Bayesian statistics, but are common to the Fisher and Neyman-Pearson points of view and which therefore characterize their union. I refer to the ideas that (1) the word 'probability' must be used only in the sense of 'frequency in a random experiment', (2) inference requires that we find sampling distributions of some 'statistics' in addition to the direct sample distribution $p(\mathrm{d}x \mid \theta)$, (3) the conclusions we draw from an experiment can depend on the probabilities of data sets which were not observed, or the psychological state of mind of the experimenter (optional stopping), (4) we can improve the precision of our results by throwing away relevant information instead of taking it into account (the procedure euphemistically called 'randomization'), (5) the attempt to dispense with prior probabilities.

Recalling the difference between the Fisher and Neyman-Pearson camps over confidence intervals vs fiducial probabilities, let's just see how great this calumny is. Given a basic sample distribution $p(\mathrm{d}x \mid \theta)$, choose two 'statistics' $\theta_1(x_1...x_n)$, $\theta_2(x_1...x_n)$ such that $\mathrm{prob}(\theta_1 < \theta < \theta_2) = P$; this defines a $100\,P$ percent confidence interval. Letting $\theta_1 \rightarrow \theta_{\min}$, the lower bound of the parameter space, we have $\mathrm{prob}(\theta < \theta_2) = P$, which is Fisher's definition (Collected works, 27.253) of the fiducial distribution of θ, based on the statistic θ_2. As we see, the deep, profound difference in basic approach is fully as great as that between Tweedledee and Tweedledum.

The difference is not in the approach, but in the perception with which it was used. Fisher, with his vastly greater intuitive understanding, saw at once something which still does not seem generally recognized by others; that all this is valid only when we are using sufficient statistics. Even in the Fisher obituary notice, Kendall (1963, p. 4) questions the need for sufficiency. My Example 5 was intended to make Fisher's point by demonstrating just what can happen when we use a confidence interval not based on a sufficient statistic. Obviously, anyone who rejects fiducial probability, but endorses the use of confidence intervals, is not doing so on grounds of their actual performance.

Surprisingly, after protesting *my* calumny of Fisher's views, we find KF (p. 380) taking a dim view of fiducial probability, saying: "If a fiducial distribution is merely a restatement of a test of significance, we see no need for it". They might better have said: "Since a fiducial distribution of θ is a simultaneous statement of *all* tests of significance concerning θ, we see no need for the separate significance tests". While we may not have an 'equal distribution of ignorance', we have a more than equal distribution of calumny.

(5) It is quite true the foundations of modern physics are in a shambles, and in this area we also have controversy arising from unsolved problems. Being deeply involved in those also, I can report that current controversial issues in physics are orders of magnitude more complicated mathematically, and more subtle conceptually, than the trivia that we are quibbling about here. Indeed, the simple facts about probability theory that I am trying to point out were seen at once by the great mathematical physicists – Laplace, Maxwell, Gibbs, Poincaré, Jeffreys. For many years I have found it a refreshing rest to take off a few hours from the problems of physics, and work out another Bayesian-orthodox comparison.

(6) Professor Kempthorne objects very strongly to my use of the term 'common sense'. May we assume, then, that he denounces with equal force Fisher's use of the term (Collected works, 26.47) in appealing, three times in one page, to common sense rather than mathematical properties, to justify his 'information' measure?

I do indeed have a very great and insuperable difficulty in reconciling quantum physics with common sense, and am on recent record as having said exactly that (Jaynes, 1973). In fact, I would note that orthodox statistics and the 'Copenhagen' interpretation of quantum theory are just two different manifestations of a single intellectual disease, closely related to logical positivism, which has debilitated every area of theoretical science in this century. The symptoms of this disease are the loss of conceptual discrimination; i.e., the inability to distinguish between probability and frequency, between reality and our knowledge of reality, between meaning and method of testing, etc.

It is true that a 'publicly agreed verdict' may well be terribly wrong. This is just what happens when the public has been misled by false indoctrination of exactly the kind that I am trying to correct here. But to throw out the notions of 'common sense' and a 'publicly agreed verdict' is to forfeit the only visible means by which this controversy could be resolved. Although the temptation is strong. I will refrain from quoting Section 17.5 of KF, entitled 'Publicly Agreed Probabilities'.

(7) We apparently agree that a statistical method should be judged by the results it gives in practice. Well and good. However, I categorically deny that "the Bayesian idea was rejected by Boole, Venn, Fisher and Neyman" on these grounds. It is just the weakness of their work that they rejected Bayesian methods on purely philosophical or ideological grounds, *without* examining their actual performance.

Since the case of Boole and Venn has been brought up, let us examine the work of these gentlemen and see for ourselves the validity of their actual criticisms, and the accuracy with which their work is reported today in the orthodox literature. I believe that Boole, like most other critics of Laplace, failed to comprehend fully his definition of probability. Since Laplace has been quoted out of context so many times in this and other matters, let us take the trouble to quote his definition in full. The first volume of his *Théorie Analytique* is concerned with mathematical preliminaries, and the actual development of probability theory begins in Volume 2. The first sentence of Volume 2 is: "The probability of an event is the ratio of the number of cases favorable to it, to the number of all cases possible when nothing leads us to expect that any one of these cases should occur more than any other, which renders them, for us, equally possible".

This definition has stated only the finite discrete case, but we know how to generalize it. The point is that Laplace defined probability in a way which clearly represents *a state of knowledge*; and not a frequency. Of course, as Laplace demonstrates over and over again, connections between probability and frequency appear later, as mathematical consequences of the theory. I claim that these derivable connections (the limit theorems of Jacob Bernoulli and de Moivre-Laplace, Laplace's rule of succession, the de Finetti exchangeability theorem, etc.) include all the ones actually used in applications.

If one has no prior knowledge other than enumeration of the possibilities (i.e., specification of the sample space), then to assign equal probabilities is clearly the only honest way one can describe that state of, knowledge. This can be formalized more completely than Laplace did, by the aforementioned desideratum of consistency: if we were to assign any distribution other than the uniform one it would be possible, by a mere permutation of labels, to exhibit a second problem in which our state of knowledge is exactly the same, but in which we are assigning different probabilities. But in this case Laplace surely considered the argument and result so obvious that he would insult the reader's intelligence by mentioning them. The only serious error Laplace made was overestimating the intelligence of his readers.

Boole (1854), not perceiving this, rejected Laplace's work on the ground that the prior was 'arbitrary', i.e., not determined by the data. He did *not* reject it in the ground of the actual performance of Laplace's results in the case of uniform prior because he, like Laplace's other critics, never bothered to examine the actual performance under these conditions, much less to compare it with alternative methods. Had he done so, he might have discovered the real facts about performance, presented 85 years later by Jeffreys. Curiously, Boole, after criticizing Laplace's prior distribution based on the principle of indifference, then invokes that principle to defend his own methods against the criticisms of Wilbraham (see several articles in *Phil. Mag*, Vols. vii and viii. 1854).

This brings up another matter that needs to be mentioned. Boole's unjust criticism of Laplace has been quoted approvingly, over and over again, in the orthodox literature, Fisher (1956) being a very generous contributor. But in that same literature, a conspiracy of silence hides the fact that Boole's own work on probability theory (Boole, 1854, Chapters 16–21) contains ludicrous errors, far worse than any committed by Laplace. Some were noted by Wilbraham (1854), McColl (1897) and Keynes (1921). See his Example 6, page 286, where by a confusion of propositions [taking the probability of the proposition: 'If X is true, Y is true' as the conditional probability $p(Y \mid X)$] he arrives at the conclusion that two propositions with the same truth value can have different probabilities. He not only fails to see the absurdity of this, but even calls it to the reader's attention as something which 'deserves to be specially noticed'. Or his solution to another problem, page 324, Equation (10), which reduces to an absurdity in the special cases $c_1 = c_2 = 1$ and $c_1 = p_1 = 1$. While Laplace considered real problems and got scientifically useful answers, Boole invented artificial school-room type problems, and often gave absurd answers. Finally, it is mathematically trivial to show that all of 'Boolean algebra' was contained already in the rules of probability theory given by Laplace – in the limit as all probabilities go to zero or unity, any equation of Laplace's 'Calculus of Inductive Reasoning' reduces to one of Boolean algebra.

Now let's turn to the case of Venn (1866), who expresses his disdain for mathematical demonstration very clearly throughout his book and its preface. Venn's Chapter 6 is an attack on Laplace's rule of succession, so viciously unfair that even Fisher (1956) was impelled to come to Laplace's defense on this issue. Fisher questions whether Venn was even aware of the fact that Laplace's rule had a mathematical basis, and like other mathematical theorems has 'stipulations specific for its validity'. He proceeds to give examples in which, unlike those of the 'great thinker' Venn, the stipulations are satisfied, and Laplace's rule is the correct one to use.

How is it possible for one human mind to reject Laplace's rule of succession; and then advocate a frequency definition of probability? Anybody who assigns a probability to an event equal to its observed frequency in many trials, is doing just what Laplace's rule tells him to do. In my Example 4, we examined Laplace's calculation underlying this rule, and learned that anybody who rejects Laplace's methods in favor of confidence intervals for the binomial, is certainly not doing so on grounds of actual performance.

I would like to plead here for a greater concern for historical accuracy, in writing on these matters. For over a century, there has been a conspiracy in the statistical literature to rewrite history and denigrate Laplace, first in the Boole-Venn manner, then by denying him credit when his principles were rediscovered (examples below). An *ad hominem* attack on Laplace (as 'a consummate politician') has even befouled the air of this Conference. I have long since learned never to accept the word of a biased source (Boole, Venn, Von Mises, Fisher, E. T. Bell, Cramér, Feller, etc.) on *any* question of what Laplace did or did not do. When working in my study,

Laplace's *Théorie Analytique* is always at my elbow; and when any question about him comes up, I go straight to the original source. It is for this reason that my judgment of Laplace differs so radically from that presented in the literature from Boole on.

Not only those who are ignorant of history, but also those who will not profit by its lessons, are doomed to repeat it. Starting with Condorcet and his omelette, those who scorned Laplace's outlook and methods – whether in science or politics – and tried to do things differently, have shared a common experience.

(A) In George Gamow's book, *The Biography of the Earth* (1941), Laplace's theory of the origin of the solar system is torn to shreds. But in 1944, Weiszäcker pointed out a few things that Laplace's critics had overlooked; and the 1948 edition of Gamow's book had a new 15 page section entitled, '*Laplace was right after all!*'

(B) Abraham Wald, in his mimeographed course notes of 1941, rejected Laplace's methods of parameter estimation and hypothesis testing and asserted that such problems cannot be solved by the principles of probability theory. During the 1940's Wald sought a new foundation for statistics based on the idea of rational decisions, which had the aim of avoiding the mistakes of Laplace; but in Wald's final 1950 book, *Statistical Decision Functions*, the fundamental place of 'Bayes strategies' is finally recognized. As it turned out, Wald's life work was to prove, very much against his will, that the original methods developed by Laplace in the 18'th century, which he and many other statisticians had scorned for years, were in fact the unique solution to the problem of rational decisions. *Laplace was right after all.*

(C) I had the same experience. In 1951, I somehow came to the conclusion that Bayes' theorem did not adequately represent the full variety of inductive reasoning, and sought to develop a two-valued theory of probability, very much like the one presented here by Shafer, except that my numbers corresponded to the sum and difference of his. I even expounded this in a Round Table Discussion at one of the Berkeley Statistical Symposiums. However, I then made the tactical error of trying to apply this theory to some real problems. At about the third attempt, the scales fell from my eyes and I saw that a two-valued theory contains nothing that is not already given by Laplace's original one-valued theory, by going to a deeper sample space. In other words, the defects that

I thought I saw in Laplace's theory were my own defects, in not having the ingenuity to invent an adequate model. *Laplace was right after all.*

Now, I don't know how many other people are doomed to follow this path – already far more man-years of potentially useful talent have been wasted on futile attempts to evade Laplace's principles, than were ever invested in circle-squaring and perpetual motion machines. But just as Lindemann's proof put an end to circle-squaring for all who could see its implications, so Cox's theorems (1946) ought to have put an end, twenty-five years ago, to these unceasing efforts to evade what cannot be evaded. The situation is described in more detail in my review of Cox (1961). This is why I can say the following to latter-day Don Quixotes:

Many of us have already explored the road you are following, and we know what you will find at the end of it. It doesn't matter how many new words you drag into this discussion to avoid having to utter the word 'probability' in a sense different from frequency: likelihood, confidence, significance, propensity, support, credibility, acceptability, indiffidence, consonance, tenability, – and so on, until the resources of the good Dr Roget are exhausted. All of these are attempts to represent degrees of plausibility by real numbers, and they are covered automatically by Cox's theorems. It doesn't matter which approach you happen to like philosophically – by the time you have made your methods fully consistent, you will be forced, kicking and screaming, back to the ones given by Laplace. Until you have achieved mathematical equivalence with Laplace's methods, it will be possible, by looking at specific problems with Galileo's magnification, to exhibit the defects in your methods.

Here are two typical examples of the kind of factual distortion that we find in the literature. KF (p. 314) quote approvingly a statement of Fisher (1956, p. 4) that: "So early as Darwin's experiments on growth rate the need was felt for some sort of a test of whether an apparent effect might reasonably be due to chance". More specifically, Fisher (p. 81) then states that the 'Student' t-test was "the first exact test of significance." Neither book makes any mention of the historical fact that Laplace developed many significance tests to determine whether discrepancies between prediction and observation 'might reasonably be due to chance' and used them to decide which astronomical problems were worth working on: a bit of wisdom that might well be noted by scientists today. Laplace also illustrates the use of these tests, including two-way classifications, in many other problems of geodesy, meteorology, population statistics, etc. As I hope to show in detail elsewhere, Laplace's significance tests were in no way inferior – and were in some cases demonstrably superior – to tests advocated in the orthodox literature today.

Likewise, both KF and Fisher denounce the use of Bayes' theorem and uphold the 'student' t-test as a great advance in statistical practice; but of course neither mentions the fact that precisely the same result follows in two lines *from* Bayes' theorem; given the data $D = \{x_1 \dots x_n\}$, the likelihood function is $L(\mu, \sigma) = \sigma^{-n} \exp(-nQ/2\sigma^2)$, where $Q = s^2 + (\bar{x} - \mu)^2$. Integrating out σ with respect to Jeffreys' prior, the posterior density of μ is $\sim Q^{-n/2}$, which but for notation is just the t-distribution. Students reading these works obtain a completely false picture of both the historical and mathematical facts about significance tests.

As a second example, KF (p. 305) consider tests of a simple hypothesis M_1 against a simple alternative M_2, on data D. The likelihood ratio in favor of M_1 is $L(D) = p(D \mid M_1)/p(D \mid M_2)$. KF note that, if M_2 is true, the expected value $E_2(L)$ is unity, and conclude that the 'Bayesian process' has bad operating characteristics. But of course, this is not the proper criterion, because it is $C = \log L$, and not L, that has equal positive and negative change for equal strength of evidence for and against M_1. The inequalities $E_2(C) \leqslant 0$, $E_1(C) \geqslant 0$ [with equality if and only if $p(D \mid M_1) = p(D \mid M_2)$ for all D] then establish what they would regard as 'good' operating characteristics. Twenty pages later, KF are back to the same problem; only now they remember to take the logarithm, represent it as an orthodox test, and have no cause to complain of the operating characteristics of the statistic C. And so the indoctrination goes on; I could cite at least twenty more examples of these tactics from recent textbooks.

Now let's come to Kempthorne's statement that "a Bayesian interval has no predictive verifiability". I suggest that the main message has totally escaped him. If the optimum confidence interval is mathematically identical with the Bayesian interval based on a noninformative prior distribution, it is a bit difficult to understand how the Bayesian result could fail to have whatever 'predictive verifiability' – or any other property – is possessed by the confidence interval.

Unfortunately, there is a serious wandering of the mind in connection with the test according to which the Bayesian will 'lose his shirt'. First we are told that the confidence interval advocate will assert at 19 to 1 that the interval contains the unknown parameter value. Now, is the Bayesian required to accept this in every case? For which confidence level is this asserted? Does the width of that interval enter into the judging of the game? It is not a matter of mathematics that an undefined claim can be sustained.

Then the game appears to change suddenly; we now learn that it is the Bayesian who is making the probability statements. It is averred that the Bayesian will lose his shirt and be consistently wrong – excuse me, coherently wrong – in a case where some individual C challenges the assertions. But now is this individual C to challenge all assertions wherever they may be? At what odds? I suggest that if Professor Kempthorne will try to back up his position by producing a specific, well defined situation instead of making assertions about undefined generalities, the mathematical situation will force him to see that his claim simply is not true.

Indeed, the contest proposed by Kempthorne has already been carried out in the Monte Carlo experiments of A. Zellner (1965) and H. Thornber (1965). The results were, in the words of H. V. Roberts (1965); "Using sampling-theory criteria, the Bayesian estimators appeared better in all examples, the margin being substantial for Zellner's experiment and modest for Thornber's". Roberts proceeds to explain why this must be so; by the time all necessary provisions for a 'fair' contest have been incorporated into the experiment, all the ingredients of the Bayesian theory (prior distribution, loss function, etc.) will necessarily be present. As Roberts concludes; "The simulation can only demonstrate the mathematical theorem".

My sixth example, on the Cauchy distribution, demonstrated (and I thought rather cogently) that the 'long-run performance' of a statistical procedure is *not* the proper criterion of its usefulness. But Professor Kempthorne simply ignores this, and continues to argue long-run performance as the criterion ('The Bayesian will lose his shirt', etc). So comtemplate this example, given by David Forney (1972):

THE WEATHERMAN'S JOB

In a certain city, the joint frequencies of the actual weather and the weatherman's predictions are given by:

		Actual	
		Rain	Shine
Predicted	Rain	$\frac{1}{4}$	$\frac{1}{2}$
	Shine	0	$\frac{1}{4}$

An enterprising fellow trained in orthodox statistics (but not in meteorology) notices that, while the weatherman is right only 50% of the time, a prediction of 'shine' everyday would be right 75% of the time, and applies for the weatherman's job. Should he get it? Which would you rather have in your city?

The weatherman is delivering useful information at a rate $I=$ (entropy of distribution of predictions) + (entropy of actual weather distribution) $-$ (entropy of joint distribution) $= (0.562 + 0.562 - 1.040)/\ln 2 = 0.123$ bits/day. As explained previously (Jaynes, 1968) this means that in the course of a year the weatherman's information has reduced the number of reasonably probable weather sequences by a factor of $W = \exp(0.123 \times \times 365 \times \ln 2) = 2.92 \times 10^{13}$. With the weatherman on the job, you will never be caught out in an unpredicted rain; with the orthodox statistician this would happen to you one day out of four.

As this example one more forces one to recognize, the value of an inference lies in its usefulness *in the individual case*, and not in its long-run frequency of success; they are not necessarily even positively correlated. The question of how often a given situation would arise is utterly irrelevant to the question how we should reason when it *does* arise. I don't know how many times this simple fact will have to be pointed out before statisticians of 'frequentist' persuasions will take note of it; but I think it is important that we keep trying.

(8) The book by Kempthorne and Folks (1971) does indeed mention the works of Good, Savage and Jeffreys, unlike so many orthodox textbooks. That is, these works are included in a list of references. This leaves to be desired only that their contents had also been noted.

(9) Here and elsewhere, Professor Kempthorne seems to regard L. J. Savage (1954) as the official spokesman for Bayesian theory; and implies that if I state anything differently from Savage, then I must not have read his book. By that reasoning, I believe we have an even stronger case for inferring that someone else has not read it. It is true that Savage, probably

more than any other person, was the one who stimulated new thought on these issues (although to me personally, the arguments of Jeffreys (1939) and R. T. Cox (1946, 1961) have always seemed far more cogent). But a great deal has happened in Bayesian statistics since 1954, and I think that at present the only thing which Bayesian statistics and Savage's personalistic theory have in common is that they both use Bayes' theorem, without apology or embarrassment. Today, very few if any Bayesians would give full support to Savage's notion of 'personalistic probability', and I am on record (Jaynes, 1968) as taking my stand with orthodox statisticians on this matter; i.e., the notion of personalistic probability belongs to the field of psychology, and not to statistics.

(10) I cannot see the point of this comment. In my paper I stated very explicitly that, while there are some differences of opinion, most would hold that the proper method for the problem is the confidence interval. I believe that is a clear and accurate statement of fact. Of course, one cannot necessarily have confidence in confidence intervals; that is just the point I thought I was making in demonstrating that there are cases in which one can have zero confidence in a confidence interval.

(11) The impelling urge to find fault rather than to understand rules the situation here. In comment No. 6, Professor Kempthorne objects to the idea of 'publicly agreed verdict', but now he apparently wishes to speak with approval of a "requirement of interpersonal validation of subjectively formed decisions". But an interpersonal validation (which would amount to a publicly agreed verdict) can only take place through the common sense judgments of different people who are all exposed to the same system of facts. I am under the impression that the comments of Bross were refuted by specific factual counter-examples, in addition to a general proof, demonstrating the opposite of what Bross claimed.

(12) Whether any particular problem should be called technically a significance test, a test of goodness of fit, an acceptance test, an hypothesis test, or a decision problem, is a matter of pedantry on which orthodox statisticians are themselves in disagreement; so why can't we just call it 'a test' and get on with the substantive issues? I believe my presentation made it clear in each case: (1) what was the problem? (2) How was it handled? And that should be enough.

(13) This comment brings to mind an older controversy, with more than one similarity to our present one. Protestant countries long refused

to accept the Gregorian calendar, in spite of its clearly superior performance. England held out for 170 years after it had been adopted by Catholic countries, leading Voltaire to quip that the British "would rather disagree with the sun than agree with the Pope". It appears that some would rather disagree with common sense than agree with Bayes.

Before getting too indignant about that high (92%) significance level indicated by the Bayesian test and denying that the evidence is that clear, let's first do that quick, short-cut calculation right. The standard error of the difference of means should be estimated not by $\sqrt{7.48^2 + 6.48^2} = 9.90$; but by (Fisher, 1958, p. 116; Hoel, 1971, p. 134):

$$\sqrt{\frac{7.48^2}{9} + \frac{6.48^2}{4}} = 4.09.$$

If this standard error were known, rather than estimated, it would correspond to a significance level, not of 92%, but of 97.5%.

(15) *'Improper' Priors.* Let me try to explain the situation. 'Complete initial ignorance' of a scale parameter σ corresponds formally to use of the Jeffreys prior $d\sigma/\sigma = d\log\sigma$. But as noted before (Jaynes, 1968), to apply this within infinite limits $(-\infty < \log\sigma < \infty)$ would not represent any realistic state of prior information. For example, if x is a measured length of some material object on the earth, we surely know that the standard error σ_x of the measurement cannot be less than the size of one atom, $\sim 10^{-8}$ cm; or greater than the size of the earth, $\sim 10^9$ cm. So we know in advance that $(-8 < \log_{10}\sigma_x < +9)$. Outside this range, the prior density must be zero.

Similarly, if x is the measured breaking stress of some structural material, we know in advance that σ_x surely cannot be less than the pressure of sound waves, ~ 1 dyne cm^{-2}, due to people talking in the room; nor greater than 10^{14} dynes cm^{-2}, which is 1000 times the tensile strength of any known material. So the prior density must be all contained in $(0 < \ln\sigma_x < 33)$. If x is a time interval measured in seconds, we can be pretty sure in advance that $(-12 < \log_{10}\sigma_x < 18)$.

Generally, thinking about any problem in this way will lead one to specify prior limits σ_{min}, σ_{max} within which the unknown value surely lies; within this interval the invariance arguments leading to the form $d\sigma/\sigma$ still apply if there is no other prior information (Jaynes, 1968). Therefore, the prior is normalizable, and we have a well-behaved mathematical problem.

Now if our final conclusions depend appreciably on the exact prior limits chosen, then obviously we should analyze our prior information more carefully than I did above, to get more reliable numerical values for σ_{min}, σ_{max}. But it just wouldn't be very intelligent to go to all that work, only to discover that σ_{min}, σ_{max} cancel out of the expressions representing our final conclusions (which might be the first few moments, or the quartiles, of a posterior distribution). So it will be good strategy to work through the solution first for general limits, whereupon the mathematics will tell us under just what conditions the prior limits matter; and when they don't.

Having thus formulated the problem, the conclusion is fairly obvious: if the likelihood function is sufficiently concentrated (i.e., if the experiment is a sufficiently informative one), then the prior limits cannot matter appreciably as long as they are outside the region of appreciable likelihood. To put it in a way somewhat crude, but not really wrong: if the amount of likelihood [integral of $L(\sigma)$] lying outside the limits ($\sigma_1 < \sigma < < \sigma_2$) is less than 10^{-6} of the total likelihood, then as long as our prior limits are still wider ($\sigma_{min} < \sigma_1 < \sigma_2 < \sigma_{max}$), the exact values of σ_{min}, σ_{max} can't make more than about one part in 10^6 difference in our conclusions. If, then, we don't worry about them, and just take the limiting form of the solution as $\sigma_{min} \to 0$, $\sigma_{max} \to \infty$ for mathematical convenience, we are committing no worse a sin than does the person who laboriously determines the proper values of σ_{min}, σ_{max}, works out the exact solution based on them – and then rounds off his final result to six significant figures. We are only getting that result with an order of magnitude less labor.

If, on the other hand, we should encounter a non-normalizable posterior distribution in this limit, the theory is telling us that the experiment is so uninformative that our exact state of prior information is still important, and must be taken into account explicitly. This phenomenon, far from being a defect of Bayesian methods, is a valuable safety device that warns us when an experiment is too uninformative to justify, by itself, any definite inferences. If someone ignores the warning, and gets into trouble with 'improper priors', what we are witnessing is not a failure, but only a misapplication, of Bayesian methods.

Finally, let us keep in mind that we are really concerned here with relative value judgments; and so if anyone attacks Bayesian methods because of the possible situation just described, fairness demands that he also takes note of what happens to orthodox methods in the same problems. Now one of the substantive factual issues illustrated in my presentation, is this: orthodox methods, when improved to the maximum possible degree, reduce ultimately to procedures that are mathematically identical with applying Bayes' theorem *with just the noninformative improper prior* about which Professor Kempthorne expresses such alarm! We saw this phenomenon in Examples 2, 3, 5 and 6. As we have just seen, this causes difficulty only when the experiment is so uninformative that our final conclusions must, necessarily, still depend strongly on our prior knowledge. The Bayesian can correct this at once by using a realistic prior, leading to the inferences that *are* justified by the total information at hand; but the orthodoxian cannot, because his ideology forbids him to recognize the existence of any prior which is not also a known frequency.

In fact, we had just this situation in the first part of my Example 3, where we took no note of the actual failure times. If all units tested fail, the test provides no evidence against the hypothesis of arbitrarily large λ. The Bayesian test (6) based on a uniform prior then yields a non-normalizable posterior distribution $p(d\lambda \mid n, r, t) \sim (1 - e^{-\lambda t})^n \, d\lambda$, which tells us that λ is almost certainly greater than $(t^{-1} \log n)$, but gives no upper limit. In this way, the safety device warns us that our prior information concerning the possibility of very large λ, remains relevant; by taking it explicitly into account, rational inferences about λ are still possible, as I showed by the maximum entropy prior.

But we saw that the orthodox ST test was, in the absence of such pathology, mathematically identical with this Bayesian test; so what happens to it? Well, this is just the case already noted where the ST test breaks down entirely, telling us to reject at all significance levels. In problems where the Bayesian cannot use the approximation of an improper prior, orthodox methods give no warning, but simply yield absurd results; and only the alertness and common sense of the user can save him from the consequences. As we see, it is the orthodoxian, and not the Bayesian, who is going to be in trouble in cases where 'improper priors' cannot be used.

Note the treatment of an almost identical problem in KF (p. 203, Equation 7.42). Here they suggest use of an estimator which estimates the mean life to be infinite if we observe one failure, to be negative if we observe no failures, and which has infinite variance unless we observe 3 or more failures! Again, I think common sense renders a rather clear verdict in this comparison. If Professor Kempthorne thinks that the Bayesian solution to this problem is open to criticism, I wonder how he would defend the solution proposed in his book.

(17) KF do indeed advocate reporting critical significance levels, and for this enlightenment over most previous treatments we can be grateful. We could be even more grateful if the enlightenment had persisted to the end of the book, where KF reproduce the same old tables, so arranged that critical levels cannot be located.

(21) In comment No. 9 Professor Kempthorne infers, from a difference in my position and that of Savage, that I have not read Savage's book. Now he infers, from a similarity between my work and Fisher's, that I have not read Fisher. These orthodox inferences – with the conclusion independent of the evidence – leave me in despair. My work was checked by another statistician for just such matters. In the first version I called y an 'ancillary statistic' in Example 6; but he objected to this on the ground that I was not using it in quite the same sense as Fisher did, so I deleted the term. Now I find myself being criticized by one orthodox statistician for having followed the advice of another.

Here is the point: Fisher (Collected Works, 27.257) held – without explaining why – that the distribution of an ancillary statistic must be independent of the value of the parameter, as expressed in his allegory of the Problem of the Nile. Presumably, this was one of the many things he saw intuitively; but whatever his private reason for this independence may have been, it is easy to see what it in fact accomplishes. To avoid a possible paradox (Barnard, 1962), we understand the conditioned probability symbol $p(\mathrm{d}\theta^* \mid y, \theta)$ to be shorthand for the limit as $\mathrm{d}y \to 0$, of the well-defined

$$p(\mathrm{d}\theta^* \mid \mathrm{d}y, \theta) = \frac{p(\mathrm{d}\theta^* \, \mathrm{d}y \mid \theta)}{p(\mathrm{d}y \mid \theta)}.$$

If $p(\mathrm{d}y \mid \theta)$ is independent of θ, then the θ-dependence of the conditioned probability $p(\mathrm{d}\theta^* \mid y, \theta)$ is the same as that of the joint probability $p(\mathrm{d}\theta^* \, \mathrm{d}y \mid \theta)$. In other words, it is fundamentally the joint probability, and not the conditioned probability, that really matters – but of course, that is just what the likelihood principle has told us all along.

With this little bit of insight, it becomes clear that mathematically,

Fisher's conditioning on ancillary statistics is just a roundabout way of restoring agreement with Bayes' theorem, without having to admit that one is using it. But conditioning is not a general method; and simple mathematics shows that if we just apply Bayes' theorem directly, there is no longer any reason for y to be independent of θ. We then have a method that works in all problems – and is guaranteed to give the same result as Fisher's, with less calculation, in cases where his conditioning is possible. I hope this excursion will clear me of the charge of not having read Fisher.

(26) Since the question is asked, I will answer by showing just how the 'neo-Bayesian prescription' works, in precisely the problem where KF deny it.

KF, p. 439, consider a standard problem of linear regression with both variables subject to error. The model is $Y_i = \alpha + \beta X_i$ with measured values $x_i = X_i + e_i$, $y_i = Y_i + f_i$, the errors e_i, f_i being independent and $N(0, \sigma_x)$, $N(0, \sigma_y)$ respectively; σ_x, σ_y unknown. We take data $D = \{(x_1, y_1);$ $(x_2, y_2); \ldots, (x_n, y_n)\}$ and from this we are to make inferences about α and β.

At this point, KF assert that 'in antithesis to the likelihood principle', the likelihood function is (1) totally uninformative, (2) ill-behaved, becoming indeterminate when $x_i = X_i$ and (3) that further assumptions are needed (about equality of several X_i, or about σ_x, σ_y, etc.) to make progress on the problem. They then suggest a method in which we partition the data points into two sets, and take the line joining their centroids as our estimate of the 'true' line.

We have here one more example – perhaps the finest yet produced – of just the Canonical Procedure that I complained about in my paper; still another time, an orthodox textbook rejects the Bayesian solution, without bothering to look at it, for patently false reasons; and gives instead an orthodox method which is far weaker in its ability to extract information from the sample.

An undergraduate in a laboratory science course does better than the proposed solution of KF, without any statistical theory at all, simply by plotting his experimental points and drawing the straight line that, as judged by eye, fits them best. He can determine the accuracy of his estimates of α, β by noting how much this line can be shifted or tilted before the fit appears appreciably worse. Furthermore, if the standard errors σ_x, and/or σ_y were unknown, he would do this in the same way whether the errors were in x only, in y only, or in both; if the ratio $\lambda = \sigma_y/\sigma_x$

were known, and/or if the errors (e_i, f_i) had a known correlation coefficient ϱ, it would make no difference in the correct data reduction procedure whatever the values of λ, ρ.

The reason why these things cannot matter is that, whether the errors are represented by a one-dimensional or two-dimensional region of uncertainty about each data point, and whatever the shape and orientation of the concentration ellipse, the component of error parallel to the estimated line contributes nothing to the error of estimation of either α or β. As common sense tells us – and the Bayesian analysis confirms – in any of these circumstances the problem of inference about α, β takes the same form, with only a single unknown error component to consider. In other words, there are *not* ten basically different estimation problems, as is implied by the elaborate KF classification scheme $(yRE \mid xCN)$, etc. If the standard errors are unknown, there is only one linear model problem.

To prove these assertions, let us just sketch the Bayesian analysis, which KF declare to be impossible. The likelihood function, which they do not even write down, is

$$L\left(\alpha, \beta, \sigma_x, \sigma_y, X_i\right) =$$

$$= (\sigma_x \sigma_y)^{-n} \exp\left\{-\tfrac{1}{2} \sum_{i=1}^{n} \left[\frac{(x_i - X_i)^2}{\sigma_x^2} + \frac{(y_i - \alpha - \beta X_i)^2}{\sigma_y^2}\right]\right\}.$$

Obviously, it is in no way 'ill-behaved' or 'indeterminate'. Now let's see just how 'uninformative' is, and whether further assumptions are needed to make progress. If we want to make inferences about α, β, then $\{\sigma_x, \sigma_y, X_1 \ldots X_n\}$ are 'nuisance parameters' that prevent orthodox statistics from making any headway on this problem. It is then interesting to see how much they deter a Bayesian.

Integrating $\{X_1 \ldots X_n\}$ out of L with respect to uniform prior, we obtain a function which depends on (α, β) only through the quadratic form

$$Q\left(\alpha, \beta\right) \equiv \frac{1}{n} \sum_{i=1}^{n} (y_i - \alpha - \beta x_i)^2.$$

Making the change of variables: $\{\sigma_x, \sigma_y\} \to \{\sigma, \lambda\}$, where $\sigma^2 = \sigma_y^2 + \beta^2 \sigma_x^2$, $\lambda = \sigma_y/\sigma_x$, the posterior distribution of (α, β) is found to be independent of the prior distribution of λ, confirming a previous remark.

Integrating out σ with respect to Jeffreys' prior, we obtain a 'quasi-likelihood' function

$$f(\alpha, \beta) \sim [Q(\alpha, \beta)]^{-n/2},$$

which, when multiplied by the prior density and normalized, gives the joint posterior distribution of α, β. This function summarizes all the information about α, β that is contained in the data; and so the optimal procedure for any inference or decision problem involving α, β – whether in the form of interval estimation, tests of any hypotheses concerning α, β, etc. – can then be found from it.

To confirm another previous remark, consider the simpler regression problem where errors are only in Y. Then $\sigma_x = 0$, $X_i = x_i$ is known, and $n + 1$ nuisance parameters drop out of the problem. The likelihood function reduces to

$$L(\alpha, \beta, \sigma_y) = \sigma_y^{-n} \exp\left\{-\frac{n}{2\sigma_y^2} Q(\alpha, \beta)\right\}.$$

Integrating out σ_y with Jeffreys prior, we get the quasi-likelihood $f(\alpha \beta) \sim$ $\sim [Q(\alpha, \beta)]^{-n/2}$, precisely the same as before. The nuisance parameters had *no effect at all* on the quality of inference about α, β.

Representing the sample means, variances, covariance, and correlation coefficient by $\bar{x}$, $\bar{y}$, $s_x^2 = (\overline{x^2} - \bar{x}^2)$, $s_y^2 = (\overline{y^2} - \bar{y}^2)$, $s_{xy} = (\overline{xy} - \bar{x}\bar{y})$, $r = s_{xy}/s_x s_y$ respectively, $f(\alpha, \beta)$ has its maximum at (α^*, β^*), where $\alpha^* = \bar{y} - \beta^* \bar{x}$, and $\beta^* = (s_y/s_x) r$. With uniform priors, further integrations yield the marginal posterior distributions of α, β: $g(\alpha) \sim [(\alpha - a^*)^2 + A^2]^{-m}$, $h(\beta) \sim [(\beta - \beta^*)^2 + B^2]^{-m}$, where $m = (n-1)/2$, and $B = \beta^* r^{-1} (1 - r^2)^{1/2} = A/(\overline{x^2})^{1/2}$. Evidently, α^*, β^* are the 'best' estimates of α, β by the criterion of any loss function which is a monotonic increasing function of the errors $|\alpha - \alpha^*|$, $|\beta - \beta^*|$, For $n > 2$, the marginal distributions are normalizable, leading to definite interval estimate statements. For $n = 3$ and $n = 4$, the (median $\pm$ interquartile) estimates of α are $(\alpha^* \pm A)$ and $= (\alpha^* \pm A/\sqrt{3})$ respectively; similarly for β. When $n > 4$, the second moments also con-converge, leading to the (mean)$\pm$(standard deviation) estimates $\alpha^* \pm$ $\pm A\sqrt{n-4}$, etc.

Thus, for example, if we need to measure β to an accuracy of $\pm 1\%$, the sample size and correlation coefficient must satisfy $(n-4) r^2/(1-r^2) > 10^4$. With a correlation coefficient $r = 0.9$, this requires $n = 2350$ measurements, while with better data, $r=0.99$, $n=208$ measurements suffice, and with $r=0.999$, only $n=25$ data points are needed. A simple analysis shows that to attain the same accuracy by the method described by KF would require at least $(16/3)=5.3$ times as many data points, if they are distributed with roughly uniform density along the line. As will be shown elsewhere, the above results can be improved a bit more by use of the invariant prior $d\alpha\, d\sin\theta$, and a similar invariant prior for X_i, Y_i.

CONCLUSION

I suppose it is possible, without actual logical contradiction, to maintain that Bayesian methods are utterly wrong, but that through a series of fortuitous accidents they always happen to give the right answer in every particular problem. However, I cannot believe that anybody will want to take that position. Now the person who, after studying the evidence given here and in the rest of the Bayesian literature, still wishes to claim that orthodox methods are superior, must realize that, if he is to avoid being forced into exactly that position, mere linguistics and ideological slogans will no longer suffice. The burden of proof is squarely on him to show us specific problems, with mathematical details, in which orthodox methods give a satisfactory result and Bayesian methods do not. My own studies have convinced me that such a problem does not exist.

Whether I am right or wrong in this belief, we now have a large mass of factual evidence showing that (a) orthodox methods contain dangerous fallacies, and must in any event be revised; and (b) Bayesian methods are easier to apply and give better results. As a teacher, I therefore feel that to continue the time honored practice – still in effect in many schools – of reaching pure orthodox statistics to students, with only a passing sneer at Bayes and Laplace, is to perpetuate a tragic error which has already wasted thousands of man-years of our finest mathematical talent in pursuit of false goals. If this talent had been directed toward understanding Laplace's contributions and learning how to use them properly, statistical practice would be far more advanced today than it is.

REFERENCES

Note: The following list includes only those works not already cited in my main presentation or Kempthorne's reply.

Barnard, G. A., 'Comments on Stein's "A Remark on the Likelihood Principle"',
    J. Roy. Stat. Soc. (A) **125**, 569 (1962).

Cox, R. T., *Am. J. Phys.* **17**, 1 (1946).

Cox, R. T., *The Algebra of Probable Inference*, Johns Hopkins University Press, 1961; Reviewed by E. T. Jaynes, *Am. J. Phys.* **31**, 66 (1963).

Deming, W. E., *Statistical Adjustment of Data*, J. Wiley, New York (1943).

Fisher, R. A., *Contributions to Mathematical Statistics*, W. A. Shewhart, (ed.), J. Wiley and Sons, Inc. New York (1950); Referred to above as 'Collected Works'.

Fisher, R. A., *Statistical Methods and Scientific Inference*, Hafner Publishing Co., New York (1956).

Fisher, R. A., *Statistical Methods for Research Workers*, Hafner Publishing Co., New York: Thirteenth Edition (1958).

Forney, G. D., *Information Theory*, (EE376 Course Notes, Stanford University, 1972); p. 26.

Hoel, P. G., *Introduction to Mathematical Statistics*, Fourth Edition, J. Wiley and Sons, Inc., New York (1971).

Jaynes, E. T., 'Review of *Noise and Fluctuations*', by D. K. C. MacDonald, *Am. J. Phys.* **31**, 946 (1963).

Kendall, M. G., 'Ronald Aylmer Fisher, 1890–1962', *Biometrika* **50**, 1–15 (1963); reprinted in *Studies in the History of Statistics and Probability*, E. S. Pearson and M. G. Kendall, (eds)., Hafner Publishing Co., Darien, Conn. (1970).

Mandel, J., *The Statistical Analysis of Experimental Data*, Interscience Publishers, New York (1964); p. 290.

McColl, H., 'The Calculus of Equivalent Statements', *Proc. Lond. Math. Soc.* **28**, p. 556 (1897).

Pearson, Karl, 'Method of Moments and Method of Maximum Likelihood', *Biometrika* **28**, 34 (1936).

Pratt, John W., 'Review of *Testing Statistical Hypothesis*' (Lehmann, 1959); *J. Am. Stat. Assoc.* Vol. **56**, pp. 163–166 (1961).

Roberts, Harry V., 'Statistical Dogma: One Response to a Challenge', Multilithed, University of Chicago (1965).

Thornber, Hodson, 'An Autoregressive Model: Bayesian Versus Sampling Theory Analysis', Multilithed, Dept. of Economics, University of Chicago, Chicago, Illinois (1965).

Wilbraham, H., *Phil. Mag. Series*, 4, Vol. **vii**, (1854).

Zellner, Arnold, 'Bayesian Inference and Simultaneous Equation Models', Multilithed, University of Chicago, Chicago, Illinois (1965).

J. G. KALBFLEISCH AND D. A. SPROTT

ON TESTS OF SIGNIFICANCE

ABSTRACT. The purpose of this paper is to discuss and illustrate by example some logical aspects of tests of significance. In particular, the importance of properly selecting the population or reference set for the test is emphasized. A dilution series example illustrates how considerations of sufficiency may sometimes be used to identify the part of the data which contains the information relevant to the hypothesis being tested.

1. INTRODUCTION

The aim of this paper is to give an elementary discussion of the logic of tests of significance. Many of the points to be made have been emphasized repeatedly in the writings of R. A. Fisher and others. Nevertheless tests of significance are either ignored or mishandled in practically all statistics textbooks, and careful consideration of the logical principles involved is urgently needed.

A test of significance is a measurement procedure whose purpose is to evaluate the strength of the evidence provided by the data against an hypothesis. The observed significance level is an index of the compatibility or consistency of the data and the hypothesis. The smaller the observed significance level, the stronger the evidence provided by the data against the hypothesis. It is a gross oversimplification to regard a test of significance as a decision rule for accepting or rejecting a hypothesis. Any decision to 'accept' or 'reject' a scientific hypothesis will certainly depend upon more than the experimental evidence. A theory which is contradicted by the data may continue to be used if no satisfactory alternative is available, or if the nature of the departures from it are judged to be unimportant for a particular application.

In the usual Neyman-Pearson formulation of hypothesis testing, the null hypothesis H_0 is to be rejected if the experimental outcome x falls in a subset C of the sample space (the critical region), and otherwise it is to be accepted. The size of the test is the probability of rejecting H_0 when it is true, and the power of the test is the probability of rejecting H_0 when it is false. C is chosen so that α has some predetermined value (such as

Harper and Hooker (eds.), Foundations of Probability Theory, Statistical Inference, and Statistical Theories of Science, Vol. II, 259–272. All Rights Reserved. Copyright © 1976 by D. Reidel Publishing Company, Dordrecht-Holland.

0.05) and then the power is maximized. This formulation of testing problems is not suitable when the purpose of the test is to weigh the experimental evidence against a hypothesis. In the first place, the size of the test should not be fixed in advance. Secondly, the consideration of power requires that there be a well-specified set of alternatives to H_0 prior to the carrying out of any test, and this is not always available. Thirdly, the significance level cannot be equated with the rejection frequency of a true null hypothesis. These objections will be discussed in the next three sections, and following this a more satisfactory formulation of tests of significance will be given and illustrated.

2. FIXING THE α-LEVEL

According to Neyman-Pearson theory, the critical region C is to be chosen to give a predetermined rejection frequency α. Whatever point of C is observed, the decision is the same: to reject H_0 at level α. However, it it will usually be the case that points in a critical region of size α will vary greatly in the force with which they contradict the null hypothesis. For example, in testing the hypothesis that the mean of a normal distribution is zero, a t-value of 4 gives a much stronger indication that $\mu \neq 0$ than does a t-value of 2, although both may lie in the 5% critical region. It is not enough to record that rejection has or has not occurred at some predetermined level; a measure of the actual strength of the evidence is required. In fact, it is standard practice in applied statistical work to report the smallest tabulated value at which the data are significant. Apart from limitations imposed by the available tables (which computers are quickly removing), this amounts to reporting the actual observed significance level of the data.

3. THE ROLE OF ALTERNATIVE HYPOTHESES

In Neyman-Pearson theory, a null hypothesis H_0 cannot be tested in isolation. The choice of the critical region is based upon considerations of power which in turn require that there be a completely specified and well-ordered set of alternatives to H_0. However in many of the most important applications, the class of alternatives to H_0 is ill-defined. Although one may have a rough idea of the type of departure that is likely to occur and take this into account when devising a test, one is usually a long way from having a well-defined parametric model.

For example, consider a test for independence in a two-by-two contingency table with cell probabilities p_{11}, p_{12}, p_{21}, and p_{22}. It is a relatively simple matter to test whether independence of row and column classifications is compatible with the data. It is much more difficult to specify what the nature of the association between the two classifications might be if indeed one exists. Two of the many possible measures of association are

$$\gamma = p_{11}p_{22}/p_{12}p_{21};$$
$$\delta = p_{11}/(p_{11} + p_{12})(p_{11} + p_{21}).$$

The former is the odds ratio; the latter arises in genetics and is equal to $2(1-\theta)$, where θ is the genetic linkage parameter in a mating of the type $AaBb \times aabb$ (backcross). If the null hypothesis of independence is true, then $\gamma = \delta = 1$. However the alternatives to H_0 are ordered differently by γ and δ, and the shape of the power curve will depend upon whether γ, δ or some other measure of association is selected. There are situations (as in genetics) where the theory is sufficiently well developed to permit the choice of an appropriate measure of association, but in many practical situations any such choice would be quite arbitrary. The natural procedure is then to determine whether there is evidence of an association. If there is not, that ends the matter – at least until further data are obtained. If an association exists, one can then undertake the difficult task of finding a suitable parametric model to explain it. A test of significance enables the building of more elaborate models to be postponed until the available data make it clear that they are needed.

In the 2 by 2 table, the class of alternatives to H_0 is one dimensional. In general it will be multidimensional, and power comparisons become even more problematical. Essentially, one must order the alternative hypotheses according to their 'distance' from H_0 and attempt to find a critical region C whose probability content is small for hypotheses 'close to H_0' and large for hypotheses 'far from H_0'. This is undoubtedly a valuable exercise when the alternative hypotheses can be placed in a meaningful order, but in most applications they cannot. When knowledge of the possible alternatives to H_0 is vague, power comparisons are arbitrary, and Neyman-Pearson theory offers little guidance in the selection of a procedure for determining whether the null hypothesis is compatible with the data.

An excellent discussion of the role of alternative hypotheses in tests of significance is given by Gillies (1971, pp. 243–254).

4. SIGNIFICANCE LEVELS AND REJECTION FREQUENCIES

It is a common misconception that the significance level should be equal to the frequency of rejection of a true hypothesis in repetitions of the experiment. Indeed, the literature abounds with papers criticizing one or another test of significance for failure to achieve preset rejection rates. Some examples are the paper of Grizzle (1967) on 2 by 2 contingency tables, that of Detre and White (1970) on the comparison of Poisson means, and that of Mehta and Srinivasan (1970) on the Behrens-Fisher problem. However, as Fisher (1959, p. 93) pointed out, one must distinguish between the strength of the evidence, which is to be measured by the significance level, and the frequency with which evidence of a given strength will be obtained. The nature of the hypothesis and experiment may be such that evidence of even moderate strength is almost impossible to obtain. The frequency with which a true hypothesis would be rejected at level 0.05 in repetitions of the experiment may then be much less than 0.05. This point is illustrated in the following examples.

EXAMPLE 1. Suppose that there are m coins and let the probability of heads for the ith coin be p_i. The hypothesis to be tested is that at least one of the m coins is unbiased,

$$H: p_1 = 0.5 \quad \text{or} \quad p_2 = 0.5 \quad \text{or} \quad \dots \quad \text{or} \quad p_m = 0.5.$$

Note that to disprove H, one must demonstrate that $p_i \neq 0.5$ for *each* $i = 1, 2, \dots, m$. It would seem logical that the strength of the evidence against H should be no greater than the strength of the evidence against any one of the m hypotheses $p_i = 0.5$.

Let each coin be tossed 10 times. If $p_i = 0.5$, the probability of 0, 1, 9 or 10 heads on the ith coin is $22 \times 2^{-10} = 0.0215$. Hence if the ith coin yielded 0, 1, 9 or 10 heads, one could claim fairly strong evidence (significance level $\leqslant 0.0215$) against the hypothesis $p_i = 0.5$. If *every* coin yielded 0, 1, 9, or 10 heads, one could claim to have fairly strong evidence against the hypothesis H, and quote an observed 'significance level' of 0.0215 (provided that at least one coin yielded 1 or 9 heads). However, the probability of obtaining evidence of this strength will be much smaller than 0.0215 whenever at least two of the p_i's are near 0.5. In particular, if $p_1 = p_2 = \cdots = p_m = 0.5$, the probability will be only $(0.0215)^m$.

It is amusing to consider the effect of attempting to construct a test with a specified rejection rate. Suppose that we require the probability of rejecting H when $p_1 = p_2 = \cdots = p_m = 0.5$ to be (approximately) 0.05. The probability of obtaining 5 heads in 10 tosses when $p = 0.5$ is $\binom{10}{5}(0.5)^{10} =$ $= 0.2461 = 1 - 0.7539$. Since $(0.7539)^{11} < 0.05$, it follows that, with probability greater than 0.95, the experiment will produce 5 heads and 5 tails on at least one coin whenever $m \geqslant 11$. One can then come close to the required rejection rate of 5% only by sometimes rejecting H when one or more of the coins shows 5 heads and 5 tails, in which case there is absolutely no evidence that H is false.

It should be noted that the difficulties encountered in this example are *not* a result of discreteness. Similar remarks apply to the problem of testing the hypothesis that at least one of m normal means is zero, H: $\mu_1 \mu_2 \ldots \mu_m = 0$. It is perhaps not clear how the significance level should be defined in such problems. However, it is clear that the frequency with which a true hypothesis would be rejected by a test in repetitions of the experiment will not necessarily be indicative of the strength of the evidence against H.

EXAMPLE 2. Let X_1 and X_2 be independent Poisson variates with means μ_1 and μ_2. The problem is to test the significance of observed values x_1 and x_2 in relation to the hypothesis H: $\mu_1 = \mu_2 = \mu$, where μ is not specified.

The problem is similar to that of Example 1 in that, under certain conditions, it will be extremely difficult to obtain evidence against H. If the experiment yields $x_1 = x_2 = 0$, one can certainly not claim to have evidence that $\mu_1 \neq \mu_2$. But the probability of this outcome is $e^{-(\mu_1 + \mu_2)}$, so that if $\mu_1 + \mu_2$ is small, the experiment will frequently yield no evidence against the hypothesis. More generally, any experiment which yields a small value of the variate $T = X_1 + X_2$ can provide, at best, only meagre evidence against the hypothesis, and this will occur with large probability whenever $\mu_1 + \mu_2$ is small. Consequently, any test of significance which adequately measures the strength of the evidence against H must have the property that the frequency of rejecting a true hypothesis at say the 5% level is much less than 5% whenever $\mu_1 + \mu_2$ is small.

Detre and White (1970) note that the conventional two-tail exact test, which is based on the conditional distribution of outcomes given T, does in fact have this property. There is no sample with $T \leqslant 5$ which leads to rejection at the 5% level, and consequently the overall rejection rate will be substantially less than 5% whenever $\mu_1 + \mu_2$ is small. This property, which is a necessary requirement of any valid test of significance, is regarded by Detre and White as a defect of the exact test, and they attempt to remedy it by enlarging the critical region. According to the test which they advocate, the hypothesis is rejected at level 0.05 whenever $|x_1 - x_2| \geqslant$ $\geqslant 1.96\sqrt{x_1 + x_2}$. Thus the sample $x_1 = 0$, $x_2 = 4$ leads to rejection at level 0.05, while the exact test gives the significance level as 0.125. The desired rejection rate of 5% is then obtained (approximately, for $\mu > 2$), but only by sometimes rejecting the hypothesis at level 0.05 when there is little evidence against it.

In summary, the failure of a test of significance to achieve a preset rejection frequency in repetitions of the experiment is not necessarily a defect of the test when the hypothesis being tested is composite. Indeed, it may be that evidence capable of simultaneously ruling out the whole range of simple hypotheses contained in H is very difficult or even impossible to obtain. In such cases, any valid test of significance will have a very small rejection frequency. Attempts to 'improve' such a test by increasing its rejection frequency closer to a preset level will render the test unsuitable for use in weighing the experimental evidence.

5. Specification of a Test of Significance

A test of significance requires two ingredients for its specification:
(a) a reference set (population) R, and a probability distribution on R;
(b) a discrepancy measure (test criterion) D.

The reference set is the set of experimental outcomes to which, for the purpose of the test, the observations are considered to be long, and upon which probability statements are based (Barnard, 1947). The discrepancy measure is a random variable defined on R which assigns to each point x in R a real number $D(x)$. The significance level of the data in relation to the hypothesis is then $P(D \geqslant d)$ where d is the observed discrepancy, the probability being computed from the distribution on R.

The purpose of the discrepancy measure is to rank the points of R according to their compatibility with the hypothesis, with points having the highest discrepancy being judged the least compatible. Two discrepancy measures which produce the same ranking of R will always give equal significance levels on the same data, and therefore define equivalent tests. It is the ordering produced rather than the actual magnitude of D which is important. A good discussion of the problem of selecting a suitable ordering is given by Kempthorne and Folks (1971).

The discrepancy measure specifies the type of departure from the hypothesis that one wishes to detect, and its choice may therefore depend upon the type of alternative hypothesis that one has in mind. If there is a well-defined and well-ordered set of alternative hypotheses, then power comparisons may be used to choose among alternative test criteria. However, as pointed out in Section 3, this will not often be the case. Usually one will have only vague notions about the type of departure that may occur, and formal power comparisons will not be possible. The choice of D then requires ingenuity and judgement, and it would be a mistake to attempt to formalize the role of alternative hypotheses in this procedure.

As stated previously, the observed significance level of the data in relation to the hypothesis is the probability of an outcome at least as unfavourable to the hypothesis as the one obtained. A small significance level can be explained in only two ways: either the hypothesis is true and an event of small probability has occurred, or else the hypothesis is false. The smaller the significance level, the greater the reluctance to accept the first explanation, and hence the stronger the evidence that the hypothesis is false. On the other hand, a large significance level indicates that, with respect to the particular test used, the data provide no evidence that this hypothesis is false. This should not be interpreted as evidence in support of the hypothesis, but merely as a lack of evidence against it. It is quite possible that a different test on the same data might subsequently show the existence of a different type of departure, and convincingly demonstrate that the hypothesis is false.

It is important to recognize the two separate ingredients of a test of significance: the population R and the test criterion D. The criterion does not of itself provide any test whatever until R has been specified; and, of course, a comparison of two criteria D_1 and D_2 must be made on the same reference set if it is to be meaningful. Thus the choice of the reference set must precede the comparison of different test criteria. For further discussion of this point in relation to the comparison of Poisson means, see Kalbfleisch and Sprott (1974a).

6. Choice of the reference set

R. A. Fisher (1959) has emphasized the hypothetical nature of probability models used in problems of statistical inference. Because most experiments are essentially unique and will not be repeated endlessly, there is generally no single well-defined long-run of experiments within which inferences must be evaluated. The experiment is imbedded in a hypothetical probability model by the statistician. Hence the reference set generally will not correspond to any sequence of actual repetitions of the experiment. In fact, the same data may be regarded as belonging to two or more different reference sets, depending upon what type of information one seeks to obtain from it.

For example, consider an experiment which involves n trials, each of which results in success or failure. The sample space would normally be taken to contain 2^n points corresponding to all possible sequences, and some sequence of, say, x successes and $n-x$ failures would be observed. If one wished to test the independence of successive trials, one would regard x as fixed and consider only the order in which the successes and failures occurred, so that the reference set would contain $\binom{n}{x}$ points. A test of significance might be based on the number of runs of like outcomes, with probabilities being calculated on the assumption that all $\binom{n}{x}$ points in the reference set were equally probable. On the other hand, it may be that independence of successive trials is to be assumed, and that one's aim is to test an hypothesized value of the common probability of success. Then only the number of successes would be considered, the order of their occurrence being ignored. The reference set would contain $n+1$ points whose probabilities were given by a binomial distribution. Thus, depending upon the type of information sought, the data may be regarded as belonging to two different reference sets, both of which are different from the sample space.

It frequently happens that the various possible outcomes of an experiment differ greatly in the amount of information they are capable of providing concerning the hypothesis of interest. In the example just considered, a sequence for which x is near 0 or n is incapable of providing much information concerning the independence of trials, but a sequence with x near $n/2$ will be much more informative. In Example 2, a sample for which $x_1 + x_2$ is small will provide little information about the equality of the Poisson means, whereas if $x_1 + x_2$ is large, even a slight difference in the means can be detected.

In problems of inference one must take into account the informativeness of the outcome actually obtained. In a test of significance, the data should be compared with, and hence the reference set should consist of, outcomes having approximately the same informativeness or precision as the one observed. Thus, in testing the independence of trials, the number of successes is held fixed. In testing the equality of Poisson means, one uses the conditional distribution given the observed total $x_1 + x_2$ (Cox, 1958; Kalbfleisch and Sprott, 1974a).

Similar principles are involved in many other problems as well: for example, in the analysis of contingency tables, the 'play the winner' rule of Zelen (1969), and the virological model discussed by Kalbfleisch and Sprott (1974b). Although results are not always as clear cut as they are in these examples, it seems that a conditional reference set is frequently appropriate (Cox, 1958).

The next section gives a more complex example illustrating the variety of reference sets available in a single problem, each appropriate for testing a different aspect of the model.

7. AN EXAMPLE

A dilution series model discussed by Roberts and Cootes (1965) will now be considered. Infective particles are assumed to be suspended in an initial solution with mean density λ per unit volume. Separate independent dilutions of this suspension are prepared, the dilution factor for the ith dilution being x_i, $1 \leqslant i \leqslant p$. Some of the diluted suspension is spread over a nutrient medium, and after incubation a count is made of the number of colonies, plaques, or pocks produced. If the ith dilution is applied to n_i different plates, one obtains n_i parallel counts r_{ij}, $1 \leqslant j \leqslant n_i$. Denote the expected count on a plate at the ith dilution by $\mu_i = \lambda_i x_i$, say. The following hypotheses are of interest:

(1) *Poisson hypothesis*: the n_i parallel counts at the ith dilution level are a sample of size n_i from a Poisson distribution;

(2) *hypothesis of homogeneity*: $\lambda_1 = \lambda_2 = \cdots = \lambda_p = \lambda$, say.

Under hypothesis (1), the joint distribution of the n_i parallel counts at dilution level i is

$$f(r_{i1}, \ldots, r_{in_i}) = \mu_i^{r_i.} \, e^{-n_i \mu_i} \bigg/ \prod_{j=1}^{n_i} r_{ij}!$$

where $r_{i.}$ is the sum of r_{ij} for $1 \leqslant j \leqslant n_i$. The total $r_{i.}$ is sufficient for μ_i, and has a Poisson distribution with mean $n_i \mu_i$. Thus the conditional distribution of the r_{ij}'s given $r_{i.}$ is

$$\frac{r_{i.}!}{\prod_j r_{ij}!} \left(\frac{1}{n_i}\right)^{r_{i.}} = \frac{r_i!}{\prod_j r_{ij}!} \prod_j \left(\frac{1}{n_i}\right)^{r_{ij}}$$

which is multinomial with index $r_{i.}$ and equal probabilities $1/n_i$, $j = 1$, $2, \ldots, n_i$. Since this conditional distribution is independent of the μ_i, it is appropriate for testing the Poisson hypothesis (1). A test may be based upon the usual goodness of fit criterion for this multinomial distribution. Since $E(r_{ij} \mid r_{i.}) = r_{i.}/n_i = \bar{r}_{i.}$, one obtains

$$D_i = \sum_j (r_{ij} - E(r_{ij})]^2/E(r_{ij}) = \sum_j (r_{ij} - \bar{r}_{i.})^2/\bar{r}_{i.}$$

whose distribution is approximately χ^2 with $n_i - 1$ degrees of freedom provided that $\bar{r}_{i.}$ is reasonably large. An overall test of the Poisson assumption may be based on $\sum D_i$, which is approximately χ^2 with $\sum (n_i - 1)$ degrees of freedom.

Assuming the Poisson model, inferences concerning the parameters λ_i will be based upon the joint distribution of the sufficient statistics $r_{i.}$:

$$f(r_{1.}, \ldots, r_{p.}) = \prod_{i=1}^{p} (n_i \lambda_i x_i)^{r_i.} \, e^{-n_i \lambda_i x_i}/r_{i.}!$$

Under the hypothesis of homogeneity $\lambda_i = \lambda$ for all i, the grand total $r = \sum r_{i.}$ is sufficient for λ, and has a Poisson distribution with mean

$\sum n_i\lambda_i x_i = \lambda \sum n_i x_i$. The appropriate distribution for testing the hypothesis of homogeneity is the conditional distribution of the r_i.'s given r, which is

$$\frac{r!}{\prod r_{i.}!}\left(\frac{\lambda n_i x_i}{\lambda \sum n_i x_i}\right)^{r_{i.}} = \frac{r!}{\prod r_{i.}!}\left(\frac{n_i x_i}{\sum n_i x_i}\right)^{r_{i.}}.$$

This is a multinomial distribution with index r and known probabilities $p_i = n_i x_i/\sum n_i x_i$ independent of λ, and will form the basis for an exact test of the hypothesis of homogeneity in small samples. In large samples, one can make use of the standard goodness of fit criterion for this multinomial distribution,

$$D = \sum (r_{i.} - rp_i)^2/rp_i,$$

which is distributed approximately as χ^2 with $p-1$ degrees of freedom.

The entire analysis is based upon a separation of the joint distribution of the counts at all p dilution levels into three factors, each appropriate for a different type of problem. A factor is first removed for testing the Poisson assumption; from what remains, a factor is removed for testing the hypothesis $\lambda_i = \lambda$ for all i; the remaining factor is the distribution of the grand total r, and contains all of the sample information concerning the parameter λ. This factorization serves to separate from one another the various types of information provided by the data, and is illustrative of a general approach to problems of inference in which considerations of sufficiency and ancillarity are used to determine the characteristics of the data which carry the information relevant to the various hypotheses of interest. First priority is thus given to the selection of the appropriate reference set for the inference problem. When this has been done, the traditional problems of selecting estimators and test statistics may be greatly simplified or even eliminated. For a similar example see Kalbfleisch and Sprott (1974b).

It is of interest to note that, in the present example, Roberts and Cootes suggested that a test of the Poisson assumption (1) should be based on the test criterion

$$\sum\sum (r_{ij} - \hat{\lambda}x_i)^2/\hat{\lambda}x_i$$

where $\hat{\lambda}$ is the maximum likelihood estimate of λ. This will not give a satisfactory test because a significant result could well be due to the failure of hypothesis (2) rather than to departures from the Poisson distribution. Considerations of sufficiency, as advocated in the present paper, lead directly to the appropriate χ^2 statistic

$$\sum D_i = \sum\sum (r_{ij} - \bar{r}_{i.})^2 / \bar{r}_{i.}.$$

as an approximation to the relevant multinomial distribution.

University of Waterloo

REFERENCES

Barnard, G. A., 1947, 'The Meaning of a Significance Level', *Biometrika* **34**, 179–182.
Cox, D. R., 1958, 'Some Problems Connected with Statistical Inference', *Ann. Math. Statist.* **29**, 357–372.
Detre, K., and White, C., 1972, 'The Comparison of Two Poisson-Distributed Observations', *Biometrics* **26**, 851–853.
Fisher, R. A., 1959, *Statistical Methods and Scientific Inference*, 2nd Edition, Hafner, New York.
Gillies, D. A., 1971, 'A Falsifying Rule for Probability Statements', *Brit. J. Phil. Sci.* **22**, 231–261.
Grizzle, J. E., 1967, 'Continuity Correction in the χ^2 Test for 2×2 Tables', *The Am. Statist.* **21**, No. 4, 28–32.
Kalbfleisch, J. D., and Sprott, D. A., 1970, 'Applications of Likelihood Methods to Models Involving Large Numbers of Parameters' (with discussion), *J. Roy. Statist. Soc.* **B32**, 175–208.
Kalbfleisch, J. G., and Sprott, D. A., 1974a, 'On the Logic of Tests of Significance with Special Reference to Testing Significance of Poisson-distributed Observations', *Information, Inference and Decision*, (ed. by G. Menges), Reidel, Dordrecht-Holland.
Kalbfleisch, J. G., and Sprott, D. A., 1974b, 'Inferences About Hit Number in a Virological Model', *Biometrics* **30**, 199–208.
Kempthorne, O., and Folks, L., 1971, *Probability, Statistics, and Data Analysis*, Iowa State University Press.
Mehta, J. S., and Srinivasan, R., 1970, 'On the Behrens-Fisher Problem', *Biometrika* **57**, 649–655.
Roberts, E. A., and Coote, G. E., 1965, 'The Estimation of Concentration of Viruses and Bacteria from Dilution Counts', *Biometrics* **21**, 600–615.
Zelen, M., 1969, 'Play the Winner Rule and the Controlled Clinical Trial', *J. Amer. Statist. Assoc.* **64**, 131–146.

DISCUSSION

Commentator: Good: I suppose the choice of the reference set is determined by selecting in some way that part of the data which does not by itself convey evidence relevant to some hypothesis of interest but in this case isn't the selection of a reference set necessarily subjective? Contingency tables are a good example, where you might assume that the marginal totals contain very little information and so can be taken as 'given' when calculating the classical significance criteria.

Kalbfleisch: In some cases, the information in the sample can be separated neatly into disjoint parts, and there is no arbitrariness in the choice of the reference set. Future research may help to identify more situations of this sort. However in most cases there will be no clean separation, and then some degree of approximation or judgment is required. Contingency tables fall in the latter category; the marginal totals usually contain *very little* information relevant to the question of independence.

Lindley: But in a very small contingency table they contain all the information.

Kalbfleisch: Yes.

Diaconis: For 50 years now we have had statistical tests of significance, and while these were admirable at the time they were invented (chiefly by Fisher) they fail entirely to consider the careful articulation of all the alternative hypotheses under test. In this case, after one has done the test of significance, one is fishing around to find out what on earth it might mean. Wouldn't it be rather more helpful to start on a more general theory which does articulate the alternatives – such as Good has provided in the special case of multi-nomials and Jeffreys' in the case of a two-dimensional array?

Kalbfleisch: It is true that it may often be helpful to articulate carefully all the alternatives, but I contend that in many situations this is next to impossible in any case and that very often it is not required; for example, if you are testing whether a set of bodies of evidence are homogeneous you do not first need to consider all the conceivable ways in which they might be inhomogeneous (though of course if the test suggests that they are not homogeneous then it is time to start thinking of exactly in which way homogeneity might fail).

Diaconis: You have said that in a test of significance whenever you get a sufficiently small value then either something is wrong or something very improbable has happened. Where do you get your notions of small and improbable from?

Kalbfleisch: These are essentially undefined. But it is important to distinguish between a measurement of the strength of the evidence, which yields the value in question, and the decision based on that measurement, which must contain some decision criterion which is not itself dictated by the measurement. In this sense, the latter criterion may indeed introduce a 'subjective' element, but the measurement itself should be objective.

Lindley: At the end of such a procedure one comes up with a figure, say 5%. What does this mean *operationally*? In the Neyman-Pearson theory it means that in the long run, if the null hypothesis is true, values like this (or more extreme) will occur 1 in 20 times. In the Bayesian theory 5% could mean odds of 19 to 1 in a gamble to win or lose an amount. Both values can be tested. How can I test Kalbfleisch?

Kalbfleisch: As in the Neyman-Pearson theory, results significant at the 5% level would occur at most one time in twenty 'in the long run' if the hypothesis were true. In selecting the reference set, one specifies the long run within which it is appropriate to evaluate the inference, and this need not correspond to all possible repetitions of the experiment.

OSCAR KEMPTHORNE

STATISTICS AND THE PHILOSOPHERS

1. INTRODUCTION

I would first like to congratulate the organizers of the present *working* conference. It is the first time that I am aware of when an attempt has been made to bring into working (and perhaps abrasive) contact philosophers, physicists, probabilists and statisticians. I am quite sure I am not alone in having views that are not at all laudatory of the philosophical world. I am not alone, I surmise, in having views that are not totally laudatory of the physics world. And I believe I am not alone in having views that are strongly critical (at best) of many parts of the probability world. Finally, as regards the statistical world I find that I have deep respect only for that part of the statistical world that is actually doing statistics and not that part which is engaged in constructing a panacea for all the problems of statisticating.

Let me in this introduction make a necessary remark. I am not accredited in philosophy, but I have read a moderate amount of what is regarded as good philosophical writings. I am not accredited in physics, though here also I have read a moderate amount. You can find individuals, some at this conference, who believe, I think, that I am stupid as regards statistics. So why should anyone listen to me? Or, rather, why should anyone entertain the possibility that I might have something worth hearing? In this connection let me state a view I have about historical writing (which is, I suggest, very close in logical nature to scientific writing). I have felt for many years that every book on history should be prefaced by a biographical statement of the background of the author, including his origin, his schooling, his religious and social values and beliefs. From such information one can make a slight stab at evaluating the author's prejudices. Just how much, indeed, of historical writing is purely and simply the choice and embedding of the historical 'facts' (many of which are not facts) in a model based on the author's set of values and beliefs? That is why history is often not helpful to the broad purposes of humanity.

Harper and Hooker (eds.), Foundations of Probability Theory, Statistical Inference, and Statistical Theories of Science, Vol. II, 273–314. All Rights Reserved. Copyright © 1976 by D. Reidel Publishing Company, Dordrecht-Holland.

The 'best' historians, it seems to me, are good novelists who present an interesting picture which we enjoy reading, as we enjoy reading 'pure' novels such as *War and Peace*. The 'history' game is not well-defined, and the successful historians seem often to be merely good entertainers.

Where does this lead? To evaluate what I write, you, the listener and reader, should be given information about me. In purely statistical circles this is usually not necessary. In the present circle, where I am almost unknown, it may aid to make three statements: (i) I have written on statistics, biometrics, and parts of genetics in a way that some scientists regard as informative, (ii) I have done a lot of homework in philosophy with particular emphasis on some modern branches, especially the existentialists, in biology, in physics, in mathematics and in statistics, and (iii) I have spent a significant proportion of the last few decades in attempting to aid scientists with their logical problems including, of course, statistical and probabilistic questions.

I am extremely doubtful that I shall say anything original in this essay. Those who have been trained in philosophy will say, "Well and good, but Plato said that more than 2000 years ago". I have been re-reading Norman Campbell (1957, originally 1919) recently and he made a quite remarkable number of useful statements and I quote some relevant ones:

If an attempt is made to introduce into any physical discussions considerations more general and more fundamental than would be appropriate to an ordinary text book, it is apt to be met with some sneer about 'philosophical' or 'metaphysical' arguments, and with a suggestion that such matters are unworthy of the attention of a serious man of science.

Because philosophers have talked nonsense there is no reason why we should follow their example: because they would not take the trouble to find out what we mean, there is no reason why we should not find out what they mean, or even what we mean ourselves.

I am not sure that the most handsome compliment that anyone could pay to my work would be to say that he knew it all before.

2. THE AIM AND STRUCTURE OF THIS ESSAY

This essay is based on my opinion that there has been almost a complete lack of communication between two groups of workers, philosophers of science and statisticians. Both groups are concerned with processes by which scientific knowledge, whatever that is, can be increased. The philosophy of science dates back for more than 2000 years, and its workers include almost all of the great philosophers of all time, from the pre-Socratics to those of the 20th century. In contrast, statistics as a formal discipline with an organized and partially systematized set of ideas and procedures dates only from the end of the 19th century. Enumeration of populations was practiced, of course, from almost prehistoric time and in a real sense must have preceded *Homo sapiens* in whatever way one wishes to define the beginning of this species. Animals, obviously, can count small populations, as mothers know their litters, and, clearly, the growth and development of organisms, in processes such as blastula formation involve in some way counting what has already happened. I state these opinions because I see throughout philosophy a restriction of discussion to the case of humans, as though ideas of knowledge enter into the picture only with the human species. That such views were natural before, say, 1850 is not surprising because informed opinion at least in Western Christian culture held that *Homo sapiens* appeared in the world by instant creation at some date in the fifth millennium B.C. and was the unique creation of God. I have no doubt that most thinking prior to 1850 or even 1900, at least in Western philosophy, was predicated on some such assumption. It is for this reason and others that are related, that a large proportion of the philosophical writing before this century must, at best, be heavily discounted and, more reasonably, be discarded as based on unwarranted *implicit* assumptions.

This lack of communication has had, I believe, most unfortunate consequences. On the one hand, statisticians, except those working purely on the mathematics of statistics, are involved in the basic philosophical dilemmas, but rarely, in my opinion, recognize the fact, and this, I believe, has led to very deep controversies in the field of statistics. On the other hand, philosophers have been as a class essentially totally unaware of what statisticians are doing, of the fact that statistics gives some suggestions with regard to the age-old philosophical problems, and of the fact that scientific workers are increasingly and predominantly using the ideas of statistics almost routinely in their endeavors to build up knowledge.

The overall phenomenon is, I think, very strange, and even ludicrous.

One can, of course, take the view that these two groups can exist side by side, involved in similar problems and not communicating, and that nothing is lost thereby. But I feel that statistics needs philosophical thinking rather desperately, and as a statistician I may, perhaps, be granted some credence. I feel also, however, that philosophy of knowledge needs statistics. The ignorance of each group with respect to the thoughts, ideas, and processes of the other group is appalling. My thesis is that both groups are losing valuable constructive ideas. I hope that conferences of the present type will be repeated frequently. I hope that it will become a completely accepted fact that a statistical education will include basic ideas of philosophy, and a philosophy (of knowledge) education will include basic ideas of statistics. I hope, furthermore, that the examination of statistical ideas and processes by philosophers will aid in the resolution of controversies with which statistics of the 1970's is plagued.

In this essay, I shall present opinions of a working statistician on some basic philosophical ideas, in the hope that the views of someone who is concerned *actively* with the accumulation of knowledge, but who has not been indoctrinated to a set of philosophical views, may be informative. I shall attempt to discuss briefly what I regard as deep philosophical issues in statistics. In doing this, I am motivated by the belief that no field of human endeavor should be its own monitor. Every field needs the evaluation of outsiders. The whole of intellectual history exhibits many examples in which a certain pattern of thought and language became entrenched to the point that unwarranted assumptions were hidden in the language and were uncovered only after decades or centuries. It would be tedious and space-consuming to give examples, but it is worth noting that even in so purely intellectual subject as mathematics this happened with the use of the Axiom of Choice by the finest mathematicians, and we may be confident, I think, that other such examples are taking place in the mathematics of today.

It may be felt that it is arrogance of the highest order for one untrained in philosophy to state criticisms of that field. This I reject, and I base my case on a modification of an age-old aphorism: A cat may look at a king, and a cat may well make some reasonable judgments of a king. Finally, I state that I shall not single out particular writers with 'page and verse', except for particular critical instances which are essential for my stream of presentation.

3. THE OVERALL PROBLEM OF KNOWLEDGE

In the past few centuries there has been a fantastic growth in science. Humanity looks to science to solve a large subclass of its problems. We may well believe that humanity has an exaggerated notion of what science can do, and this exaggeration has led to the deep doubts on the value of scientific work which have been voiced by people in all walks of life. The doubts are justified if science claims too much and it is surely the case that highly unreasonable claims have been made and are being made continually. Along with a loss of faith in many human processes, there has developed a deep distrust of the nature and results of science, a distrust that many scientists ignore. Part of the cause is that science is outside morality: it can be applied to immoral ends, and has been applied to immoral ends. But a rejection of science is not admissible. It has had many successes and even the most anti-scientific humans rush to science or applied science (e.g. the medical doctor or the garage mechanic) with some of their problems.

The basic problem seems to lie in the natures of the concept of knowledge and the concept of truth. The literature on these topics would fill a small library, and it would be arrogant and foolish to claim to describe adequately these matters in a few lines. History tells us clearly that this cannot be done. But rarely in the literature does one find a recognition of some elementary facts. For instance, the concept of truth, outside of formal logic in which truth is merely one value in a two-valued logic, which could be replaced by a neutral symbol such as Δ, is not operational at its ultimate limit. Can anyone say of any proposition about the world of actual or potential experience that it is true? I search my mind and find no such proposition. But, of course, there are many propositions in which I have very high belief. A problem is, I surmise, that if all humanity were forced to replace the proposition (assertion), 'It is true that it is raining', by the proposition (statement), 'I have extremely high belief that it is raining', so much scepticism would be engendered that reasonable human relations would be impossible.

When we turn to 'knowledge', the difficulties increase exponentially. I will not list all the books I have tried to understand, and will say only that while I have found almost all to be informative, I have not found a presentation that satisfies me even slightly. And one of the reasons for

this is obvious. The only way I can *explain* the matter to myself is to interpret and use words, and I am not perfectly confident that I know what I
mean by the words that I use, let alone what others mean by the words
they use. I can be fairly confident in simple cases, of course. I think, for
instance, that I use the word 'water' in a sense that is nearly definite, un-
altering, and consistent. But when I read any writer I encounter huge
difficulties, and this arises not only with general literature, but critically
with the philosophy and nature of science, because meaning is then critical. In fact, the difficulties increase in this area.

Let me give some examples. Wittgenstein (1922) *starts* with:

"1. The world is everything that is the case".

"1.1. The world is the totality of facts, not of things".

"1.2. The world divides into facts".

Perhaps I am being obtuse, but I suggest that if the proverbial man-in-
the-street entered a philosophical meeting and made these pronouncements
he would be met by derision (and justifiably, I believe). Yet the work from
which these quotations are taken is widely regarded as a work of genius
(and perhaps it is). It is happenstance that I give these particular quotations and I intend no derogatory implications. As another example, Ayer
(1956, p. 34) says:

I conclude then that the necessary and sufficient conditions for knowing that something
is the case are first that what one is said to know be true, secondly that one can be sure
of it, and thereby that one should have the right to be sure.

At a superficial level this statement has great appeal, but we may reasonably ask: What does it mean to say something is true?; What does it
mean to be sure of something?; 'What gives one the right to be sure of
something?' Clearly, the last of these questions is utterly crucial and
without a presentation of some basis for a 'right to be sure', Ayer's statement is void of content. I am not alone, of course, in raising such questions. And, of course, the whole book discusses questions such as these.
Unfortunately, however, it does not give any operational answers.

This leads to a point on which I am deeply critical of and disappointed
in the philosophers. The great bulk of the literature consists of posing
questions and discussing answers, but my own reading has not led me to
any discussion of what constitutes a proper question and what constitutes
an answer to a question. This matter I judge to be truly philosophical.
A question is meaningless to me unless I am aware by implicit assumption

or tacit agreement what are possible forms of an answer. I go further and suggest that the atom in philosophical discussion is an ordered pair (q, a) consisting of a question and an answer, and that the ordered pair is not meaningful unless a class a, say of possible answers is given. It is appropriate, perhaps, to state in relation to all the implicit criticisms that are contained in what I say, that I do find the later Wittgenstein informative, and perhaps him alone. To make another comment, the proliferation of meta-language, meta-meta language and so on does not, I believe, aid the cause of science at all. On this route, we seem to be involved in one sort of infinite regress if we seek the ultimate justification – the 'right to be sure'.

The main thrust of what I am saying or trying to say is, I believe, that language in a generalized sense is at the root of the whole business. This is obviously a well-worn cliché, but the consequences seem to be not at all well perceived by workers in the knowledge area. Examples abound and I will give only a few. For me, an exemplar case was the hope to reduce all human reasoning to the processes of two-valued propositional logic, which is nothing but the construction of complicated strings of symbols from simple strings of symbols by a set of rules. The methods for generating complex strings were a remarkable human invention, but the implementation was exceedingly tedious, and we can see now is a task for which a modern computer is well suited except for the problem of deciding which strings shall be called theorems. The deep problem is, of course, to establish some correspondence of the strings of symbols to aspects of human experience.

The problem becomes critical in all systems of axiomatization. I have for some decades, after initial exposure to the lectures of Dirac, been totally mystified by the processes of quantum mechanics. It was deeply interesting to me to be exposed at the present conference to a large amount of questioning of the nature of the logic, to the point at which there is no agreement at a level that is basic in 1973 on what this nature is. Rather obviously, some thinkers in the area have acquired one language that they believe, while others who have, clearly, thought extensively about the same area have not acquired this language and in fact reject it. My own reactions were initially that I was being given an almost meaningless string of statements with a prescription which I could follow only blindly. But curiously the whole process led to verifiable predictions. I had hoped in the ensuing years that the area would have been 'fixed up', but I find that recent presentations follow the same route in that one is given axioms and one has to follow a set of rules. The whole affair is one of magic, which is meaningful only to those who have crossed a bridge of blind acceptance. I saw no discussion of the possibility that the whole affair is not just a gigantic brain-washing job, until the present conference. I shall refer to this matter in a later section.

A second example which is of deep relevance to the knowledge area is the axiomatization of probability and of statistics. In the probability area the first deep effort was, I believe, that of Keynes (1921). This work is still, it appears, highly regarded by many philosophers of science. On the other hand, the work has had no perceptible effect on the application of ideas of probability by the statistical world. I suggest to the philosophers who favor this work that they reconsider their position. Quite apart from the question of whether the axiomatization is appealing, which is largely an aesthetic evaluation, I suggest that the book gives no guidance as to how any of the processes that are suggested in the abstract may be applied to concrete questions of the real world. This is like having a book on how to knit which merely talks of the theory of knitting. A second example in the probability-statistics area is the axiomatization of L. J. Savage (1954) which has been taken as the basis for neo-Bayesian statistics as presented, for instance, by D. V. Lindley. I do not have the space to go into a dissection of this system; I have never been able to accept it. I was glad to hear at the present conference an evaluation of this system by P. Suppes. He was very critical, I believe. The formulation of Savage is essentially in terms of game theory and rational behavior. A fairly recent book by Howard (1971) appears to bear on the whole issue and contains a theorem "to the effect that to be rational in two-person games is usually to be a sucker". It says, I think, that the sure-thing principle is stupid. In giving these statements, I am not asserting that I accept Howard's presentation and that Savage's axiomatization is stupid. But I do state with complete confidence that there are serious thinkers, with some claims to competence, who reject the Savage formulation. Hopefully, then, it will not be stated that it has never been seriously challenged. The moral seems clear that we should be exceedingly wary of accepting any axiomatic system.

But all this sort of discussion reduces to nearly nothing in face of the progress of science and some technology. There is a deep mystery which

I have not seen addressed in the literature. From analysis of what are thought to be primitive, elemental, and necessary ideas we can become convinced that progress is simply impossible, but obviously progress is being made. The situation is ludicrous. It can be explained only, I think, by postulating that all the analyses that have been made are not merely confused, but are downright wrong. It is an elementary fact of logic that we must accept that if p implies q and q is false, then p is false. Here, I may be accused of inconsistency, but I see no way of avoiding it.

Perhaps the appropriate response to all such discussion is to say that philosophy is a useless subject except for those participating in it. This does seem, indeed, to be the general response of the scientific world, a response with which I disagree. It is a rare curriculum in any branch of science which demands that its students study some philosophy. I have not seen a discussion by a philosopher of this almost universally occurring appearance [may I say 'fact?']. What happens in science [restricting the term somewhat] is that students are taught how to observe (not an easy task, always, as, for instance, the observation of the stages of meiosis), and how to organize observations into a coherent system of related observations. They are also taught in the elemental science of physics, and to a lesser extent in other physical sciences what theories have been developed.

I feel that the field of philosophy of science has not addressed itself to the fundamental question: Why is philosophy of science an activity which most scientists support only in the abstract? I recall a reported conversation between X, a scientist, and Y, a famous philosopher of science, which went somewhat as follows: X said to Y "When I have a mathematical question, I go to a mathematics colleague, but when I have a philosophical problem, I do not go to a philosophical colleague". At this, Y became rather upset. But I believe this is what happens.

I close this section with one comment. It would be fascinating to accomplish the miracle of time and language of having Kant alive at the present time. He would find that his total belief in the correctness of Newtonian mechanics placed him alone and in conflict with experience. He would find a considerable degree of disbelief with regard to his basic ideas. He could be told, I think, that there really are no analytically true propositions which are other than linguistic and empty, as, for example, the proposition 'X is the mother of her son'. He would be challenged to produce one true synthetic a priori proposition. I find in looking over

college curricula that his name usually appears in the offerings of philosophy departments. I wonder how many expositions in books and by
teachers of philosophy take cognizance of the very deep questioning and
rejection of Kant's exposition, and teach their students so. I am alarmed
at the amount of brain-washing that is current.

4. STATISTICS AS APPLIED PHILOSOPHY OF SCIENCE

In contrast to previous statements, it is a demonstrable fact that working
scientists go to statisticians for aid in analyzing their problems and their
data, and compel their students to take courses in statistics. For some
years in the 1960's, according to *Science Citation Index*, the book most
widely referenced in scientific literature was Snedecor's 'Statistical Methods'. I have spent perhaps 10 years of my whole working life talking to
scientists about what data should be collected, how data should be collected, what analyses of data are 'reasonable', and what conclusions may
be drawn. I have colleagues who spend three-quarters of their lives engaged in such activities. The clientele of these colleagues return to them
again and again for discussion and aid. Statisticians are used widely in all
noisy sciences.

It may be questioned whether the activities of statisticians should be
classified as philosophical. I believe so, but, obviously, I may be wrong.
So I shall try to give support for the proposition.

As background, I shall give my impression of the class of books on
philosophy of knowledge. They start with general discussion, and then
give basic Aristotelian logic, and often a discussion of causation like that
of J. S. Mill. After one or two hundred pages they get into induction and
probability. The well-informed book will give a theoretical discussion of
various ideas of probability, probability as relative frequency, logical
probability, probability as degree of belief, and then ends. But one rarely
finds any application of the ideas except to the classical problems of the
induction literature such as the sex ratio.

In contrast to this, on the basis of probability as relative frequency
there were extensive developments in statistics in the period beginning
from the late 19th century, including estimation, tests of significance,
tests of hypotheses, the theory and application of sampling, the theory
and application of the design of investigations. During the same period

the ideas of likelihood were formulated by R. A. Fisher and have been worked on extensively. But it is the rare book on philosophy of science that took account of this development. Keynes (1921) presented his axiomatic development of logical probability but then wrote a presentation of statistical methods appropriate perhaps to the year 1900. Ramsey (1926) wrote on the foundations of probability but when it came to discussing how the ideas could be used said 'For all this, see Fisher'. In more recent years Braithwaite (1953) took cognizance to a moderate extent of some of the developments in statistics and Hacking (1965) of the idea of Fisher's likelihood. Bunge (1967) gives considerably more than mere recognition to the stream of work. On the other hand the recognition by others who are regarded as leaders in philosophy of science is either miniscule or zero. Even in the cases in which there is recognition, I see essentially no recognition of the design of experiments as formulated by R. A. Fisher with the notions of randomization, replication, and blocking.

It is extremely interesting to me that one of the earliest modern efforts in statistical ideas was the development of the χ^2 goodness of fit test, in 1900 by K. Pearson and that this is a procedure directed to the question of whether data support a statistical model. For instance: Is a ratio of 10 successes to 5 failures consonant with a probability of $1/2$ of success? This seems to me to be strictly a matter of confirmation of a model, but a moderate search of the modern philosophical confirmation shows no recognition of the existence of the Pearsonian idea and of the many developments on the same lines over the ensuing 70 years.

That the main thrust of statistical methods is towards the ever-present questions of the philosophy of science and even the philosophy of knowledge is evidenced, I think, by the fact that statistical thinking of the 20th century gives suggestions on the answering of questions such as the following:

What is a random sample?

Do the data support such and such a model?

What values of a parameter are reasonable in the light of such-and-such data?

What are useful ways of analyzing data?

How does one make a judgment of a probability?

The language in which these questions are addressed is not the language of the philosopher, though, of course, the suggestions are based on Aristotelian logic and mathematics. I wish to insist that these are questions of applied philosophy of knowledge.

It is, perhaps, worth interjecting the view that in almost all disciplines application precedes theory – or if one wishes, theory grows out of successful application. And, I think it should also be stated that validation of a theory comes *always* from application of that theory, primarily by way of making predictions and observing that actual observations agree in some respects with predictions. It seems reasonable to require then that a theory of philosophy of knowledge should be validated likewise. But any validation would require application and I have not found any philosophical book that suggests how the theory may be applied and gives examples of application.

5. A COMMENT ABOUT PHILOSOPHY OF SCIENCE

The great bulk of the philosophy of science has been dominated by physical science and even in that context with physical science prior to 1900. I wish to express the view that physics and thereby philosophy of science missed essential parts of science and was dominated by a hypothesized sub-class of situations. The class of situations may be characterized by the following:

    (i) the measurements were simple;

    (ii) the process of measurement could be regarded as perfect;

    (iii) the variables studied could be envisaged by a not-too-great a stretch of the imagination as satisfying perfectly rather simple mathematical relations such as

$$PV = \text{constant}$$
$$S = kt^2 .$$

There were some minor problems to be sure. Measurements were not perfect, so a theory of errors was developed, a theory which has, it appears, been discarded in this century. [I refer to 'proofs' of the normal law of error]. But there was rarely if ever any discussion of what errors of measurement might really mean and the mode of accommodating them was to compute a probable error. A result was then quoted as say 2.71 ± 0.03 with no discussion of what the meaning of this result was intended to be. And, unfortunately, this still seems to be the case in the general area of physical measurement.

A second characteristic of physical science was that it was possible to prepare specimens that were so nearly identical as regards what could be measured that they could be regarded as being identical and that physical specimens did not change over time. What little variation that did occur was 'swept under the rug' by means of a probable error calculation. So when there was a matter of determining the effect of a stimulus such as an increase in temperature there was no difficulty of the sort that occurs in determining the effect of a stimulus on a specimen, such as a human subject, which would not remain constant over an experimental period. Examples of this sort permeate biological and psychological investigation. A simple example that I use to get across the problem to a student who has had no experience in anything organismal is to ask how he would make an experimental test of the comparative merits of two text-books for a semester course, and to make matters easy, let the area be basic freshman calculus, or to take something even easier, a book on elementary mathematical logic.

I have not found in a rather strong effort any mention of a problem of this sort in a book or the philosophy of science. I have not found in any philosophical book on the logic of discovery or confirmation any suggestions on how one might address such a problem. But this is surely a practical problem of knowledge, and one which almost anyone is aware of. I wonder if a problem of this sort would be called a non-scientific problem.

The problems of elementary physics deal, it seems, with situations in which there is not in a real sense an interaction of the object being treated and the stimulus to which the object is subjected, in the sense that the object can be returned to its original status by a reverse stimulus. This is not to say that there are no such things in elementary physics like irreversible reactions, but a casual introspection suggests that reversible stimuli are the usual occurrence.

So I state the view that an attempt to understand the nature of science by means of classical physics and classical chemistry will fail because it is impossible (nearly always) to achieve two features: (i) the existence of two objects which are so nearly identical that they can be regarded as identical, and (ii) the existence of a reverse stimulus that can, colloquially speaking, undo the effect of a stimulus.

It is a fact that R. A. Fisher (1935) developed a procedure by which progress towards knowledge and towards confirmation of hypotheses could be made, and this is described in his 1935 book *The Design of Experiments*. It is the approach of this book which dominates the processes of accumulation of knowledge, and the confirmation and rejection of hypotheses in what may be termed noisy experimental sciences. I hazard the view that by failing to take cognizance of this area of the knowledge endeavor philosophy of science has failed. I also hazard the guess that just as the methodology of parametric statistical models is now creeping into the literature of philosophy of science, some 40 years after it had become commonplace in the actual practice of accumulating knowledge, so in the next decade or two will the problems addressed in the design and analysis of experimental investigations be recognized in the philosophy of science.

6. The general setting

In the months prior to preparing this essay, I was led, rather by happenstance than deliberate choice, to fairly extensive reading of modern existentialist writing. I have formed the view that much of existential thought is highly relevant to the problem of knowledge, a view which I find interesting because I have not met a book on philosophy of knowledge which takes any note of existentialist thinking. A central theme of this philosophy is conveyed by the phrase 'Life is absurd'. Just what it means to those who state it is rather obscure but it makes sense to me if it is interpreted to mean that life is not perfectly predictable. Humanity is exposed all the time to happenings for which there is no rational complete explanation. *Homo sapiens* has been searching for a panacea or a prescription by which all of life can be justified, explained, and predicted. The problem of justification as indicated by the question "What is the meaning of life?", I shall not discuss, except to say that in my opinion the first matter is to decide on the nature of the question and what sorts of statements may be considered as answers, an aspect which is not in my experience dealt with convincingly.

It is when we turn to 'explanation', that what may be termed absurdity appears. Explanation is always explanation in terms of concepts that are regarded in the discourse as being unnecessary of explanation, as every dolt knows. We can always push the explainer to the wall and eventually he will have to say something like "I don't know" or "That is the way things are," or "That is the way God made things". The existentialists seem to me to be saying just this: Existence is; existence precedes essence; the essence of anything is the totality of its appearances. I think that this outlook contains a valuable message for all the philosophy of knowledge, and philosophy of science, a message which the knowledge philosophers seem to have ignored. Interestingly enough, Sartre wrote a mindcracking philosophical work trying to explain or rationalize the business, – i.e., the absurdity, an effort which on the face of the axiom seems foredoomed to failure. It is interesting to me, also, that for every writer who develops a vast edifice of supposedly rationalist thinking, there appears a horde of apologists, by which I mean, workers who spend their lives putting the best face on the exposition and being most careful not to discuss the huge cracks in the edifice. In Sartre's case, two writers whom I regard almost totally as apologists are Simone de Beauvoir and Hazel Barnes. In Camus' case, the apologist appears to be Germaine Brée. It is only by chance almost that I stumbled onto Alfred Stern (1953) on Sartre, and Conor Cruise O'Brien (1970) on Camus. Do professors in these topics present the student with the seriously made destructive criticisms? The student reads the edifices and wonders if he is a dolt; he needs open presentation, and not brain-washing. On Heidegger, Stern (1953) says:

Heidegger surrounds his ideas with "that sort of luminous intellectual fog" which according to Mark Twain, "stands for clearness among the Germans".

Camus wrote in *The Myth of Sisyphus*, "I want everything or nothing". Stern asks what is new about this and mentions Pastor Brand of Ibsen. Stern also says:

In my opinion Kirkegaard was the prototype of a philosophical absolutist and of a frustrated rationalist.

If we recognize that we cannot have a world deducible with logical necessity, a world of absolute certainty, we have to accept the fact wisely and not to adopt the attitude of a child crying because he cannot have the moon.

But, as the critics of existentialism indicate clearly, there are some 'truths' in what the existentialists say or imply. A 'truth' for me is that there cannot be a complete explanation of life and hence of science. My impression is that they reach this truth by completely inadequate reasoning, usually deeply emotional and non-rational. Another truth for me is that the Platonic essences which have dominated rationalist thought for millennia simply do not exist. That knowledge of the real world can be built up by a pure thought (whatever that is) is a stupid and dangerous Platonic residue. The best example of this attempt I have seen is the attempt to derive the whole of cosmology and particularly of the number 137 by Eddington, a physicist-astronomer of renown, and justifiably so. It is interesting to read Gamow (1961) on this topic. It is also interesting to read Stebbing (1937) on philosophy and the physicists.

Another example of importance in the area of probability and statistics is the axiomatization of Savage. Another 'truth' for me is that rationalism can achieve only some state of partial coherence in its explanation of the real world, a partial coherence which I cannot describe and which I judge to be the task of the philosophy of knowledge and science to elucidate.

It is interesting, and I believe totally relevant, that philosophers of knowledge have not contributed to what humanity describes as its knowledge. These are 'hard' words and I do not state them lightly. Furthermore, I challenge anyone to document any case of deep invention or discovery in science which can be traced to the treatises on philosophy of knowledge. Indeed, the converse seems to be the case, that scientists uncover implicit assumptions which a philosophy of knowledge should have pinpointed, e.g. the assumptions of Newton mechanics. This should not be the case. In the area of probability and statistics as applied to science, it must be agreed surely that an exemplar case was that of Mendel and genetics, one which makes one raise one's eyebrows at anyone, no matter how philosophically well-trained and well-educated, who pursues the question of determinism.

I close this section with the affirmation, then, that modern existentialism has ideas of deep relevance to the theory of knowledge, even though we may well reject totally some of its dominant themes, such as the role of authenticity.

7. OBSERVATION

The beginning of the whole business, at least with regard to human science, is observation, which is ultimately sensory perception. But even here words are misleading and carry their loads of conventionalism. You will note that I said human science. Why did I insert 'human?' Because science consists, partly, at least, of observation, organization of observations into a structure, and a choice of action after observation. And then we must recognize that this is not an attribute only of the human species. Most biological species have ability to recognize dangerous situations and to make avoidance moves. Should we say that lower biological organisms have scientific ability? I do not see why not. The knowing of humans is much more sophisticated. A programmed computer knows matters: it can check it all and find out when mistakes have been made. The whole approach to these questions seems to me to have been dominated by some theological views of *Homo sapiens* as being not merely unique in mental abilities because it is so far above other species, but a unique creation of a divine hand endowed with qualities that are different in kind from all other life.

So what is observation? In the simplest cases, it seems to be no more than photographing. The analogy is not bad. A photograph picks up only certain details. What it sees is determined by the seeing or filtering aspects and what it records by the recording apparatus. In the same way, what a human observes is determined strongly by the seeing and the filtering aspects and what is recorded by the recording apparatus, features that have evolved presumably by natural selection. The processes are subject to all sorts of biases and errors. There is not a perfect reliability of observation. Repetitions that should agree perfectly do not agree. It is only with the simplest type of presence-or-absence observation in science that 'error' of observation can be ignored. What do the philosophers of science say about observation? I suggest that you, the reader, examine the works and form your own conclusion. When one turns to measurement of a supposed continuous variable, the matter becomes very obscure. If we turn to the physicists as exemplars of scientific method what do we find? A century ago, it was, it seems, a favorite activity of some rationalist minds to *prove* a law of error, the normal or Gaussian law. Now we recognize that those proofs were utterly fallacious, we do not bother to teach them. This whole problem has not been faced in the philosophy of science. In many books I see no recognition of error of measurement at all. In others, I see in books on philosophy of science statements such as that a measurement was 2.71 ± 0.03. What is the status of such a result computed from data? Is it merely a summarization of a number of separate observations? Is it appropriate to regard modern statistical methods as giving a validation or justification of the procedure? When we turn to statistics we see that the field is currently engaged in a deep controversy about this 'simple' matter. I have asked physicists what they mean by such a statement and have failed totally to receive a statement that has content *for me*. Is the matter important? Perhaps not to a practicing physicist who follows a procedure and obtains an intuitive feeling for it. But I insist that it is a problem of the philosophy of science and I was very pleased to see that Stebbing (1931) many years ago referred to the sloppiness of analysis in the area. This is surely a problem which we may hope that philosophers would address themselves to, because if we do not understand our observations, we have nothing. It is obvious that observation in the context of human science is a personal (or subjective) activity. A lot of philosophy seems to be based on the false axiom. I see: therefore I have a fact – or a proposition, or I see, therefore I know. Take, for instance, the first sentence of Keynes's (1921) book on probability, "Part of our knowledge we obtain direct". I read this and I gag instantly. What is being said here?

It seems rather clear in the elementary context, at least, that observation is based on a belief of the persistence over at least a short period of time of an entity that is being observed: an assumption, it seems, that bothered some pre-Socratics more than 2000 years ago. The assumption seems reasonable, of course, if one is observing the gross characteristics of a physical object, such as a golf ball, but not one applicable to entities that are changing rapidly. In fact, whether a set of observations repeated over time can be regarded as repeated observations on a fixed object is a deep question which modern statistical theory attacks, though whether successfully on philosophic criteria I do not say. Related, of course, are the questions of space and time, and on these I have found philosophical writings in general at best deeply obscure. I make exception in the cases of Borel (1922) and Reichenbach (1927). It is also of interest to note that the theory of relativity came, it seems, out of precisely a consideration of space and time. There are, however, still great obscurities, it seems, evidenced by the clock paradox. We may be sure. I think, that the situation has not been resolved adequately in the language and thought of our times.

What is the status of measurement in more difficult cases? It seems completely clear that observation in general is an interaction of an observer and an object being observed. It is obvious that there may be considerable and complex interaction. My colleague, C. P. Cox, reminds me that even a juvenile bird-watcher is aware of the interaction. So should we be surprised really by the Heisenberg Uncertainty Principle, as relating to observation? Do we really have a good theory for this interaction? And when we turn to areas such as animal and human behavior what happens? What is the nature of observation in psychiatry? How much bona-fide observation did Freud make, and to what extent was he finding what his inner psyche wished to find or what his observation techniques would necessarily bring out? But, even in stating this view, I do not wish to denigrate Freud. He was dealing, surely, with cases that exceed by far in complexity any that physical science has tackled.

Instead, as a statistician, let me point the finger of criticism at the basic presentations of theoretical statistics with which our students are brainwashed. It is supposed in these presentations that *it is known* that observations are distributed independently around a true value according to a Gaussian distribution with zero mean and certain variance. This assumption leads to very interesting mathematics, some nice well-formed formulae, if you like. But when does anyone *know* that the assumption is correct? Never, I assert. In advanced work, we follow this by the assumption that the distribution of errors is symmetrical. The question and answer remain the same.

An example of observation misinformation and misuse occurs in so-called theoretical statistics, in which it is assumed that a continuous random variable can be measured exactly. Some of you may think I am talking hot air – that is your prerogative. But I state my opinion that if anyone here can observe a continuous random variable without error I congratulate him. I wonder how he records his observation. I certainly do not have time to listen to him to hear what it was. Observations are, I assert (and not merely believe), grouped by some peculiar process that I do not know. So anyone who does statistical manipulations in mathematics with the assumption that continuous random variables are perfectly observed is, I believe, in a quite unreal world. You may ask: Does it really matter? I answer that if one performs certain operations, e.g. averages, it matters little. But, turning to a technical matter of statistics, if one wishes to test goodness of fit by taking logarithms of the sample spacings, then the mathematics goes to pot. Like the quantum field theory people, no one can handle infinities. In this connection I mention also the matter of fiducial inference. Fisher stated that it was impossible without actual observation of continuous random variables. I believe (perhaps erroneously) that I showed, Kempthorne (1966), that a sequence of increasingly fine measurements, in terms of a diminishing grid, at *no* stage enables fiduciating, that is, the making an exact fiducial inversion. Fisher, who was a very great man, a genius in my opinion, never faced this elementary fact. And then what about our friends from Toronto and Waterloo with structural inference? If one takes the stance that data, even on a presumed underlying continuous variable, are grouped, then I surmise that the group structure and the right and left Haar measures, etc., etc., go out the window. I believe these workers have not faced an elemental philosophical aspect of the process of observation. At a more elementary level, what is the likelihood function? I believe it is

$$\frac{\text{Prob}\,(\text{Data};\,\theta)}{\underset{\theta}{\text{Sup Prob}}\,(\text{Data};\,\theta)}.$$

in which the divisor is inserted to give uniqueness. Somewhere along the line, the likelihood function for observations on a continuous random variable was written down as

$$\prod_i f\,(x_i;\,\theta)$$

where $f(x_i;\,\theta)$ is the probability density. I believe this is an error at an obvious (but philosophical) level.

But let me not merely knock my mathematical statistical friends. Who, after all, should have thought out the observation business? The philosophers, I believe. And have they? I think not, but the philosophical members of this conference may say: You have not read $X\,Y\,Z$! To which I reply that I have done a fair amount of reading of the supposedly leading individuals and I find *nothing*.

It is impossible, I believe, to make sense of the process of observation without a concept of repetition and without a historically obtained distribution of 'error of observation'. And hence without a concept of probability, a concept of probability that is based on relative frequency. But my neo-Bayesian friends will ask "What population of repetitions? You are making a subjective choice". And I will reply "Of course, the population out of history that I have chosen is a subjective one, done by me on the basis of imperfect analogy, but checked as far as I can by the use of ideas of statistics". The outright Bayesian states something like the following: I am just working on beliefs. But when questioned about the origin of the beliefs, he gives a quasi-existential answer of the form "Your beliefs are!" The whole process reduces then to a formal manipulation of mathematical variables that are supposed to represent beliefs, without, it appears, any questioning of whether beliefs are justified.

The moral I draw from this section, as indeed from all the previous ones, is that the philosophers need the ideas of statistics and the statisticians need deep philosophical analysis of their ideas and procedures. We need philosophers who address themselves to the broad purposes of humanity just as we need statisticians with the same overall aim. There is no limit to the number and variety of intellectual problems that can be invented, and they are often beautiful. But at this juncture of the history of humanity purely aesthetic desires seem somewhat immoral.

8. THE ORGANIZATION OF OBSERVATIONS INTO A SCHEME

The process here is again rather obvious, when it is not obfuscated by the pursuers of Platonic essences. One tries to make sense of the observations, to reduce observations to general relationships, which are, unfortunately, called laws. Now a law is a statement of what must happen, or in societal affairs, a prohibition. But there is no law in science of this nature. The words of humanity have been chosen very badly. The reasons are, I suppose, rather obvious. That the world was created by a Divine creator, as a huge clockwork mechanism which ran according to definite rules incorporated into the clockwork by God. The depths of humanity's attachment to such a concept have been fabulous, and humanity is still carrying the burden. At the end I give several quotations from Feynman, with a few others. I do not want to single out these individuals but they happened to be on hand, because I became re-interested in modern physics after some decades.

There is, I maintain, a huge difference between two propositions:
  (i) Nature unfolds according to a set of laws which are hidden from us and discoverable, and
  (ii) We construct a model of Nature which must by the nature of the

concept of model consist of regularities which we discover by our choice of observation and analysis of observation.

In the second proposition, the so-called laws of science are generated by human mental activity. They are not, like gold, lying hidden under the soil waiting there for us to discover. We can readily imagine, I think, another Earth with a physical history the same that of our own Earth in which a different language, not translatable into any of ours, and different laws have been developed to describe regularities of observations there.

Consider the laws which we are given in basic physics and, in particular, Newton's law: $F = ma$. What was m in this equation? We were told that it was 'the quantity of matter'. It is only after some introduction to relativity theory that we find that the law and the answer are both wrong. I would ask those who have studied much to read the basic books on physics with a deeply questioning mind. My own experience was to realize that a very remarkable story with some aspects of corroboration had been built, but that it was necessary to accept the mode of presentation rather blindly or one would not make headway.

The fact is, of course, that the building up of a theory such as that of physics or chemistry is a boot-strap operation. Why is there an absence in the philosophy of knowledge and the philosophy of science of recognition of and accommodation to the boot-strap process? Why is there absence of recognition of the obviosity of a process of model building that never leads to a final model, in spite of what some leading scientists say, perhaps in careless moments. I give illustrations later.

What really seems to underlie the whole matter is a process of learning, which does not come about by building complicated well-formed formulae from simple ones by the processes of classical logic.

A persistent theme in this essay is that philosophy of science has been dominated by physical science and particularly elementary physics. Why is it that one finds biologists so rarely represented in conferences on philosophy of science? I found it of interest to look over indexes of authors of several books on the philosophy of knowledge and science in search of names of leading biological scientists. The result with the exception of Bunge (1967) was appalling. In many treatises, I found not a single biologist referred to. There were the standard names of Galileo, Kepler, Newton, Einstein, with a few others. I use a principle of induction to lead me to the view that those philosophers I examined consider only a fraction of science, and a special one at that. You, the reader, perform your own examination, and then form your own inference. Perhaps I am missing my target. Perhaps what is called biological science is not science. Interestingly to me, a very early great biologist was Aristotle.

It seems to be a deep-seated human attribute to believe that the model which is currently supported is the final truth. Kant certainly so accepted Newtonian mechanics, and there appear to be physicists who think that Schrödinger's equation is the final truth (see later). But for most of us, we use the induction that no law has withstood challenge indefinitely, and we should guess that at best Schrödinger's equation will be found to be an approximation to a better law. At worst, of course, it may be discarded. Our real problem lies in the fact that we must steer a path between two poles:

(i) that our best model is perfect, and

(ii) that our best model is merely a boot-strap model and therefore not true and therefore not worth considering.

Each pole is to be avoided and not merely by asserting (i) as some physicists do, or by asserting (ii) as the verbal, poetic, literary and café philosophers do.

I was interested in Stern's characterizing Sartre as an absolutist. It seems to me that he is not alone. I regard some of our works in the foundations of statistics as absolutist. You, the reader, are to accept a few rather simple and intuitively appealing axioms, and it is then shown that you *must* behave in such and such a way. You must behave so as to maximize expected utility, though no one has succeeded in making the notion of utility operational. I have referred above to the recent questioning of the sure-thing principle, by Howard (1971), a principle that earlier I had thought to be unassailable. I am of the opinion that accepting a few seemingly innocuous assumptions is a dangerous action, and that anyone who considers that the whole process of learning from experience can be encompassed by a few axioms and their logical consequences should be treated with ridicule. This does not happen because we are all searching for a panacea, including myself who is completely convinced that there is none.

9. DISCOVERY

The term 'discovery' has been used widely in philosophy of science, e.g., the use of the phrase 'The Logic of Scientific Discovery' as a book title by Popper (1959). I find myself quite ill at ease in the whole discourse. I have no doubt of there being facts of discovery but I think it is a mistake to base one's ideas on classical physics in a discussion, and I suggest that a consideration of the growth of what we call biological knowledge or biological science, if these phrases may be permitted, is much more informative.

It has been said that there is no observation without a hypothesis. If by this one means that one uses the hypothesis that there are real objects which can be observed and that our observational process means something, I cannot object. The first part, it seems, is necessary because otherwise what is called observation would be a complete waste of effort. The second part that our observations mean something is a matter for investigation. We need only think if the making and interpretation of Rorschach tests to visualize part of what is happening. I am inclined to to view that one can just observe and then attempt to see a pattern in what one observes.

It is interesting to glance over the development of the picture of a plant or animal cell over the past. There is, I believe, no useful logical theory behind the process that can be formalized. The process was only one of looking at cells with increasingly better equipment and gradually the picture of the 1970's was evolved. There is no question of the picture being final. It is sure that new methods of observation will be developed of the sort: Subject tissue to operations O, place in instrument I, project light of frequency f, etc., etc. There is no need for a formal logic of the discovery of mitochondria and their role in growth and development. There seems to be little more from a logical or philosophical viewpoint than that one individual states: If you follow such and such a protocol of preparation and observation, you will see such and such. There is, of course, a huge amount of what we call originality in the development of observation processes, and there seems to be little possibility of formulating any logic of this originality. Clearly, the originality consists of having deep knowledge of what has been observed in the past and of the 'laws' or generalizations obtained from past observations, and of the opening of the mind to possibilities that have not been thought of before. Perhaps the biggest 'discovery' of the past 30 years was that of the DNA model by Watson and Crick. The sequence of ideas in this is certainly fairly clear at least in hindsight, though to understand that sequence involves for most of us deep study for several years. This case is interesting because there were clearly many workers in the chase, and there were elements of happenstance which led to the discovery by particular individuals. I cannot avoid the view that the philosophical writings on discovery bear essentially no relation to the sequence of ideas in this case.

I have had a moderate amount of contact with the development of genetics before the molecular biology era. The process of discovery of Mendel's laws is described quite well in many basic college biology texts. An attempt to describe the actual process would lead to the sort of presentation given in such texts and it would be foolish and useless to give the recapitulation that is possible in an essay such as the present. But it does seem worth expressing the opinion that a large component in the development was the use of analogy, often of particular analogies that were quite ludicrous in the context of knowledge and belief of the time. In Mendel's case, the analogy was to the results of coin-tossing. Interestingly, the same sort of analogy is used widely to motivate the probabilistic models of elementary modern physics. In the latter case the analogy seems reasonable but in the former case, and at the time of Mendel, the analogy was mind-boggling. I can imagine a treatise by a very broadly educated individual on the use of analogy in science, but I know of none. And whether there can be any logic of thinking by analogy seems to me to be very moot.

A dominant aspect of scientific theorization of the past century, in contrast to previous eras, is the use of probability models in which a deliberate decision is made to take the position that one will accept without further explanation repeatable patterns of irregularity in what is observed. Because of our indoctrination in the notion of cause and effect, we wish to say that the pattern of irregularity is 'due to something', and so we say it is 'due to chance'. The linguistic convention is found useful, clearly. But a more accurate statement is to say that the irregularity is not explainable, simply that *it is*. Saying that something is 'due to chance' is no more, I think, than saying that we do not have any picture or model of causality in the classical sense. So we admit our ignorance but do not wish to say so, and, instead, say that what happened is due to chance. I am not referring here to the deliberate use of a validated chance mechanism such as in the drawing of a random sample, or the randomization of a comparative experiment.

It is interesting to speculate how the neo-Bayesian movement can accommodate the development of phenomenona such as Mendel's laws or quantum physics. The notion of science as being a mapping from an observation space to a defined decision space does not seem at all viable. The posterior probability of a model must be zero for any model with zero prior probability. I would like to see a discussion of the whole matter by a neo-Bayesian. Can one make the processes conform to application of the principle of maximizing expected utility? I surmise not.

One important part of the whole process is the formulation of hypotheses or expectations on the basis of analogies. The next question is then one of whether the obtained data are in agreement with the hypothesis, the problem of 'goodness of fit' of the model, not confining the idea to probabilistic models. My reading of the neo-Bayesian literature suggests that the workers simply do not admit that there is a problem of assessing goodness of fit of a model. One aspect of this which has become very popular in the past 2 decades is what is called residual analysis, that is determining whether the residuals from a fitted model are reasonable. We are told that Kepler used precisely this type of analysis and rejected one model because a residual was 'too large'.

This type of data examination has clearly been of great importance in scientific discovery. On the basis of the model or picture of reality of a given time, an observation is simply too aberrant. The scientist then attempts to produce similarly aberrant observations at will, and then modifies his theory, and analogies to accommodate the newly discovered repeatable phenomenon. The interplay and feed-back between modelling and observation has been discussed by many statisticians and there should be no need to add yet one more representation. However, my reading suggests that a process that is accepted as a matter of course by scientists and consulting statisticians is not given a significant place by the philosophers of science. It seems somewhat as though two languages have been developed with no intercommunication at all. This view is part of my reasoning for choosing the title 'Statistics and the Philosophers'.

10. VERIFICATION, CONFIRMATION, AND CORROBORATION

Here matters become very deep. It seems at times as though philosophers make a basic distinction between discovery and verification. I think I have the view that the dichotomy is fallacious in general. There seems to underlie much philosophical writing that a word has an essence, given to it by some deity, and that *logical* analysis will discover this essence. That this is fallacious, I take to be one of the basic messages of Wittgenstein (1958, 1959). The word'discovery' is used with many connotations which cannot be combined or abstracted into one essence. We have the discovery of gold, the discovery of Golgi bodies, the discovery of Mendel's laws, the discovery of the Schrödinger equation, and so on.

In many instances, discovery is the finding of a real object for which search was being done. There is no question of verification. Of course, it may be thought that one has discovered gold but only has discovered pyrites – but in that case one has not discovered gold. In the case of discovering Golgi bodies, the observer sees a regularity in pictures of different entities of a kind, and he has not discovered them without a verification. Discovery and verification are, it seems, inseparable. In the case of Mendel's laws, a statistical regularity was observed by data examination and that regularity was confirmed by obtaining new data. Here the discovery was the recognition of a pattern of regularity in existent data, and the confirmation was achieved by the use of the Pearson χ^2 goodness of fit statistical test, which is explained at length in many, many statistical books and is described in many basic biology books. It seems that workers in this area were not at all bothered by paradoxes of confirmation which have occupied a significant portion of philosophical literature of the past decade. I suggest to the philosophers that there is something strange for them to explain – that confirmation was going on and was somewhat effective decades or even centuries before the subject became so deep and unresolved. The discovery of the Schrödinger equation is obviously a different type of discovery, and it is not in my area of competence and knowledge to understand it. A rational view is that all theories have been corroborated but not that they have been verified. Indeed, it seems that the term verification should be avoided. I think one obtains simplicity in terms of a concept of discovery and a concept of verification only by perverting the words and the facts, and losing the essence of science.

On the matter of falsifiability, it would seem that much of what is being said in the philosophical literature was given decades ago in statistics under the simple assertion that one cannot prove a null hypothesis, and one can only disprove one. The route has been the use of statistical tests of one sort or another. There has entered also from statistics the concept of there being observation schemes of differing sensitivity. Is one, qua philosopher, to take the view that all this type of methodology is not worth considering as it is not being well-grounded philosophically?

One deep problem is that the literature on theoretical statistics and on philosophy seems to take a most untenable stand with respect to probabilistic models. Most of the time it assumes that data arise from a known class of parametric models, and the task is then to form a judgment about the values for the parameters. Nothing, it seems, is farther from the facts. The classes of models are obtained by data inspection and the use of analogies. They are never known. Indeed, it is almost always the case that the class of models is itself an intellectual construct. So from the viewpoint of the theory of knowledge of the real world, any statistical process which takes as its initial point that one knows the distribution function apart from the values of some parameters, is fallacious. It is downright wrong. Deep philosophical discussions as to what one should do with such cases, is not completely forcing in the building of knowledge of the real world, simply because it ignores the foundation.

The only case that I know of for which one knows the model is that of sampling at random, i.e., with a validated randomizer from a finite population, which in the case of a dichotomous observation leads to the hypergeometric distribution. Whether at a deep philosophical level one can say that one *knows* the proper model even in this case is probably moot, though the world of investigation ignores the philosophical difficulties, and, I think, wisely so.

A summit of the corroboration problem is the paradox of the ravens. A deep discussion is beyond my ability, but I assert, rather dogmatically, that the observation of a white shoe does *not* lend support to the proposition that all ravens are black. There has to be a deep error in logical analysis to reach the opposite conclusion. Good (1960, 1961) has written on this from a neo-Bayesian viewpoint and reaches the two possible answers from two population models and sets of assumptions. It seems that if we follow Good's 1960 argument, and if the number of black non-ravens is greater than the number of ravens, then observing a black non-raven lends support to the hypothesis that some ravens are non-black.

The matter is of tremendous importance for the processes of science and technology. Consider two regimes for treating humans, one being usual diet $(\sim R)$ and the other, usual diet plus substance $X(R)$. Let the observation be classified as no increase in weight (B) or increase in weight $(\sim B)$. Then we have the 2×2 table:

	Increase	No Increase
Treatment X		
Not treatment X		

The supporters of the idea that observing a non-black non-raven supports the hypothesis that all ravens are black, must consider, I think, that the observation of no increase without treatment X supports the hypothesis that treatment X produces an increase. The conclusion is ludicrous, and I can only express the view that I am glad experimental science has not accepted it. A human being wishes to gain weight. He consults his doctor, who asks him: Have you tried eating sand? The patient says no. So the doctor says "Your not having eaten sand and not having increased supports the hypothesis that eating sand will result in your gaining weight; so why not try eating sand". The argument may sound very appealing in terms of ravens and blackness, but, like all logical arguments, does not depend at all on the situation or the names that are used, and changing the names is informative.

The only response that can be made on the matter is that logical analysis has broken down, not in a minor way, but totally. Even a Bayesian analysis needs, it appears, the inclusion of assumptions that cannot be substantiated, and it does seem to need the notion of random sampling, which can be given substance for most statisticians only in terms of repeated sampling, which is outside neo-Bayesian frame.

My reaction to the problem is that one cannot talk in the abstract about the support for a hypothesis given by an observation. One can consider rules for calculating a statistic, which is an element of some space of elements, which is a function of the observation and the hypothesis and the alternative hypotheses (whether precisely or vaguely defined). One must then obtain the operating characteristics of use of the rule in defined classes of repetition, and choose a rule which has a good (or the best, if this is possible, but I doubt that it is possible) operating characteristic. This has been the outlook of 20th century statistics of Fisher, Neyman-Pearson and all the ensuing workers until the neo-Bayesian argumentation was brought forward. I am perplexed that I see essentially no attention to any of these streams of thought to the matter in philosophical literature.

If one views the matter as a 2-way table

	B	$\sim B$
A	p_{11}	p_{12}
$\sim A$	p_{21}	p_{22}

in which there is an enumerable large population of entities (e.g. e^N, where N is the number of elementary particles in the universe), and the p_{ij} are the relative frequencies of the different possibilities, the hypothesis that is advanced is that p_{12} is 0. Consider a random member from this population. Then it may be thought that an observation in any cell except $(A, \sim B)$ supports the hypothesis that p_{12} equals 0, because $p_{12} = 1 - p_{11} - p_{21} - p_{22}$. But the hypothesis $p_{12} = 0$, is a composite hypothesis. One is interested in the hypothesis $p_{12} = 0$ with regard to the alternative hypothesis that $p_{12}/(p_{11} + p_{12})$ is not zero. It seems then that no observations on $\sim A$ individuals bears on the hypothesis with regard to this alternative. I do not wish to be assertive here, because the statistical theory of testing composite hypotheses is in great difficulties. I am inclined to the view that deep logical analysis will result in some resolution of the difficulties.

11. CAUSALITY, DETERMINISM, AND INDETERMINISM

These matters loom large in most literature on the philosophy of science, but as a practicing consulting statistician I have found the literature almost totally uninformative, and, as I have indicated before, it seems that practicing scientists ignore this literature also. I tend to the view that the bulk of the discussion is a sort of hang-over from the fundamentals of Judeo-Christian theology.

For the practicing scientist, the concept of cause is used, except in literary science writing, only for the case in which one can imagine interfering with the system, e.g. by altering temperature, and insofar as we are concerned with altering our environment or improving our condition, this is the only notion of causation that is of any use. It seems unquestionable that the notion of causation has been used very effectively to generate useful models. It seems clear, also, that notions of probability must be introduced (Suppes, 1970).

On the matter of determinism versus indeterminism, it seems also that there has been a huge literary output which is of little relevance to the human condition. The notion is useful in models for problems involving a small number of physical bodies which interact with each other in a simple way. But the notion seems useless in the large because a computer to store the initial conditions and to pursue all the consequences of a deterministic physical world would have to be huge, bigger, I suspect, than the universe. When one turns to the biological world, the deterministic model falls flat, for the simple reason that judgments and choices must be made on the basis of incomplete information. Without a forcing logic which all organic life uses there cannot be determinism. I ask, "What has all the fuss been about?"

12. SAMPLING INFERENCE

I have expressed the view, and I am not alone, particularly on account of Dr Godambe, who is here, that the basic problem of inference is that of forming an opinion about a finite population. Let me describe the problem for our philosophical and physical friends. I have 20 opaque envelopes. I have 20 small pieces of paper. On each piece of paper I write a number, e.g. 10^{-23}, 2^{137}, 3, 5, e, etc. I do this very carefully so that the number will be hidden. I insert each piece of paper into a separate envelope. I do this out of the room. I walk in. I lay the envelopes out in a line on this desk. I tell you that your task is to form an opinion about the set of numbers in the envelopes with the proviso that you are permitted to examine 5 envelopes. You may react that this is a false problem. If you should do so, I would have to reply: "False or not, it is a problem which humanity faces". It is simply useless for a philosopher or logician to say that the problem is not well-formulated.

It would be most interesting to have educated individuals who have not been exposed to (or should I say brain-washed by?) the teachings of 20th century statistics react to this problem.

It would be most interesting to have one of our most ardent neo-Bayesians, such as D. V. Lindley, who is present, to address the problem.

At one time Lindley informed us that he would draw 5 from the population 'at haphazard'. He thought, obviously, that this was a meaningful statement, but unfortunately to many, very many, statisticians the statement has very obscure content. This alone serves to illustrate a crucial point in the whole business. Language, at worst, is a purely personal matter, but the essence of rationalism which I hope we all support, is to attempt to use common language or to explain in common language what one's special language means. How many have made themselves dizzy trying to understand Heidegger's personal language? See also Stern on Sartre and Sartre's language.

It is interesting that governmental and business organizations all over the world have adopted the procedure of 20th century statistics (apart from the Bayesians), which is to use random sampling, that is, a tested and 'verified' randomizer. I would like philosophical analysis of the whole problem and process.

An attempt to use the ideas of parametric statistics except in an approximative sense, seems doomed, because there are in the problem 20 parameters. A random sample of 5 may well be sufficient in the technical sense, but it seems clear that it is inadequate from a parametric model viewpoint. I do not have a formal definition of inadequacy.

13. EXPERIMENTAL INFERENCE

Fisher, whom many of us regard as being about 1 Gauss in stature, proposed in his book *The Design of Experiments* (an arrogant title to be sure, but Fisher was rather bright, so we should forgive him) the problem of the lady tasting tea. I wonder if any philosopher has examined this piece of writing. I see several philosophers picking up Fisher's ideas on likelihood, but not this. Recall the problem. The lady asserts that she can tell whether the milk or the infusion of tea was placed in the cup first. How does one form a partial opinion about the correctness of her claim? I am reminded of ideas of electrons with inner works, etc. This is, I believe, a beautiful 'Gedanken' or 'thought' experiment and I would like our physicists to address their minds to it. Interestingly enough, I think this problem can be reasonably well formulated as a contest or a game between the lady and humanity.

This is not the time or the place to describe Fisher's analysis of the problem and his proposed solution. Read him, but keep a salt cellar handy because there are some obscurities, I believe, in his writing.

This problem is interesting because a pure scientific approach would, I surmise, involve an examination of the lady's psychophysical process. We may hazard the guess that humanity will have a fairly reasonable model of this process in a few centuries, but by then both the lady and we shall be dead. Of course, you may say "What a stupid problem! Defer it for a few centuries". But this misses the boat entirely. Gaining knowledge is a bootstrap operation, and we who are alive now need ideas to help us in our finite lives. We try to understand electrons at first without considering their internal workings and a partial, probably fallacious understanding of the electron as a permanent single entity lies at the beginning and before any attempt to understand its inner workings.

I believe the problem is a reasonable one and I leave it with you.

Now let us consider a 'simple' earthly problem. Someone asks: "Does a daily supplement of 8 ounces of orange juice improve the growth of children of London, Ontario?" If you deny that this is a reasonably posed problem I must part company with you. I will assume you admit it as meriting consideration. Now I ask: "What would a philosopher do about the problem?" Also I ask: "What would the physicist say?" Perhaps he would try to apply physical science ideas and suggest that we find pairs of identical children to only one of which we give the supplement. I would then say: "That is not a bad idea, except that it is useless. You find me two children who are identical and I will accept your solution. But I defy you to do so!" The notion of complete experimental control which pervades classical physics is simply not operable. And the notions of quantum physics are hardly applicable, I surmise. I shall not discuss a suggested approach (not *the* solution). I leave the problem with the reader. The type of problem is critical to the hopes and aims of humanity and I have yet to see a philosopher address it.

14. PROBABILISTIC INFERENCE

This has been the bane of philosophy, of scientific method and of statistics for centuries. We have seen the ups and downs of Bayesian inference. It was published by Price (and one hopes Bayes forgives him), bought hook-line-and-sinker by Laplace, rejected by Boole and Venn, and others, rejected after a brief initial acceptance by Fisher, and then resuscitated by Jeffreys, Savage and others. If one can learn from history at all, we shall see again its rejection. But hope springs eternal and it will be revived again by young people in another 50 years and so on.

Some views are presented by Kempthorne (1971) and Kempthorne and Folks (1971) that are regarded as useful by some. I have nothing to say in general.

But there is a particular matter which must be raised. Jeffreys and others coined the idea of improper prior distributions. Hacking voiced the views of many of us by asking how you could combine something that was not a probability with a probability and get a probability. It was all rather obvious. However, the use of improper priors has been pursued avidly. There has recently come to my attention a paper which I have not had time to digest by Dawid *et al.* (1973). My impression is that these workers give examples of inconsistency that result from the use of improper priors. Perhaps one should not be surprised because playing with infinities is a tricky business. There is a strong suggestion that the whole improper prior distribution gambit should be thrown out. It is possible that the process can be justified as a matter of approximation, but no such justification has been given, it seems.

15. The Processes of Reasoning in Physics

Along with study of the philosophy of science, which presumably gives indications of how science should be pursued, it is useful to examine how science is actually pursued or at least how the progress is presented. Some months ago, I was led to an examination of presentations of basic physics, and was pointed towards the famous lectures of Feynman (1965). I have been bothered about quantum physics for decades and read the lectures with the hope of understanding physics and understanding the development process. I was interested in trying to decipher the logic of the development that was presented. During my reading, I came across many statements which were hard to understand or to reconcile with any truly logical process. I also came upon statements from other sources which seemed to me to round out the picture. I now give a subset of the statements which struck me very forcibly.

I ask the reader to consider what picture this set of statements gives.

I suggest that the reader tries to make sense of the statements as he proceeds. I shall comment later.

Quotations from Feynman *et al.* (1965)

1–4	We should say right away that you should not try to set up this experiment. This experiment has never been done in just this way. The trouble is that the apparatus would have to be made on an impossibly small scale to show the effects we are interested in. We are doing a 'thought experiment',....
1–7	Here is what we see:
1–9	No one has ever found (or even thought of) a way around the uncertainty principle. So we must assume that it describes a basic characteristic of nature.
1–11	Quantum mechanics maintains its perilous but still correct existence.
2–1	... that would mean that the probability of finding a particle is the same at all points. That means we do not know where it is – it can be anywhere – there is a great uncertainty in its location.
2–6	So we now understand why we do not fall through the floor.
2–7	... – but never mind, we have the *right* way out, now!
2–7	... in a situation where there are many electrons it turns out that they try to keep away from each other.
2–9	The tiniest irregularities are magnified in falling, so that we get complete randomness.
8–6	*We do not really know* what the correct representation is for the world.
8–6	... Does a proton have internal parts?... We don't know.
8–6	The *main problem in the study of the fundamental particles today* is to discover what are the correct representations for the description of nature.
8–6	The question of what *is* a fundamental particle and what *is not* a fundamental particle... is the question of what is the final *representation* going to look like in the ultimate quantum mechanical description of the world.
11–5	Of course, nature knows the quantum mechanics,... – so maybe you will not be so unwilling to take the quantum formula, Eq. (11.13), as the basic truth.
16–13	In principle, Schrödinger's equation is capable of explaining all atomic phenomena except those involving magnetism and relativity.
21–19	I wanted most to give you some appreciation of the wonderful world and the physicist's way of looking at it, which, I believe, is a major part of the true culture of modern times. (There are probably professors of other subjects who would object, but I believe that they are completely wrong.)

With regard to 1–7 above, this refers to the two-hole experiment that has never been done. See also Gamow (1961, p. 263) who says:

Dirac's relativistic wave equation is too complicated to be discussed here, but the reader may rest assured that it is perfectly all right.

I now comment. I suggest that there is a strong element of irrationality or incoherence in the set of statements. We are told that quantum mechanics is correct and that we know the right way. But then "we do not really know". Then we are told that "Nature knows the quantum mechanics" and that Schrödinger's equation can in principle tell us everything. Gamow assures us that Dirac's equation is "perfectly all right". But surely one may wonder. The writing is full of images and is like a work of art rather than of science.

The picture that these quotations present is not consistent. At times, the view is expressed that we know exactly what the correct representation or model is, and then, at another place, that we do not.

I think that in many respects the quotations show the actual nature of scientific thought and progress. One knows nothing, one gets observations, one then finds a 'nice' model which stands up to various tests, and one feels that one has the ultimate answer; one then finds that the model is inadequate, one searches for a better model, and having found one, one then thinks one has the ultimate answer, only to realize later that one does not. And so on.

That this should be the nature of scientific progress is inevitable, I think. The basic question facing the philosophy of science is to construct a rationale for this seemingly illogical process. The process is, I think, rational in the large, though some of the views expressed are clearly not logical, and are often completely inconsistent. The task of developing a logic for science is obviously extremely difficult in the view of this nature. It does not help, however, to distort the process to force it into a coherent logical structure. Parenthetically, we may wonder how the student is to make sense of the whole affair. But the fact is that he does.

ACKNOWLEDGMENTS

I am indebted to B. C. Arnold, C. P. Cox, C. E. Fuchs, R. Fuchs, W. A. Fuller, R. N. Giere, and G. Meeden for critical comments on a draft of this essay. The defects are, of course, my responsibility.

Statistical Laboratory, Iowa State University; *Ames, Iowa*

REFERENCES

Ayer, A. J.: 1956, *The Problem of Knowledge*, Macmillan.
Borel, E.: 1922, *Space and Time*, Reprinted 1960. Dover.

Braithwaite, R. B.: 1953, *Scientific Explanation*, Cambridge Univ. Press.

Bunge, M.: 1967, *Scientific Research II*, Springer-Verlag.

Campbell, N. R.: 1919, *Foundations: of Science The Philosophy of Theory and Experiment*, Reprinted 1957, Dover.

Dawid, A. P., M. Stone and J. V. Zidek: 1973, 'Marginalization Paradoxes in Bayesian and Structural Inference, *J. Roy. Stat. Soc. B.*, **35**, 189–233.

Feynman, R., R. B. Leighton and M. Sands: 1965, *The Feynman Lectures on Physics*, Vol. III, Addison-Wesley.

Fisher, R. A.: 1935 (rev. ed., 1960), *The Design of Experiments*, Oliver and Boyd.

Gamow, G.: 1961, *Biography of Physics*, Harper Torchbooks.

Good, I. J.: 1960, 'The Paradox of Confirmation', *Brit. J. Phil. Sci.* **11**, 145–149; 1961, **12**, 63–64.

Hacking, I.: 1965, *Logic of Statistical Inference*, Cambridge Univ. Press.

Howard, N.: 1971, *Paradoxes of Rationality*, MIT Press.

Kempthorne, Oscar: 1966, 'Some Aspects of Experimental Inference', *J. Am. Stat. Assoc.* **61**, 11–34.

Kempthorne, Oscar: 1971, 'Theories of Inference and Data Analysis', in T. A. Bancroft, (ed.), *Statistical Papers in Honor of G. W. Snedecor*, Iowa State Univ. Press.

Kempthorne, Oscar and J. L. Folks: 1971, *Probability, Statistics and Data Analysis*, Iowa State Univ. Press.

Keynes, J. M.: 1921, *A Treatise on Probability*, reprinted 1962, Harper and Row.

O'Brien, C. C.: 1970, *Albert Camus*, Viking Press.

Popper, K. R.: 1959, *The Logic of Scientific Discovery*, Basic Books.

Ramsey, F. P.: 1926, 'Truth and Probability', in R. B. Braithwaite, (ed.), *The Foundations of Mathematics and Other Logical Essays*, Kegan, Paul.

Reichenbach, H.: 1927, *Space and Time*, reprinted 1958, Dover.

Savage, Leonard, J.: 1954, *The Foundations of Statistics*, Wiley.

Snedecor, G. A. and W. G. Cochran: 1967, *Statistical Methods*, first edition, 1937, Iowa State Univ. Press.

Stebbing, L. S.: 1931, *An Introduction to Modern Logic*, reprinted 1961, Harper Torchbooks.

Stebbing, L. S.: 1937, *Philosophy and the Physicists*, reprinted 1944, Penguin.

Stern, A.: 1953, *Sartre, His Philosophy and Existential Psychoanalysis*, Liberal Arts Press.

Suppes, P. C.: 1970, *A Probabilistic Theory of Causality*, North-Holland, Amsterdam.

Wittgenstein, L.: 1922, *Tractatus logico-philosophicus*, Paul, Trench, Trubner, London.

Wittgenstein, L.: 1958, *The Blue and Brown Books*, reprinted 1965, Harper Torchbooks.

Wittgenstein, L.: 1959, *Philosophical Investigations*, Macmillan.

DISCUSSION

Commentator: Kyburg: Professor Kempthorne's opening reference to existential absurdity is a fitting introduction to a statement of existential dispair; it seems to me that he suffers more from *Angst* than from honest doubts. He takes seriously certain questions and problems raised by professional philosophers, but then throws up his hands because the answers aren't obvious. If the answers were obvious, the questions wouldn't be very good ones.

For example, Professor Kempthorne wants certainty in observation. He looks Professor Godambe in the eye, and says: "I do not really *know* that Professor Godambe is there; I may, after all be dreaming." (Descartes' First Meditation.) Of course, in any reasonable sense of the word, he knows perfectly well that Professor Godambe is before him; but having raised the question in a relatively technical philosophical form, he expresses frustration at the lack of a common sense answer.

Professor Kempthorne goes on to make a point that strikes me as extremely important: that observations occur on a grid, or occur in granular form: we do not ever observe the value of any continuous random quantity, but only that it falls in a region of a size dependent on our techniques of observation. This is an important epistemological point. But it should not be taken as another plaintive skeptical cry; in that form it is pointless, for no grid is coarse enough to yield the deep Cartesian certainty that Professor Kempthorne seems to seek.

Causality, too, comes in for some hard knocks. Like Hume – but not so wholeheartedly – Kempthorne is skeptical of the notion of causation. He finds it particularly absurd that something can be "due to chance". (But for the Greeks there was no problem in regarding certain events as due to chance; they even had a goddess for it: Tyche.) But here again we have a relatively good idea of causation that functions relatively usefully in science: we know the cause of yellow fever; we are seeking the cause or causes of cancer; we say intelligibly and perhaps even truely that whether a given atom undergoes a given energy transition is, on the other hand, a matter of chance. Metaphysical concepts of causality are something else again: they raise different questions, questions that seem to call for technical philosophical answers.

One way in which philosophers attempt to answer certain questions is to make precise and explicit what is implicit in common sense judgements. Professor Kempthorne deplores the wasted effort that goes into producing axiom systems and "strings of well formed formulas". (Somewhat inconsistently he admires Gödel's famous results on arithmetic and the foundations of mathematics – results that were obtained in and are entirely about the world of well formed formulas.) But this is one way (and I think the best way) of making things precise. Were it not for the formalization of mathematical argument, we would not know nearly so much *about* mathematics as we now know, and we might not even know so much mathematics.

Kempthorne makes a plea for tolerance and flexibility. That is not a difficult plea to go along with. But flexibility and tolerance have their limits; a man who wishes to regard four as the sum of two and three would not and should not be tolerated in a scientific laboratory. There is no point to being broadminded, unless we are willing at least to grope toward some grounds for distinguishing between the better and the worse alternatives. To say that a certain thing is a matter of taste is not to say that one shouldn't attempt to cultivate *good* taste, and to develop criteria for taste.

"The Professional philosophy of Knowledge has contributed nothing to the knowledge business." Surely Professor Kempthorne is not lamenting the admitted fact that few philosophers of knowledge have discovered the causes of mysterious diseases or enunciated important physical laws. The only way in which an epistemologist could conceivably contribute to the knowledge business is by instilling the appropriate views toward observation, experiment, inference, and the like, in those whose business is acquiring knowledge. We may not yet know exactly what these appropriate views are. But it is certainly true that these philosophical vies have had and do have a profound if indirect effect on the way in which scientists approach their work. Kempthorne, for example, has obviously been profoundly affected by Hume, Peirce, Descartes, and assorted existentialists, among others. So it is not that philosophy has made no contribution; it is that the contribution is inadequate, so far.

I know of no philosopher who would disagree with this latter point.

Kempthorne: It is quite impossible, and possibly not worthwhile if it were possible, to convey in 10 or 20 pages my views about the *Angst* of present times. However, as a man in the street without deep training in philosophy, with my limited reading, I am distressed by the tower of Babel which one finds in philosophy and particularly in the philosophy of science. A theme which runs through my essay is that the philosophers of science have failed us rather badly. I know of no sustained piece of writing by a philosopher of science which should be strongly recommended to scientists, particularly those who wish to obtain a feel for structure in scientific activity.

Is there a structure and if so, has it been exposited? I think there is, and I also think it has *not* been exposited. I think that statistical methods as exposited by statisticians with biological, psychological, physical, engineering, economics and other orientations have taken up the task *at a working level.* This statistical method activity is a phenomenon of this, the 20th century. I see statistical consultants interacting with scientists over much if not all of the scientific spectrum. These workers discuss the design of studies, methods of analysis, and formulation of conclusions. I see essentially no recognition of this activity in philosophical circles. I see no discussion in philosophical circles of important questions of humanity, such as that of how we build up a body of informed and partially justified beliefs to formulate programs to aid the Angst's of everyday life of humanity. I see discussions of the "infamous ravens" which seem to be merely clever pieces of writing, possessing no force at all for real problems.

I do not throw up my hands because "the answers aren't obvious". I throw up my hands because of two opinions I have which I think to be observational facts: (a) scientists do not use philosophers of science to give them clues about the structure of science, (b) scientists do use statisticians to attempt to bring out structure of their activity. My *Angst* is deepened by the occurrence of wide diversity of opinion on the nature of statistical methods, and subsequently, on the nature of scientific inference, whatever that is. I am encouraged, however, by the existence of a few philosophers, of whom Dr. Kyburg is one, who attempt some integration of statistical methods and philosophy of science.

I do not recall my wanting certainty in observation, because I realize that it is impossible. So the world of facts is not clear. Of course I believe totally that Dr. Godambe is there in front of me. I simply do not believe that I have the mind and imagination to produce the London conference within my mind. However, I find philosophical writings on this sort of thing most obscure and fruitless. We must admit generally that observations may have error: statisticians have realized this for decades and have words and numerical procedures which they believe combat problems.

I do not seek Cartesian certainty but I do seek an acceptable philosophical basis for the uncertainty which I believe totally to exist.

As regards causality, a long essay requiring weeks of effort, would be involved. My point is the simple one that causality cannot be inferred except with the comparative experiment. The different roles of experimentation in the comparative sense and of mere observation, e.g. of planetary motion, seems again to be a matter on which philosophy of science, speaking generally, has taken little cognizance. Has any philosopher of science looked at Fisher's Lady-Tea-Tasting experiment, or on the role of randomization in comparative experiments? It appears that those philosophers of science who have become acquainted with some of statistics have ignored this totally, and regard the Neyman-Pearson theory of accept-reject rules for parametric hypotheses as the essence of statistics. The empirical fact is that this theory is only a part of statistical practice, and perhaps only a small part in the scientific, as opposed to the technological, sphere. I think I see in scientific writing a wide use of notions of causality that cannot be sustained.

As regards something being "due to chance," I stay with my original criticism. To say that an atom disintegrates because of (*due* to) chance is perhaps a useful colloquialism. The fact is that we have been unable to find a structure in such disintegration except a *descriptive* one, that such disintegrations follow a Poisson process, say. I believe that to say such disintegration is 'due to chance' is pure 'hocus-pocus' at a philosophical level. This is not to deny that considerable progress has been made in science by assuming the existence of an all-powerful God tossing a coin, for example, to decide whether the product of the mating which produced me would be male or female, or whether one of the tires on my car will burst on my next highway trip. So I would like to see a technical philosophical discussion of the nature of *causation by chance*.

As regards well-formed formulae, I admit a profound lack of understanding, not of what a well-formed formula is, though even that is obscure. How is it that well-formed formulae with apparently no real-world content are able to help us in building a good picture of the real world, as obviously they do? What is the basis for establishing a correspondence, thought to be perfect, between the elements of a well-formed formula, and aspects of the real perceptible world? And if the correspondence is not perfect, how does one, philosophically, make peace with an imperfect correspondence? The punchline in all such questions is possibly that the building of a model for the real world is a bootstrap operation of very deep complexity. I wonder if the 'bootstrap' nature is recognized in *any* philosophical writings. I wonder if, with all the discussion of languages and meta-languages, peace has been made with the fact that we have to work with the one and only one incomplete and growing language (the union of existent languages) that we have.

I do not lament that philosophers have not found the cause of cancer, for example. I lament that philosophy of science, as developed over the centuries, seems to have no contribution to the formulation of a structure of investigation which will perhaps lead to a cure for cancer.

My overall Angst is, then, that I feel that philosophy of science could be, but fails to be, a contributor to scientific processes.

It is not inappropriate, perhaps, for such a view to be stated at a conference on philosophy of science.

HENRY E. KYBURG

STATISTICAL KNOWLEDGE AND STATISTICAL INFERENCE

I

In order to set the stage for what follows, I should like to explain, if I can, why philosophers sometimes worry about the sun's rising rather than about techniques for randomized sampling. Ordinary people ordinarily take themselves to know some things, to have reasoned opinions about others, and to be in ignorance of yet others. They take other people to know some things, to have reasoned opinions about others, and to be in ignorance of yet others. But they *also* take other people to be immensely pig-headed about some things, and to be irrationally opinionated about others. You recall Shaw's conjugation: *I* am firm, *You* are stubborn, *he* is pig-headed.

Now it is commonly held that some of the things we know are things we have seen with our own eyes. There is no disputing about them – at least not often. Some of the things we know are logical and mathematical truths. It is possible to dispute about them, but there exists a framework, within which the dispute can be conducted, that renders the dispute constructive and useful rather than destructive. Some of the things we know are facts about the world that go beyond the data of observation. Some of the things about which we have justified opinions are things that go beyond the data of observation, and some merely go beyond the data of past observation. About both these kinds of things it is possible to dispute. In neither case is it clear what the framework should be for resolving the dispute.

Why do philosophers worry about the sun's rising tomorrow? Partly, of course, just because it is annoying – the desire to irritate other people (for their own good, of course) has been an unlovable characteristic of philosophers since Socrates. But partly also because it is such a clear case of a bit of knowledge we all have. With luck, if you can explain *how it is* that you know that the sun will rise, you will gain some insight into how you know things in general; and armed with this insight you may be able

Harper and Hooker (eds.), Foundations of Probability Theory, Statistical Inference, and Statistical Theories of Science, Vol. II, 315–352. *All Rights Reserved. Copyright* © 1976 *by D. Reidel Publishing Company, Dordrecht-Holland.*

to overcome Shaw's conjugation. That is, you may be able to establish a framework in which two rational people, who share the same evidence, can argue constructively about whether they know that p – for example, that the sun will rise, but of course also much more interesting and controversial things.

Now it might be maintained that this is impossible – that it is just part of human nature, or a basic fact about the world, or a profound metaphysical truth that there is no such framework. Or it might be maintained that we don't know anything, *really*. Either of these claims strikes me as wholly dogmatic and unwarranted. The claim that no such framework exists strikes me as both true and challenging.

This problem is one problem that has concerned philosophers who have thought about induction. When it is expressed as the problem of how you know that the sun will rise, I can see why a statistician might find it boring and irritating. When it is expressed as the problem of how you know that the ratio of male babies to female babies is roughly one – something else we all know – it is not clear that the statistician can afford to be totally unconcerned.

How about justified opinion or belief? If you have a thoroughly tested and well made die, and you are about to roll it in a perfectly normal way, you are certainly entitled to the opinion that it will not yield an ace. More exactly, you are justified in having a degree of belief of roughly 5/6 that the die will fail to yield an ace. Again this is an opinion that we all hold. Again, with luck, if you can explain *how it is* that that is the correct opinion, you will gain some insight into what justifies opinions in general; and armed with this insight you may be able to establish a framework in which two rational people, who share the same evidence, can argue constructively about whether a given opinion is justified.

Now it might be maintained that this is impossible – that it is part of human nature, or a basic fact about the world, or a profound metaphysical truth, that there is no framework in which disputes concerning the justifiability of opinions can be carried on constructively. Maybe the best we can do is agree to disagree and go on about our business. Again, to assume this strikes me as negative and dogmatic.

This is another problem that has concerned philosophers who have thought about induction. Surely this is a problem that has also concerned some statisticians – at least in particular cases. To be sure, philosophers tend to approach things more generally and more abstractly. But that is something one just has to put up with, like their tendency to use irritating examples.

On the other hand, I must confess that philosophers have tended to be too ignorant of real statistics. I don't think that to do philosophy you must in principle deal with the complex, or even with the realistic; but in this particular case the problems change qualitatively as we get beyond the only case philosophers have tended to consider – the binomial case. But there are signs that this is changing.

II

Against this background, then, let me sketch some of the hopes (and possibly prejudices, but not, for heavens' sake, presuppositions!) that lie behind the approach to probability and inference that I have been pursuing.

Given a certain body of data, and a certain language, there is a set of statements in that language that may be construed as the body of knowledge based on that data. More precisely, we should speak of a body of knowledge of a certain level, for one of the necessary conditions that a statement must satisfy to get into that body of knowledge is 'having a high probability' relative to that body of data. I propose that there is a set of necessary and sufficient conditions for acceptance into a body of knowledge of a given level.

From this it follows that if two people speak the same language and are willing to pool their data, the same set of statements will constitute justified knowledge (of a given level) for each of them. This doesn't mean that as people they will agree about what they know – what it means is that in principle any dispute should be resolvable, in terms provided by the necessary and sufficient conditions for acceptance.

Probability will be defined relative to a body of knowledge. That is, I construe probability as a function from ordered pairs consisting of a set of statements which is a body of knowledge, and a statement, to the set of intervals included in $[0, 1]$. Now the set of data statements is a perfectly good body of knowledge, and probabilities relative to this body of knowledge exist – that is what leads to the bodies of knowledge of various other levels. But we also have probabilities – more interesting ones – relative to these other bodies of knowledge.

Since probability is defined explicitly, it follows that two people who share the same body of knowledge will, in principle, be able to agree on the probability of any given statement. (Recall that that probability will be an interval, so that the least odds they are willing to offer on the statement may still differ.) Now since two people who are willing to share data will (in principle) be able to agree on a common body of knowledge of a given level, they will – if they can settle on a level – agree (in principle) on what opinions are justified. This is to say, if they disagree, there is a set of principles – the necessary and sufficient conditions for acceptance and the definition of probability – which provide a framework within which the dispute can be resolved.

This means that if you and I agree to share data, and can settle on an acceptance level, representing what for the moment we are both willing to call practical certainty, we will accept the same statements – including statistical hypotheses – into our body of knowledge; and relative to that body of knowledge we will assign the same probabilities to statements – including statistical hypotheses.

III

The most important mechanism involved in this program is statistical inference; but the program itself will throw some light on statistical inference. There is no need to belabor the fact that there is no approach to statistics that has won the universal support of statisticians. There is one approach, the Subjectivistic Bayesian approach, that purports to provide a global interpretation of probability, in which any legitimate statistical procedure will find its justification. But as yet this approach is swallowed whole hog by relatively few statisticians, though there are many who adopt it in part and on occasion. The kicker is the word 'legitimate'. There are circumstances, in decision theory for example, in which it can be shown that any admissible solution to a problem is some sort of a Bayesian solution: in these special cases every legitimate solution is a Bayesian solution. But there are other problems in which objectivistic techniques and Bayesian techniques do not coincide: mixed or randomized experiments, for one; optional stopping for another. And even the reduction of objectivistic solutions of a decision problem to Baysian solutions is not without its problems; the reduction works both ways. If every admissible decision function is the Bayes solution for a certain assignment of prior probabilities, so every Bayes solution is a strategic choice of a decision function for some sort of a strategy.

Many statisticians are put off by the arbitrariness of the Bayesian approach. This arbitrariness would be alleviated by a logical measure version of Bayesianism (such as that of Carnap) but no logical measure theory of probability has been offered which even begins to do the job that statisticians require of it. But, of course, other statisticians are put off by the rather haphazard collection of criteria that are devised by objectivist statisticians to justify their procedures: if no Uniformly Most Powerful test exists, maybe there is a Uniformly Most Powerful Unbiased test; perhaps we should seek a most powerful invariant test; and so on.

Under these circumstances it does not seem implausible to try to start over from the beginning with the basic concept of probability.

IV

I shall define probability as a purely logical concept, applicable to sentences. It will be defined relative to a body of knowledge. Thus, 'The probability of S is p relative to the body of knowledge K' will be taken to be perfectly analogous to 'S is derivable from premises R'. Such a statement, if true, is a truth of logic; if false, logically false. In the one case, as in the other, to be precise we must specify the language we are speaking. We shall construe a body of knowledge K as a set of statements of the language L; probability we shall take to apply to statements of L.

The unit of statistical knowledge will be taken to be a set of statistical distributions: The distribution of the random quantity X in the population P is one of those in the set of distributions $\{F: D(P, X, F)\}$.

We shall say that the probability of the statement S is the interval (p, q) just in case the following conditions are satisfied:

(i) S is known to be equivalent to a statement of the form $X(a)\in b$, where X is a (real or a vector valued) random quantity, b is a borel set of appropriate dimension, and a is a closed individual term.

(ii) a is, relative to what we know, and relative to having an X-value in b, a random member of some reference set R.

(iii) It is known that the distribution of X in R is one of the distributions of the set F.

(iv) The supremum of the measures that distributions in F assign to b is q, and the infimum is p.
More formally:

    (i)   '$S \leftrightarrow X(a) \in b$'$\in K$

    (ii) $RAN(a, P, X, b, K)$

    (iii)   '$D(P, X, F)$'$\in K$

    (iv) $\sup \{\int_b df : f \in F\} = q; \quad \inf \{\int_b dF : f \in F\} = p.$

The problem here is the notion of randomness. I assume that $RAN(a, P, X, b, K)$ holds just in case P is the appropriate reference class (if we know K) for evaluating the probability of the statement $X(a) \in b$. This is not a definition, of course, but is intended to give an intuitive idea of what is going on. Some philosophers have offered characterizations of reference classes: Reichenbach suggests that it is the smallest reference class about which we have adequate statistical information. Salmon suggests that it should be the largest homogeneous reference class. I offer a rather more complicated definition – indeed, too complicated to present here – but it is essentially built along the same lines. The point is that what the correct reference class is depends on what we know and don't know. Two natural principles suggest themselves:

PRINCIPLE I: If a is known to belong to both X and Y, and our statistical knowledge about X and Z differs from that about Y and Z – i.e., if one interval is not included in the other, then, unless we happen to know that X is included in Y, Y prevents X from being an appropriate reference class for '$a \in Z$'. If neither is known to be included in the other, then each prevents the other from being an appropriate reference class.

PRINCIPLE II: Given that we want to avoid conflict of the first sort, we want to use the most precise information we have; and therefore among all the classes not ruled out by each other in accordance with Principle I, we take as our reference class that class about which we have the most precise information.

    We cannot obtain a definition of randomness from these two principles, however. Surely we should take into account different object names, just as we should take into account different random variables and different Borel terms. Thus if $\ulcorner x' \epsilon y' \urcorner \epsilon w$ and $\ulcorner z \langle x \rangle \epsilon b \leftrightarrow z' \langle x' \rangle \epsilon b' \urcorner \epsilon w$ and $\sim \ulcorner y \subset y' \urcorner \epsilon w$, then we should also require that a strongest statement about y', z', and b' not be stronger than or different from a strongest statement about y, z, and b. But this is no longer workable. Let us suppose that we know that x belongs both to y and to y', that $\ulcorner z \langle x \rangle \epsilon b \leftrightarrow z' \langle x \rangle \epsilon b' \urcorner$, that a strongest statement about y, z, and b differs from a strongest statement about y', z', and b', and, finally that $\ulcorner y \subset y' \urcorner$. There seems to be no reason why we should not have $RAN_L(x, y, z, b, w)$; but there is.

Let '$\mathscr{P}_1(t)$' denote the one-membered subsets of t, 'z_s' the function whose domain is $\mathscr{P}_1(\mathscr{D}z)$ and whose value for every object $\{x\}$ in that domain is $z \langle x \rangle$, and 'y_s' $\mathscr{P}_1(y)$. We then know that the unit set of x belongs to y'_s and the stongest statistical statement we know about y', z', and b, is just exactly as strong as the strongest statistical statement we know about y'_s, z'_s, and b, and thus a strongest statement about y'_s, z'_s, and b will also differ from a strongest statement about y, z, and b. But of course we need not (and generally will not) have $\ulcorner y \subset y'_s \urcorner$ in our rational corpus w. Thus we would not, after all, have $RAN_L(x, y, z, b, w)$.

The way to save the situation is obvious: $\ulcorner y \subset y' \urcorner$ entails $\ulcorner y_s \subset y'_s \urcorner$, and, naturally, our statistical knowledge about y_s, z_s and b agrees precisely with our statistical knowledge about y, z, and b. Thus rather than requiring that the set y itself be a subset of any set concerning which our statistical knowledge is different, we need only require that there be *some* subset such that our knowledge about it agrees exactly with our knowledge about y.

But even this requirement is a bit too strong. Suppose that y has the form $y' \times y''$, and x is of the form $\langle x', x'' \rangle$. Now consider a new quintuple $\langle y''', x''', z''', b''', s''' \rangle$ such that we know that x''' belongs to y''', that $z'''(x''') \epsilon b'''$ if and only if $z(x) \epsilon b$, and the strongest statistical statement we know about y''', z''', and b''' *differs* from the strongest statistical statement we know about $y' \times y''$, z, and b. There are circumstances under which we do not want y''' to prevent the randomness of x (i.e., of $\langle x', x'' \rangle$) in $y' \times y''$, even though there is no subset of y''' which matches $y' \times y''$. They may be characterized as circumstances in which there is a term y'''' (which may be y''' itself) such that we know that $\ulcorner y'''' \subset y''' \urcorner$ so that y'''' is known to be a subset of y''', and there are terms x'''', z'''', b'''', and a statement s'''', such that we know that $\ulcorner x'''' \epsilon y'''' \urcorner$, that $\ulcorner z'''' \langle \langle x'''', x'' \rangle \rangle \epsilon b'''' \leftrightarrow$

$\leftrightarrow z \langle x \rangle \in b^\neg$ and the strongest statement we know about $y''' \times y''$, z''', and b''' is exactly as strong as the strongest statement we know about $y' \times y''$.

An example will help to make this clear. Suppose that y is a sequence of trials of a two-stage experiment: one of two urns, U_0 and U_1, is selected, and then a ball is drawn from that urn. We may know that in the long run, urn U_0 is selected 1/4 of the time. If we know the frequency with which black balls are drawn from each of the urns, we may also know that the frequency with which urn U_0 is selected on those occasions when one ball has been drawn and has been black is 1/3. Clearly the appropriate reference class for the probability that U_0 was selected is the cross product of the set of selections of an urn, and the set of drawings of a single ball which turns out to be black. We must be sure that the set of selections of an urn, even though the frequency with which urn U_0 is selected in this set differs from 1/3, does not prevent the pair ⟨selection of an urn, draw of a black ball⟩ from being a random member of the cross product.

The concepts of probability and randomness have the following properties:

Every probability is based on *known* statistical proportions.

The definition of randomness is such that if $X(a) \in b$ is known to be equivalent to $X'(a') \in b'$, $RAN(a, P, X, b, K)$, and $RAN(a', P', X', b', K)$, then the measure of b in P is just the same as the measure of b' in P'. Thus, every statement has exactly one probability; probability is a function.

Given any finite set of statements s_i with probabilities (p_i, q_i), there exists a coherent probability function P satisfying the axioms of the probability calculus, such that $p_i \leqslant P(s_i) \leqslant q_i$.

If the probability of s is (p, q), the probability of $\sim s$ is $(1-q, 1-p)$.

The poor behavior of some preference rankings may be accounted for by interval probabilities rather than – as some economists have suggested – interval utilities.

Sometimes the probability may be a degenerate interval $[p, p]$; sometimes the probability may be utterly useless because it is the whole interval $[0, 1]$.

SIMPLE EXAMPLE: Suppose we have a bent coin, of which we know that it falls heads a third of the time or two thirds of the time, but we don't know which. Let S be the statement that the coin falls heads on the next toss. Let z be 'the next toss of the coin'. Let X be the characteristic function for heads ($X(y)=1$ if and only if y is a toss that yields heads, ($X(y)=0$ otherwise). b is $\{1\}$. We suppose that 'the next toss of this coin' is, relative to what we know, and with respect to having an X-value of 1, a random member of R, the set of tosses. D consists of the two two-point distributions corresponding to the two hypotheses, say d_0 and d_1. The supremum of the measures that members of D assign to $\{1\}$ is 2/3; the infimum is 1/3. Thus the probability of S is the interval (1/3, 2/3).

We begin now from the simple and natural view that statistical inference is a way of assigning probabilities to statistical hypotheses. This sounds Bayesian, and in some measure it is. In the simplest case the relevant family of distributions has but a single member, and the inference is classically Bayesian.

V

We have a coin drawn from an urn of coins. The coins all behave binomially when tossed, but with different parameters p representing the measure of heads; p we take to be known to be distributed in the urn according to a beta distribution with parameters 3 and 8. We toss our coin nine times, and get six heads. By Bayes' theorem, applied directly, we obtain a posterior beta distribution for p with parameters 9 and 11. This yields probability statements regarding p, as well as probability statements regarding the frequency of occurrence of heads in finite sequences of future tosses of the coin. The next toss, for example, is a random member of the set of trials consisting of the selection of a coin and the tossing of it ten times, of which the first nine yield 6 heads, with respect to yielding heads on the tenth toss. What we have supposed to be in our body of knowledge, including the results of the nine tosses, yields distribution for heads on the tenth toss in this reference class in our body of knowledge with parameter:

$$q = \int_0^1 p \, df_\beta (p \mid \rho, v).$$

Similarly, the coin is a random member of the set of coins in the urn that are tossed nine times and yield six heads, with respect to having a p-value in any borel set b. Thus the probability that the p of this coin lies between r_1 and r_2 is

$$\int\limits_{r_1}^{r_2} df_\beta(p \mid \rho, v).$$

This is all just the way the Bayesians would have it. Of course, we cannot really take it all literally, if only because the beta distribution is continuous and the number of coins in an urn is finite. (Our degree of belief concerning p may be distributed continuously, even if we are speaking about a single coin, and no urn enters at all; but I have required that all probabilities be based on *known* statistical distributions, and not on degrees of belief, even if such things exist.) What we might plausibly know is that of the large number of coins in the urn the proportions exhibiting various proportions of heads are close to the corresponding proportions in a beta distribution.

VI

It has been observed that posterior Bayesian probabilities are often only mildly affected by variations in prior probabilities. For example, my prior distribution of belief concerning the p of a certain coin may be quite different from yours; but our posterior distributions of belief, after having observed a few hundred tosses together, will in the ordinary course of events be very nearly the same. This phenomenon, the coming together of posterior distributions generated by diverse prior distributions and given evidence, often enables the subjectivist statistician to deal with two problems simultaneously. He has, as already remarked, the difficult problem of determining the prior joint distribution of his client (or himself); to do this in detail seems an extremely difficult job, and handbooks for budding subjectivist statisticians usually suggest that knowledge of the deciles, or even the quartiles, of the prior distribution is as far as it is worth trying to go, unless there are special features in the distribution. Furthermore, the statistician is often concerned with a number of people whose prior distributions of belief are different; he would like to be able to say something to all of them, and indeed sometimes to force them into near (or near enough) agreement. If the exact shape of the prior distribution function doesn't matter much, then we do not have to spend much effort finding out about that exact shape; and the posterior distribution for a number of people whose prior distributions vary will turn out to be about the same.

But what is it for a prior distribution to be 'gentle'? What does it mean to say that its 'exact' character doesn't matter 'very much'? Without rather precise answers to these questions, we are not in a position to advise anybody about anything. The theory of stable estimation is the Bayesian theory that provides answers to these questions. It has an extremely useful counterpart in the present theory. Edwards, Lindeman, and Savage have provided a set of assumptions under which the argument will go through. Suppose θ is the parameter whose posterior distribution is sought.

There is a set B of possible values of θ such that:

(3) It is practically certain that $b \in B$, given r, on the assumption that $f_T(\theta)$ has a constant value in B where it is not zero; i.e., for small α:

$$\int_{\bar{B}} f_T^* (\theta/r) \, d\theta \leqslant \alpha \int_{B} f_T^* (\theta/r) \, d\theta,$$

$$\text{where} \quad f_T^* (\theta/r) = \frac{f(r/\theta)}{\int f(r/\theta) \, d\theta}.$$

(4) The prior frequency function $f_T(\theta)$ must not change much in B – i.e., for some number k and some small number β, we have for all θ in B,

$$k \leqslant f_T(\theta) \leqslant (1 + \beta) \, k.$$

(5) The prior frequency function $f_T(\theta)$ must be bounded by a multiple (which may be as large as 100 or 1000) of k, i.e., $f_T(\theta)$ is nowhere very much larger than its relatively uniform values in B.

They give a famous example, in which, feeling out of sorts and a bit feverish, you take your temperature with a good thermometer. They show that your posterior distribution of temperature is quite insensitive to the exact shape of your prior distribution; but they do assume that your prior distribution for temperature is precisely defined.

Now let us abandon the unrealistic and demanding assumption that we have a precisely defined prior distribution for our temperature. Are we worse off? Not at all. While we may not have accumulated a great store of knowledge concerning the general relationship between the way we feel and the temperature we actually have, we do have some knowledge about it – enough to narrow down the possible distributions to a set S, each member of which satisfies the assumptions made by Edwards, Lindeman and Savage. We may then go through exactly the same argument as the subjectivist goes through, to conclude, for example, that whatever member of S be true, the frequency with which θ will lie in C will differ from the number calculated on the basis of a uniform prior by less than 0.068; that for any C included in B the difference between the artificial and the true measure of C will be less than 6%; and so on. These measures bound the measures which yield the probabilities we are interested in, subject to conditions regarding randomness analogous to those mentioned earlier. That is, if my temperature today is a random member of the set of my temperatures on days when I feel as I do today, and observe a thermometer reading of 101.0, then the *probability* – the regular, epistemic probability – that my temperature falls in a certain interval C falls within the intervals calculated.

Thus the theory of stable estimation, though developed to aid the subjectivist statistician confused about his prior distribution, is splendidly suited to provide bounds on the epistemological probability interval, just so long as the conditions on randomness are satisfied. Indeed, in many respects, the theory of stable estimation seems to fit the present approach to probability even better than it fits the subjectivist approach.

VII

There is really nothing more to Bayesian inference (in principle) from our point of view than we have discussed in these few examples. The major points of difference between our form of Bayesian inference and the subjectivistic form are (1) we require that the prior distribution be based on statistical knowledge, rather than mere opinion; and (2) we do not require that a single prior distribution be employed, but rather allow that our prior knowledge may be represented by a whole family of distributions. Where the standard subjectivist analysis claims that the posterior distribution of a parameter is largely determined by the likelihood function when the prior distribution is 'gentle', our analysis would claim that the family of posterior distributions in our body of knowledge will be narrow and determined largely by the likelihood function when the family of prior distributions includes none that is significantly 'ungentle'. Furthermore, for the posterior distribution to determine probabilities, requires that certain conditions regarding randomness be met. To make a probability statement about the population parameter, given knowledge of the outcome of a sampling test, we require that the population be a random member of the set of populations whose sampled subsets are such as that we have observed, with respect to having a population parameter in a specified Borel set. For example: a selected coin subjected to a test consisting of n trials, r of which result in heads, is to be a random member of the set of all coins subjected to a test consisting of n trials, of which r produce heads, with respect to having a long run frequency of heads in (say) the interval (0.5, 0.6). Alternatively, we may require the complex object consisting of the selection of a coin and the performance of n tosses of the coin, to be a random member of the cross product of coin selections and the set of n-fold sets of tosses yielding r heads. To make a probability statement about the likelihood that a future set of m trials will result in s heads, we may require that the selection of the coin, the test consisting of n trials, of which r result in heads, and the performance of m subsequent trials, be a random member of the set of all selections of a coin, followed by the performance of n trials, r of which result in heads, followed by m subsequent trials, with respect to having s heads turn up on the subsequent trials. These conditions regarding randomness are complex, but they lie at the very heart of the application of probability as a guide to the future. In the subjectivistic account they are concealed by subjective judgements; in the account offered here they are objective in that they depend explicitly and formally on the body of knowledge, real or hypothetical, that is being considered.

VIII

Let us now look at fiducial inference. Suppose that we know that X has a normal distribution in R, and that that is all we know about X. Thus the set of distribution functions one of which is known to apply to X is the set of *all* normal distribution functions:

$$\{\Phi(m, \sigma): m \in \mathbb{R} \wedge \sigma \in \mathbb{R}^+\}$$

(Properly speaking, what we know is that the distribution of X is 'very close to' some normal distribution.) From this it follows that in R^n,

$$t = \frac{\bar{X} - m}{s}\sqrt{n-1}$$

has a t distribution with $n-1$ degrees of freedom, where s^2 is the sample variance and $\bar{X}$ the sample mean. If we compute t from a sample k of n members of R, and if k is a random member of R^n with respect to having t-values in the appropriate Borel sets, we have:

The probability that $-t_p \leqslant t(k) \leqslant t_p$ is $(1-p, 1-p)$, for any p, where t_p is the p value of t for $n-1$ degrees of freedom.

By the equivalence condition (p. 112), since we know that $-t_p \leqslant t(k) \leqslant \leqslant t_p$ if and only if

$$\bar{X}(k) = t_p \frac{s(k)}{\sqrt{n-1}} \leqslant m \leqslant \bar{X}(k) + t_p \frac{s(k)}{\sqrt{n-1}},$$

we obtain: the probability that m falls within these limits is also $(1-p, 1-p)$. This seems to be an unnaturally exact probability – but recall that we cannot plausibly suppose ourselves to *know* that X has a normal distribution, and thus there is some degree of approximation in our result. Again, note that the relevant roughness of our approximation increases with the specificity of the hypothesis with which we are concerned. Having observed that $X(k) = 50$, and $s(k) = 5$, we can ask for the probability that m is less than 10, or that m lies between 50.000 and 50.001. To answer these questions sensibly, however, we must know much more about the details of the set of hypotheses we regard as possible – i.e., about their detailed deviations from normality. We say that the family of distributions is 'roughly' normal, and if we are considering the probability that m lies between 40 and 60, the 'roughness' may only be relevant to the fifth decimal place, which is only the fifth significant figure. But if we ask one of the other questions, the 'roughness', though still perhaps relevant to the fifth decimal place, now represents a difference in the first significant figure, or indeed may preceed the first significant figure by a couple of decimal places. Furthermore, note that m itself is not a random quantity distributed in any reference sequence. Thus the insights of Neyman and others who refuse to regard fiducial probabilities as 'real' probabilities are to a certain extent vindicated. Nevertheless, although m has no distribution in this instance on the view of probability presented here, we may obtain probability statements about m that are just as good – and just as soundly based in statistical knowledge – as probability statements about t. It is this indirect statistical basis that provides for the possibility of sampling indefinitely 'to demonstrate the correct frequency'.

A number of conditions must be met before the fiducial inference will go through. Some of these conditions are reflected in a natural and simple way in our epistemological system. One such condition concerns 'recognizable subsets'. The random quantity t has as its domain the set of n-sequences of items from the domain of x. Since t has the same distribution for *every* set of n-sequences from every normal population, the derived distribution "is applicable to the enlarged set of all pairs of values t, $[m]$ obtained from all values of $[m]$. The particular pair of values of $[m]$ and $[t]$ appropriate to a particular experimenter certainly belongs to this enlarged set, and within this large set the proportion of cases satisfying the inequality $[9]$ is certainly equal to the chosen probability P. It might, however, have been true... that in some recognizable subset to which this case belongs, the proportion of cases in which the inequality was satisfied should have some value other than P. It is the stipulated absence of knowledge *a priori* of the distribution of $[m]$ together with the exhaustive character of the statistic $[t]$, that makes the recognition of any such subset impossible, and so guarantees that in this particular case,... the general probability is applicable". (R. A. Fisher, *Statistical Methods and Scientific Inference*, p. 55.)

There are a number of comments that may be made on this passage. First, while it is true that the distribution is applicable to the enlarged set obtained by considering a set of populations, with a normally distributed random quantity defined on each of them, and, corresponding to each population, a set of n-sequences constituting samples from that population, it is not *necessary* to consider any population other than the domain of x. The particular n-sequence we obtain as a sample may well be a random member of the set of n-sequences from R. With respect to having a t value less than 2.787, relative to our body of knowledge; if so, the probability is 0.99 that its value is less than this, and thus the probability is 0.99 that m lies within the limits mentioned.

We may *also* consider the enlarged set referred to by Fisher. In this enlarged set, comprising a set of populations, in each of which a certain quantity is, we know, normally distributed, we have several random quantities to consider. The mean of the normally distributed quantity in each population is now a perfectly good random quantity, whose domain is the set of populations under consideration. The random quantity t is also a perfectly good random quantity – though its meaning changes from population to population, since each population is the domain of a different random quantity x – and in each population it has a student's distribution with $n-1$ degrees of freedom. What we require, in order that the argument go through from our point of view, is that the n-sequence sampled be a random one with respect to having an absolute t-value less than 2.787. We must, of course, know which population we are sampling, but *if* the pair (population, sample) is a random member of the set of pairs of populations and n-samples from them, we may convert a distribution for t into *something like* a distribution for m in the particular population sampled, just as we did before. This condition of randomness will be met just in case we know of no set of terms (k, b, c, d) such that: we know $c(k) \in d$ if and only if $|z^*(a)| \leqslant 2.787$; k is known to belong to b; and the frequency with which objects in b have c-values in d is different from 0.99; and there is no proper subset of b in which a corresponding frequency *is* 0.99; and there is no product set of which b is a projection, in which a corresponding frequency is 0.99. Now this clearly includes Fisher's stipulation that there be 'no recognizable subset leading to some value other than P', but it is equally clearly rather more broad. It is furthermore quite inadequate simply to stipulate that the inference will be valid just in case we have no knowledge of the distirbution of m among the populations.

Fisher goes on to say, "...had there been knowledge *a priori*, the argument of Bayes could have been developed, which would have utilized all the data, and which would in general have led to a distribution *a posteriori* different from that to which the fiducial argument leads. Bayes' method in fact calculates the distribution of m in particular subset of pairs of values $[t, m]$ defined by $[t]$, and to which therefore the observation belongs". (*Ibid.*, p. 55.)

It is not generally the case, however, either that we are in total ignorance regarding the distribution of m in a population of populations, or that we have knowledge of the exact distribution of m in order to apply Bayes' theorem in a straightforward way. And Fisher's suggestions do not give us any way of dealing with the common instances in which we have some, but by no means complete, knowledge of the distribution of m. Our own suggestions with regard to this question will be mentioned in the following section.

Within a classical framework, Robert Buehler takes the question of relevant subsets seriously, and provides explicit definitions of 'relevant' and 'semirelevant' subsets. The considerations he adduces are presented in a classical frequentist framework, and therefore it is difficult to make connection to bodies of knowledge. Nevertheless he correctly sees that what is at issue is a "generally accepted notion of an 'appropriate reference set' for inferences".

IX

Confidence intervals also receive a reasonable representation from our point of view. They also represent the most basic form of inference. Consider any finite set S, and any property P. (There is no reason to suppose that any set of empirically given objects is infinite.) We know *a priori* (on the basis of set theory alone) that there is some rational number r such that the proportion of S's that are P's is r. Suppose that we are totally ignorant of r – i.e., suppose that the set z of distribution functions for the characteristic function X_P of P in S includes all possible two point distributions.

We propose to examine a sample of N, and make an inference regarding r. Now for every value of r, we can compute two numbers, ε_r and δ_r, such that a proportion c of all possible samples will have a ratio of P's in the interval $(r-\varepsilon_r, r+\delta_r)$, and further such that the sum of these numbers is a minimum. In Figure 1 these points are represented by points inside the elliptical area.

We now argue thus: For every value of r, c of the samples of N will have the property that they fall inside the ellipse. In particular, for the true value of r, c of the samples will fall inside the ellipse. This does not yet lead us to the probability that our sample falls in the ellipse: we need one further item: that our sample be a random one (i.e., a random member of the set of all subsets of N), with respect to having this property, relative to what we know. There are a number of ways in which this could fail. Given a prior distribution on r, for example, it may be that the membership of the pair consisting of a selection of r and the selection of a sample in some appropriate space yields a different (or more accurate) interval for the parameter r. This does not render the claim false that the proportion with t lying between $r-\varepsilon_r$ and $r+\delta_r$ is c – for that is true of every r. But it may render it irrelevant, just as the knowledge that Jones is a coal miner renders irrelevant the actuarial fact that he is a 40 year old white male. Alternatively, we may know that the sample at hand comes from

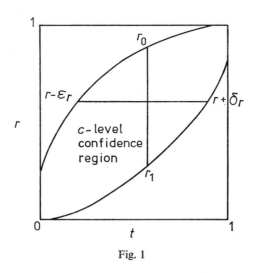

Fig. 1

some subset of the set of all possible samples in which the frequency of the property we're interested in differs from its frequency in the general population S. (E.g., if we are concerned with the proportion of strawberries in a basket that are rotten, a sample drawn from the top layer will not be a random member of the set of all equinumerous samples with respect to revealing this proportion.) It is possible that the sample itself will reveal its own lack of randomness – e.g., if we find P's occurring more and more frequently as we draw the sample.

But *given* the satisfaction of the randomness condition, the probability is about c that our sample falls in the ellipse – i.e., that the relation between r and the observed t is such as to be represented by a point in the ellipse. This is just to say that the probability is about c that the true value of r falls between r_0 and r_1, in the diagram.

There is an important philosophical observation to be made here: to reject a sample as a random member of the set of all N-membered samples is to *know* something that puts that sample in a special class about which we have empirical knowledge. Thus in that state of total empirical innocence that epistemologists like to imagine, and statisticians seem to abhor, the sample *must* be a random member of the canonical reference class.

It is often asserted of the confidence interval analysis, as it is of the fiducial analysis, that if we have knowledge of a prior distribution for the parameter, the confidence (or fiducial) analysis no longer applies. But on the analysis I have offered (as in real life) we always have *some* prior knowledge, even though it may be vague. What counts is not whether there is *any* way of construing the population P as being drawn from a superpopulation P^* (we can always do that), but rather whether there is any way of doing so which will yield information about the frequency of Q in P, given the observed sample. The resolution of the problem lies in the fact (which emerges from the definition of randomness) that if we have *enough* prior information so that the Bayesian posterior interval actually *differs* from the confidence coefficient (rather than merely including it) the Bayesian form of inference takes precedence. But we do not require that there be no way of assigning the population about which we are making an inference to a set of populations.

In general one may expect that confidence sets will be the appropriate yield of statistical inference in the absence of much knowledge, and that, as knowledge of a subject matter increases, conditional Bayesian intervals will become more often relevant. The question of when Bayesian intervals based on prior known distributions supercede confidence intervals is a question which cannot even be raised in any of the alternative approach to statistical inference: Bayesians and those who offer logical interpretations of probability will find themselves obliged to say 'always', except when the confidence analysis is construed as a cheap and easy algorithm; frequentists will essentially find themselves obliged to say 'never', since they say that only when prior distributions are *known* are the Bayesian techniques appropriate; but it is never the case that a prior distribution is 'known' with perfect accuracy. The practising statistician will of course use that most precious commodity, judgement; but it is surely advantageous to replace judgement by reasoned argument when that is possible.

X

As a final illustration of the theory I have been describing, let us consider
the problem of Decision Making. In the simplest case, we have two possible states of nature, and we can compute the expected loss under each
possible state of nature of each decision function. The space of decision
functions is convex, and may be illustrated in Figure 2.

A number of doctrines have been proposed for choosing from among
the possible decision functions (represented by points along the curve).

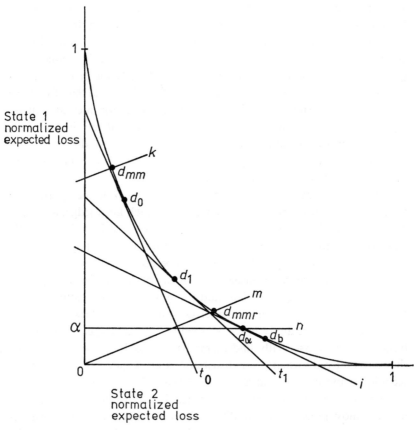

Fig. 2

The Bayesian procedure (which, according to subjectivists can always be applied) takes account of the fact that the relative probabilities of the two states are well determined; this fact yields a ratio of posterior probabilities that determines a support line, such as j. The point (points) at which this support line touches the convex space of decision functions represents the desired decision. Other procedures that have been advocated by statisticians of objectivist persuasion are minimax, minimax regret, and perhaps a significance level procedure. A minimax procedure is represented by a line of positive slope and intercept (determined by the values or costs involved) on our picture, such as the line k. The point at which that line intersects the decision curve is the decision chosen. A minimax regret line is represented on our picture by a line m of the same slope as the minimax line, but with an intercept at the origin; the point at which it intersects the decision line is the decision function chosen. A significance level approach would consist of choosing a significance level alpha at which to test the hypothesis that state 1 obtains; the intersection of the decision curve and the horizontal line n then represents the decision chosen on this basis.

Under the analysis that has been offered here, the hypothesis that S_1 obtains will have a probability of (p_1, q_1) and the hypothesis that S_2 obtains will have a probability of $(1 - q_1, 1 - p_1)$.

These probabilities will be the upper and lower bounds derived from *known* distributions. No 'degree of belief' falling outside these intervals can be regarded as 'rational' or defensible. Consider the supporting lines corresponding to prior probabilities for S_1 of $r: p_1 \leqslant r \leqslant q_1$. Clearly only those decision functions lying on one of these support lines, bounded by t_0 and t_1, can be regarded as rational. Equally clearly there will be a number of such functions. (Someone whose degree of belief in S_1 did not fall in the interval (p_1, q_1), although his body of knowledge was such as to support the hypothesis to that degree, would be irrational; he might be inclined to adopt a decision function beyond the bounds of rationality. Could he be motivated to adopt one of the acceptable – rational – decision functions? Only if he could be motivated to be rational. That raises knotty problems.)

As I see it, this is as far as *rationality* takes you: it constrains you to choose a decision function between d_0 and d_1. But you are still faced with the choice of a specific decision function. A number of ways of making that choice might present themselves.

(a) One might use the support line corresponding to an average of p_1 and q_1.

(b) One might use a minimax procedure, provided the minimax line intersected the boundary of the set of decision functions between d_0 and d_1.

(c) One might use a minimax regret procedure, subject to the same constraint that the minimax regret line intersect the boundary at a per missible place.

(d) One might use a minimax regret procedure restricted to that portion of the decision space between d_0 and d_1. That is, one might take the expected losses under d_0 and d_1 to represent the unavoidable losses of the problem, and, graphically speaking, use a minimax regret line proceeding at the appropriate slope from the vertex of perpendiculars intersecting d_0 and d_1.

The variations and alternatives seem endless, and the choice among them seems appropriately a matter of taste. Why not? With the restrictions imposed by the support lines we have used up the *rational* constraints. Just because we should have sound opinions and justifiable knowledge doesn't mean that we have to end up making the same decisions.

University of Rochester

DISCUSSION

Commentator: Giere: Perhaps you would care to indicate how your theory differs from say, that of Carnap?

Kyburg: Well it differs in one very obvious way: My theory is formulated within a language rich enough to do real science instead of just in a monadic predicate calculus.

Giere: In Carnap's case the origin and form of the metric is quite obvious, where does it come from in your case and what is its form?

Kyburg: Even if we begin only by accepting logical truths and observation statements, even in that body of knowledge, we have some knowledge of distributions of values for physical quantities and it is this knowledge of distributions which generates the metric. And that is why I need an epistemological definition of randomness: One wants to be able to talk about the situation in which the item in question does not belong to a recognizable subset in which the frequency is known to be different.

Good: This matter of recognizable subsets seems to me to be crucial to your theory and is a long standing problem for statistics, a problem which statisticians almost ignored before about 1960 and still have not given enough attention. Consider, e.g. a typical case in medical diagnosis where you are entering discrete information about the indicants of patients into a contingency table of perhaps 100 dimensions. By the time you take a small sample of about a billion patients you find that most of the cells are empty. Thus there is nearly always a recognizable subset that will reduce your sample size to zero. This is an ancient and important problem in actuarial science and in medical diagnosis.

Kyburg: I approach the problem somewhat abstractly and a priori. I set down a definition which seems to me to work and to yield appropriate answers and which has as a consequence that for any given statement there always is a reference class which has the property Good refers to, but I haven't done any work on the question of how to construct that reference class in any particular case. One remark is to the point, however: If one reference class gives us an estimated range for the probability of $0.5 \rightarrow 0.6$, while another reference class gives us the range $0.3 \rightarrow 0.9$, then my theory is going to select the former reference class because it gives us more precise knowledge. With regard to cells that wind up having fewer and fewer things in them, we know less and less about the frequencies. The subsets are recognizable, but don't interfere with useful probabilities.

Garner: I think it would be more appropriate if you chose to deal with problems other than the simple student or binomial problems. For example the Fisher-Behrens problem or the problem of the weight mean with regard to the analysis of Box and Tiao who showed that the differing suggested fiducial distributions of Yates and Fisher correspond to a marginal or a conditional posterior probability distribution respectively. I'm not suggesting that a fiducial statistician would accept your explanation, but I think it would be a more discerning test of your formulation than the present exposition by way of binomial distributions.

Kyburg: Well, I was after something uncontroversial on purpose.

Suppes: What relation does your notion of randomness bear to the standard notions, e.g. that of Kolmogoroff-Martinoff?

Kyburg: None. My definition of randomness is not meant to capture the frequentist notion of randomness.

Giere: On your conception of randomness you could have a random individual and a random set of individuals whereas on the classical conception you could only have a randomly *selected* individual for randomness is characteristic of the method of selection and not of the individuals thus selected. Similarly, on your view you could not fail to know whether a given individual, or set of individuals, was random or not – you would only need to sit down and consult your knowledge, whereas on the classical view it is an empirical question whether the given individual has been selected at random or not.

Kyburg: Yes, thank you.

Teller: What impact does the redescription of background knowledge have on your recognizable subsets, random individuals and so on? Would, for example, re-parameterizing the data with a 'stretched co-ordinate' change these things?

Kyburg: Merely stretching a co-ordinate would not change anything. Changing the predicates in the language would change things – we have to assume that we are operating with a fixed, and settled language. For we have to construct a canonical list of reference classes with reference to the descriptive resources of that language.

Teller: You mean to say that my language might differ from yours but that we might restrict ourselves to saying only those things which were logically equivalent to one another and that, even so, you might come out with different numbers at the end of your calculation than I would come out with at the end of mine?

Kyburg: Yes, that could happen. And I think that it ought to happen. After all, one of the things that is happening when scientists propose sophisticated theories is an alteration in the language which is being used. Among other things, we presumably want criteria for better scientific languages. One of those criteria might well be that in one language we get many more probabilities which are very much closer to zero or one than we do in other such languages. I think that this is the kind of thing you are pointing to and that it is exactly one of the problems in the foundations of science that we should be investigating.

Good: It does seem to me that this feature of your system re-introduces subjectivism through the back door by making our results relative to the languages we speak, for we might all have our own personal languages. This somewhat undermines your aim of obtaining more general agreement which was the original motivating spirit behind the construction of your theory. Indeed, I should like to construct a language which would more nearly match my own subjective probability judgments! (Cf. my *Probability and the Weighing of Evidence*, London, 1950, p. 48.)

Kyburg: I do think that we could have criteria for better scientific languages. We can get interpersonal agreement insofar as we can get people to agree to talk the same language. And we can even share a common metalanguage in which I can agree that the probabilities that you assign in your language are correct for that language and you can agree, conversely, that the probabilities I assign in my language are correct for my language. But over and above this I do believe that we can have criteria for choosing to use one scientific language rather than another.

I. J. Good: I think Dr. Kyburg's paper exemplifies a point I made yesterday that people tend to identify the Bayesian's position with the use of *sharp* probability estimates.

Kyburg: It is not the sharpness of the probabilities that I take to be crucial, but their subjectivity. I find myself forced to abandon sharp probabilities, not in order to distinguish myself from 'Bayesians', but in order to base all probabilities on known, objective, measures.

Critical Commentary

I. *Levi: Kyburg's Frequency Principle*

Given a language L based on predicate logic, let the urcorpus UK be the set of logical truths, set theoretical truths and mathematical truths expressible in L together with other sentences in L which one may wish to regard as incorrigible or immune to revision.

DEF. 1. A *potential corpus of knowledge* K expressible in L is a set of sentences in L satisfying the following conditions:

(i) K contains UK.

(ii) K is deductively closed.

DEF. 2. A *potential credal state in L relative to K* is a set B of real valued functions $Q(h; e)$ where h is any sentence in L and e is any sentence in L consistent with K satisfying the following conditions:

(i)   If K is consistent, B is nonempty.

(ii) B is convex.

(iii)   Every Q-function in B is a probability function such that if K and e entail h, $Q(h; e) = 1$.

DEF. 3. A *potential confirmational commitment* for L is a function $C(K)$ taking as arguments potential corpora (expressible in L) and as values potential credal states (relative to the corpora which are the values).

I assume that a rational agent X is committed at time t to a confirmational commitment $C_{X,t}$, corpus $K_{X,t}$ and credal state $B_{X,t}$ satisfying the following condition:

Principle of total Knowledge: $C_{X,t}(K_{X,t}) = B_{X,t}$.

In addition, I assume that confirmational commitments satisfy a further condition. Let K_1 and K_2 be potential corpora of knowledge such that K_2 is the deductive closure of e (consistent with K_1) and K_1. Let B_1 be a potential credal state relative to K_1 and B_2 be a potential credal state relative to K_2.

DEF. 4. B_2 is *the conditionalization of B_1 relative to K_1 and K_2*
if and only if (i) for every Q_1 in B_1 there is a Q_2 in B_2 and (ii) for every
Q_2 in B_2 there is a Q_1 in B_1 satisfying the following:
If f & e is consistent with K_1, $Q_2(h; e) = Q_1(h; f \& e)$.

The additional condition on confirmational commitments may be
stated as follows:

Principle of Confirmational Conditionalization: If K_2 is obtained from K_1
by adding e (consistent with K_1) and forming the deductive closure, then
for every potential confirmational commitment C, $C(K_2)$ is the conditionalization of $C(K_1)$ with respect to K_1 and K_2. If K_2 is inconsistent,
$C(K_2)$ is empty.

In virtue of the principle of confirmational conditionalization, potential
confirmational commitments can be defined by specifying the set of
probability measures $C(\text{UK})$ – at least insofar as potential corpora of
knowledge are obtainable from UK by adding a sentence e (in L) and
forming the deductive closure. I shall call the measures in $C(\text{UK})$ P-
functions to distinguish them from the Q-functions in $B = C(K)$ for other
potential corpora.

It is clear that the set $C(\text{UK})$ must satisfy the following conditions:

Confirmational Consistency: $C(\text{UK})$ is nonempty.

Confirmational Convexity: $C(\text{UK})$ is convex.

Confirmational Coherence: Every P-function in $C(\text{UK})$ is a probability
measure relative to UK.

DEF. 5. For a given confirmational commitment C, $c(h; e) =$ the set of
real numbers r such that there is a P-function in $C(\text{UK})$ according to
which $P(h; e) = r$.

DEF. 6. For given K and credal state B relative to K, $Cr(h; e) =$ the set
of real numbers r such that there is a Q-function in B according to which
$Q(h; e) = r$.

$c(h; r)$ and $Cr(h; e)$ will be interval valued functions owing to the
principle of convexity.

In the special cases where $C(\mathrm{UK})$ contains exactly one P-function, $c(h; e)$ can be viewed as a real valued probability function. It then corresponds to what Carnap called a "credibility function". $Cr(h; e)$ then corresponds to what Carnap called a "credence function".

Unlike Carnap, de Finetti, Savage and other strict bayesians, I do not assume that for an ideally rational agent $C(\mathrm{UK})$ is single membered. To the contrary, I suppose that even an ideally rational agent should not, under normal circumstances, adopt a single membered confirmational commitment.

Corpora of knowledge can be partially ordered with respect to strength so that UK is the weakest potential corpus and the inconsistent corpus is the strongest.

Similarly, potential confirmational commitments can be partially ordered with respect to strength. The strongest would be the empty set of P-functions. (I have ruled this out of consideration by the principle of confirmational consistency.) The weakest would be the set of all P-functions allowed by our principles. I shall call this confirmational commitment CIL(UK). The P-functions in CIL(UK) are the 'logically permissible' P-functions. If we consider a potential commitment $C(\mathrm{UK})$ stronger than CIL(UK), the set of logically permissible P-functions in $C(\mathrm{UK})$ are the P-functions which are 'seriously permissible' according to $C(\mathrm{UK})$.

I shall call those principles which impose necessary conditions on logically permissible P-functions, principles of 'inductive logic'. This accords to a good degree of approximation to Carnap's own characterization. Thus far, only one principle of inductive logic has been introduced – to wit, the principle of confirmational coherence.

The question arises as to whether the principle of coherence constitutes a 'complete' inductive logic IL in the sense that it constitutes a set of necessary conditions for logical permissibility which are jointly sufficient. 'Coherentists' like Savage and de Finetti have answered in the affirmative. Others, like Carnap, have supposed that other principles could be defined.

Reichenbach, Fisher, Hempel, Hacking, Kyburg and Salmon have formulated principles which may be construed as regulating the credal states rational agents should adopt for hypotheses about the results of 'trials' on objects or systems given knowledge of the statistical probabilities or objective chances of obtaining outcomes of these kinds on such trials. Hacking calls his principle "the frequency principle", Carnap called an analogous principle a principle of "direct inference" and Reichenbach construed his principle as a criterion for assigning 'weights" to hypotheses. Hempel construed his criteria as regulating the legitimacy of "statistical syllogisms". Most of the authors mentioned have differed with one another in some detail or other concerning the precise formulation of such a frequency principle.

On one formulation, a statistical syllogism contains two "premisses". One of these is a statement of the form 'The chance (statistical probability) of an event of kind R resulting on a trial of kind T'' is equal to r'. This is the 'major premiss'. The 'minor premiss' is of the form 'e is an event which is a trial of kind T'. The conclusion is the assignment of a degree of probability r to the sentence 'e results in an event of kind R'.

In the context pertinent to the current discussion, the premisses of the statistical syllogism are taken to be contained in X's corpus of knowledge at a given time and the probability assignment based on these premisses is construed as an assignment of a degree of credence to the appropriate sentence. Consequently, a 'frequency principle' which specifies the conditions under which such assignments of degree of credence are legitimate serves as a restriction on Q-functions which are permissible relative to given corpora of knowledge and, hence, via the principle of confirmational conditionalization, on the logically permissible P-functions eligible for membership in potential confirmational commitments.

One of the problems which bedevils the adequate formulation of a frequency principle is the fact that X will often include in his corpus information not only to the effect that e is a trial of kind T but also information that e is a trial of kind T'. On some occasions this extra information may prove no bar to basing an assignment of degree of credence to the hypothesis that e results in an event of kind R. But on other occasions, it appears that such information will prove relevant. The challenge to advocates of frequency principles is to formulate conditions under which such extra information is relevant in this way and when it is not. (This problem is customarily but misleadingly called "the problem of the reference class".)

Among the authors I have cited, only Henry Kyburg has seriously attempted to come to grips with the full panoply of situations where a question arises as to which description of the trial event should serve as the minor premiss for a statistical syllogism. Hence, in discussing these matters, only his position can be described with any confidence. For the most part, one must speculate as to how the other authors cited would respond to the questions which Kyburg has attempted to answer with some thoroughness.[1]

Nonetheless, one can identify (at least tentatively) two positions concerning the problem of the reference class which have been adopted:

(1) According to one view, one should always base an application of the frequency principle on the strongest known description of the trial event. Thus, if X knows that e is a trial of kind T and of kind T' he should use as a minor premiss 'e is both a T and T'' and he should appeal to his knowledge of the chance of obtaining an R on a trial which is both T and T' in assigning a degree of credence. Sometimes restrictions are placed on the predicates which may occur in descriptions of trial events. Controversies as to whether such restrictions are appropriate and what they should be will play no important role in the subsequent discussion and will be glossed over here.

(2) According to other authors (including Kyburg and, I think, Reichenbach), there are some occasions where one should not base the statistical syllogism on the strongest known description of the trial event.

Of course, advocates of the first position do concede that on many occasions one does not have to appeal to the strongest known description of the trial event. X may know that e is both T and T' and also know that the chance of obtaining an R on a trial which is both T and T' is equal to the chance of obtaining an R on a trial which is a T. In that case, the description of e as not only a T but a T' is stochastically irrelevant information and X may, even for those who endorse the first approach, base his use of the frequency principle on knowledge that e is a T. The difference between the first view and the second is that according to advocates of the second view there are occasions when X may base his use of the frequency principle on knowledge that e is a T even when he does not know that the information that e is a T' is stochastically irrelevant.

Suppose, for example, that on the basis of actuarial data X knows that 90% of American males aged 30 survive to 31. He knows that Jones is picked at random from American males aged 30. Moreover, he knows that

Jones is diabetic. However, he does not know what percentage of diabetic American males aged 30 survive an extra year. As I understand Reichenbach's view, since X lacks "reliable statistics" concerning diabetic American males aged 30 but has reliable statistics concerning American males aged 30, the latter characterizes the "narrowest reference class for which he has reliable statistics". Hence, it seems that, for Reichenbach, X should assign a degree of credence of .9 to the hypothesis that Jones will survive to 31. This is so even though X does not know whether being diabetic is stochastically relevant information. It would appear that Reichenbach endorses the second approach. Unfortunately, Reichenbach does not clearly confront situations of this sort so that one cannot say with confidence what his position would be. There is no doubt about Kyburg. He does explicitly endorse the second point of view.

My contention is that the second point of view as developed by Kyburg and adumbrated by Reichenbach is inconsistent with the conditions previously imposed on confirmational commitments. Since each of the principles introduced has as strong a rationale as any principle which might be introduced in this extremely controversial problem area, the presence of contradiction with these principles constitutes a serious objection to the Kyburg-Reichenbach view and lends substantial support in favor of the first approach to the frequency principle.[2]

If this conclusion is correct, the applicability of the frequency principle for the purpose of obtaining fairly definite assignments of degrees of credence to hypotheses is limited; for in real life, we often have extra information about trial events without having any knowledge as to whether this extra information is stochastically irrelevant or not. To be sure, if I toss a coin known to be fair in Philosophy Hall at Columbia, I do assume that the information that the coin is tossed in Philosophy Hall is stochastically irrelevant and no bar to basing an assignment of a degree of credence to the hypothesis that the coin will land heads on the chance that the coin will land heads on a toss. The frequency principle does, indeed, have application in this case. But I would be quite reluctant to assume stochastic irrelevance of extra information in many other cases including many important applications of the frequency principle in scientific inquiry. Kyburg has recognized this fact about frequency principles and his move toward contradiction is part of an effort to extend the scope of applicability of inferences from statistical knowledge to outcomes of trials. That his effort fails is an important fact, therefore, about the scope of applicability of an adequately formulated frequency principle.

Let us suppose that X knows that a given urn contains 1,000 balls which are different from one another in the following respects: Some are black and others are white. Some are labelled with numeral '1' and some with the numeral '2'. In addition, X knows that at time t the contents of the urn were thoroughly mixed and one of the balls drawn by a blindfolded agent in a manner such that the chance of his selecting any given ball is equal to the chance of his selecting any other. Call a trial on the urn (i.e., selecting a ball 'at random' from the urn) of this sort a trial of kind T_1. The statistical probability of obtaining a black ball ('1' labelled) on a trial of kind T_1 will equal the percentage of black ('1' labelled) balls in the urn and X knows this. Let this information be d.

Case 1. Suppose that in addition to the information d, X knows that the percentage of black balls in the urn is $50(e_{.5})$. If his corpus consists of the deductive consequences of the urcorpus UK, and $e_{.5}$ & d his degree of credence for the hypothesis h that the draw will yield a black ball should equal .5. On this particular point, Kyburg and all the other authors cited before would agree.

This means that every logically permissible P-function should be such that $P(h; d \& e_{.5}) = .5$ and for e_r asserting that $100r\%$ of the balls in the urn are black, $P(h; d \& e_r) = r$. ('r' is an L-determinate designator for a real number as, at least Kyburg and Hacking have explicitly insisted, and any charitable reader of the other authors cited should understand.)

Case 2. X knows d and, in addition, knows that either $e_{.4} v e_{.6}$ is true but does not know which disjunct is true. According to Kyburg, relative to this corpus of knowledge X should assign a degree of credence equal to the interval from .4 to .6 to the hypothesis h.

Kyburg's analysis of Case 2 implies commitment to the view that every logically permissible P-function must be such that $P(h; d \& (e_{.4} v e_{.6}))$ takes some real value in the range from .4 to .6. This approach is congenial with the analysis of Case 1. According to the analysis of Case 1, every logically permissible P-function must be such as to satisfy the following two conditions:

(1) $P(h; d \& e_{.4}) = .4$
(2) $P(h; d \& e_{.6}) = .6$

Observe, however, that the principle of confirmational coherence implies that

(3) $P(h; d \& (e_{.4}ve_{.6})) = .4x + .6(1 - x)$ where
        $x = P(e_{.4}; d \& (e_{.4}ve_{.6}))$

Since Kyburg imposes no restrictions on the values of x other than those implied by the principle of confirmational coherence, it follows that they may take any value between 0 and 1. This means that the left hand side of (3) may take any value between .4 and .6.

Kyburg also takes one further step. He assumes that a rational agent should endorse the weakest confirmational commitment CIL(UK) consonant with inductive logic. Hence, $c(h; d \& (e_{.4}ve_{.6}))$ should equal the interval [.4, .6].

In this respect, Kyburg follows Carnap in endorsing a necessitarian position according to which a rational agent should endorse CIL as his confirmational commitment. He differs from Carnap in that, according to Carnap, a complete inductive logic would be so powerful as to single out a unique logically permissible P-function and, hence, a unique real valued 'logical' confirmation function. Kyburg's logical confirmation functions are interval valued. To repeat, this difference reflects, on my analysis, a disagreement concerning what constitues a complete inductive logic. Yet, both Kyburg and Carnap are necessitarians.

Those authors who reject necessitarianism might agree with Kyburg's analysis of Case 2 with respect to which P-functions are logically permissible. They would, however, hold that X might use a convex set of P-functions which restricts the range of 'seriously permissible' P-values for h given $d \& e_4$ to some subinterval of the unit line or to some real value (which may be viewed as a degenerate interval). Hence, they would not say that X is obliged to assign h the degree of credence [.4, .6] in Case 2 but say, instead that his degree of credance must be restricted to some subinterval of the interval.

Case 3. In addition to d, X knows $e_{.5}$ (as in Case 1) and that the trial of kind T_1 yielded a ball labelled '1'. (t_2). Moreover, he knows that the percentage of balls labelled '1' which are black is .9 $(f_{.9})$. Let any event be a trial of Kind T_2 if it is a trial of kind T_1 yielding a ball labelled '1'. The chance of obtaining a black ball on a trial of kind T_2 is .9. It follows from d that every trial of kind T_2 is of kind T_1 but not conversely. Hence, the information t_2 together with d entails that the strongest description of the trial event known by X to be true of that event is as a trial of kind T_2. For this reason, Kyburg (and everyone else) agrees that X should base his assignment of a degree of credence to h on that description (or "reference class' as it is misleadingly called). Hence, the degree of credence to be assigned to h is .9.

Case 4. In addition to d, X knows that $e_{.5}$ and $f_{.9}vf_{.95}$. Moreover he knows t_2. Once more, Kyburg would recommend basing the degree of credence assigned to h on the description of the trial as a T_2. In this case, Kyburg would favor assigning the intreval from .9 to .95 to h.

Case 5. In addition to d, X knows that $e_{.5}$ and $f_{.4}vf_{.6}$. As before he knows t_2. In this case, Kyburg would not recommend adopting the narrowest known reference class – i.e., the strongest known description T_2 of the trial event. The reason is that the interval from .4 to .6 spans the 'interval' from .5 to .5 (i.e., the point value .5).

Observe that in this situation, not only does Kyburg recommend that X ignore the narrowest known reference class even though X does not know it to be stochastically irrelevant but also in spite of the fact that he knows that the information that ball is labelled '1' is stochastically relevant. (For reasons cited previously, it appears that Reichenbach would agree.)

It is apparent that Kyburg is committed to the view that every logically permissible P-function satisfies the following condition:

(4) $P(h; d \ \& \ e_{.5} \ \& \ t_2 \ \& \ (f_{.4}vf_{.6})) = .5.$

It is also apparent that confirmational coherence and (4) imply that every logically permissible P-function satisfies the following pair of conditions:

(5) $P(f_{.4}; d \ \& \ e_{.5} \ \& \ t_2 \ \& \ (f_{.4}vf_{.6})) =$
                  $= P(f_{.6}; d \ \& \ e_{.5} \ \& \ t_2 \ \& \ (f_{.4}vf_{.6})) = .5.$

Close scrutiny of this result and its import will disconcert many although it falls short of establishing an inconsistency. What is disconcerting is how effective Kyburg's frequency principle seems to be in enabling one to obtain a numerically precise 'prior' probability for $f_{.4}$. (It is this feature of his approach which enables him to use his version of the frequency principle to generate fiducial arguments without an extra "principle of irrelevance" of the sort invoked by Hacking. I shall not, however, elaborate on this here.)

Before proceeding further, it may be useful to offer an explicit statement of Kyburg's frequency principle.

Kyburg's frequency principle: A: X should base his assignment of a degree of credence to the hypothesis that *e* yields an outcome of kind *R* on the strongest known description *T** known to be true of *e* unless there is a weaker description *T** X* knows to be true of *e* meeting the following conditions:

(1) *X* knows that the chance of a *T*** in an *R* is a value falling in a subinterval of the interval $[r, \bar{r}]$ where the strongest information he knows about the chance of a *T** being an *R* implies that **r** is the least value of that chance and $\bar{r}$ is the largest value

(2) For every other description *T*, *X* knows to be true of *e*, either *X* knows that all events of kind *T** are events of kind *T* (but not conversely) or $[\mathbf{r}, \bar{r}]$ is a subinterval of the narrowest interval in which *X* knows the chance of *T* yielding an *R* to fall.

When this condition is met, $Cr_{x,z}$ (*e* results in an event of kind *R*)= $= [\mathbf{r}, \bar{r}]$.

Review of the previous cases should indicate how Kyburg's rule operates to exempt Case 5 (but not the first four cases) from the prima facie obligation to use the strongest known description of the trial event.

Case 6. The situation resembles Case 5 except that *X* knows $f_{.4}vf_{.5}$. Once more, the interval from .4 to .5 spans .5. Kyburg would recommend assigning *h* the degree of credence .5. That is to say, we obtain the following condition on logically permissible *P*-functions:

(6) $P(h; d \ \& \ e_{.5} \ \& \ t_2 \ \& \ (f_{.4}vf_{.5})) = .5.$

From this by reasoning entirely parallel to what has been used before, we obtain the following result:

(7) $P(f_{.5}; d \ \& \ e_{.5} \ \& \ t_2 \ \& \ (f_{.4}vf_{.5})) = 1.$

This result does not yet contradict our principles; but it comes close. I have not required that *P*-functions be such that $P(x; y)=1$ only if *y* and UK entails *x*. My reason for avoiding this condition of 'regularity' is that in situations where an infinite number of hypotheses exclusive and exhaustive relative to the corpus of knowledge and each consistent with it are considered, it is often desirable and necessary to assign 0 Q-values to some of the alternatives. However, no such necessity obtains in our case. It would seem plausible to suppose that $f_{.5}$ be assigned P-value 1 conditional on the information specified in (7) only if that information entailed $f_{.5}$. It does not.

Notice by the way that on a suitable variant of Case 6, we obtain

(8) $P(f_{.5}; d \& e_{.5} \& t_2 \& (f_{.5}vf_{.6})) = 1.$

Case 7. This is like Cases 5 and 6 except that X suspends judgement between $f_{.4}, f_{.5}$ and $f_{.6}$. We must conclude that every logically permissible P-function satisfies the following condition:

(9) $P(h; d \& e_{.5} \& t_2 \& (f_{.4}vf_{.5}vf_{.6})) = .5.$

The left hand term of (9) equals the left hand side of (6) multiplied by $P(f_{.4}vf_{.5}; d \& e_{.5} \& t_2 \& (f_{.4}vf_{.5}vf_{.6}))$ plus $P(h; d \& e_{.5} \& t_2 \& f_{.6}) \times$ $\times P(f_{.6}; d \& e_{.5} \& t_2 \& (f_{.4}vf_{.5}vf_{.6})).$

This means that the following holds:

(10) $.5 = .5y + .6(1 - y)$

where

$$y = P(f_{.4}vf_{.5}; d \& e_{.5} \& t_2 \& (f_{.4}vf_{.5}vf_{.6}))$$

From this it follows that $y = 1$. It is easy to see that by reiterating the reasoning of Case 6 once more, we obtain:

(11) $P(f_{.5}; d \& e_{.5} \& t_2 \& (f_{.4}vf_{.5}vf_{.6})) = 1.$

Clearly, we can elaborate on this example by adding more disjuncts to the statistical assumption about the percentage of balls labelled '1' which are black. We might indeed allow as disjuncts all statements f_r where r ranges over all numbers represented by three placed decimals from 0 to 1. In this latter case, we have a situation where X knows nothing about the percentage of balls labelled '1' which are black. He completely lacks 'reliable statistics'. In this case, it seems apparent that Reichenbach would favor the Kyburg approach and base his assignment of credence on $e_{.5}$. For him as for Kyburg, the result is the same. X is to assign a degree of credence equal to 1 to the hypothesis that 50% of balls labelled

'1' are black – counter to the clear intent of both Kyburg and Reichenbach and counter to all presystematic judgement.

Thus far we have not obtained a strict contradiction. That requires consideration of one final case:

Case 8: This resembles case 5, 6 and 7 except that X suspends judgement between $f_{.4}, f_{.55}$ and $f_{.6}$. Once more, Kyburg would favor violating the injunction to use the narrowest known reference class. The degree of credence for h should, according to him, equal .5.

(11) $P(h; d \& e_{.5} \& t_2 \& (f_{.4}vf_{.55}vf_{.6})) = .5.$

By reasoning entirely parallel to prior reasoning,

(12) $P(h; d \& e_{.5} \& t_2 \& (f_{.4}vf_{.55})) = .5$

(13) $P(f_{.4}vf_{.55}; d \& e_{.5} \& t_2 \& (f_{.4}vf_{.55}vf_{.6})) = 1.$

(14) $P(f_{.4}vf_{.6}; d \& e_{.5} \& t_2 \& (f_{.4}vf_{.55}vf_{.6})) = 1.$

(13) and (14) imply that the following holds:

(15) $P(f_{.4}; d \& e_2 \& t_2 \& (f_{.4}vf_{.55}vf_{.6})) = 1.$

It is easy to show by the calculus of probability that (15) is incompatible with (11) and the following condition implied by Kyburg's approach:

(16) $P(h; d \& e_2 \& t_2 \& f_{.4}) = .4.$

If we avoid such violations of the injunction to use the strongest known description of the trial event, we escape the difficulties described above. But as Kyburg has pointed out, we confront other problems. Even though we may have precise statistical knowledge of the chance that an A is a B, we may know that e belongs to a narrower reference class and only have vague knowledge of its chances. Indeed, this will often be the case. We cannot appeal to knowledge of stochastic irrelevance to justify ignoring the extra information. If we do take the extra information seriously, we shall be committed to numerically indeterminate degrees of credence often represented by the entire interval from 0 to 1. This will happen in embarrassingly many situations in real life where we feel entitled to reach more definite conclusions.

Kyburg is right in pointing to the problem and stands alone among authors who have formulated frequency principles in attempting to come to terms with it in a thoroughgoing fashion. I have argued that his own solution to the problem is untenable. In my opinion, we should endorse a frequency principle requiring use of all the stochastically relevant information about the trial event. What then should we do short of giving up in despair?

There are two approaches worth mentioning:

(i) One can retain necessitarianism as Kyburg does and supplement the frequency principle and principle of confirmational coherence with additional principles of inductive logic. Hacking, in effect, does just that by introducing a special 'principle of irrelevance'.

(ii) One can abandon necessitarianism and allow rational agents to adopt confirmational commitments stronger than CIL while restricting inductive logic to the frequency principle and the principle of confirmational coherence.

My own view is that a suitably qualified version of the second approach is preferable to the first; but that is a matter which cannot be elaborated upon here.

There is, of course, one other response one can make to these remarks. One can reject some of the restrictions imposed on confirmational commitments. There is some evidence in Kyburg's writings that he is prepared to surrender the principle of confirmational conditionalization which lies at the core of my argument. Once more, space does not permit exploration of this issue in detail. Observe, however, that confirmational conditionalization regulates the range of confirmational commitments x is rationally free to adopt at some time. It does not obligate x to retain the same confirmational commitment over time and, hence, does not imply that the revision of credal states with changes in knowledge is always controlled by conditionalizing arguments. Kyburg's arguments against conditionalization are often cogent when addressed to the use of conditionalization as a principle regulating changes in credal states but they misfire when they are directed against confirmational conditionalization.

Columbia University

NOTES

[1] Kyburg's principle was published in his *Probability and the Logic of Rational Belief*, Wesleyan University Press, Middletown, Connecticut (1961). For a fairly straightforward formulation of Kyburg's frequency principle, see his *Probability and Inductive Logic*, New York: MacMillan (1970) p. 81.
[2] An outline of the considerations favoring these principles is given in my 'On Indeterminate Probabilities' *Journal of Philosophy* **71** (1974), 391–418.

D. V. LINDLEY

BAYESIAN STATISTICS

My purpose in presenting this paper is to explain the basis of the Bayesian argument in statistics as I see it and, as a by-product, to convince you that it is the only satisfactory approach to statistics amongst those currently available. There is nothing new in the paper: it provides merely a brief review of the subject.

Reasons for adherence to the Bayesian viewpoint fall into two categories: firstly, those of a foundational and theoretical character; secondly, those of an operational and practical form. The former category has to be split into two parts corresponding to decision theory and to statistics proper. We discuss these in turn.

1. FOUNDATIONS OF DECISION-MAKING

The generally accepted model for decision-making for a single decision-maker is that of a space D of decisions d, and a space Θ of events θ; the problem being to select a member of D which is, in some sense, best, not knowing which θ obtains. An ordered pair (d, θ) is termed a consequence and it is the relative attractiveness of different consequences that, amongst other things, affects the choice of a decision.

Having set up this structure the next stage of the argument consists in formulating axioms that would seem to describe sensible behaviour in selecting a member of D. Let me informally discuss two axioms that are usually included in some form or other. If we are to select a decision, d_1 say, as best, then this choice means that there is no other $d \in D$ which is strictly preferred to d_1. Let us write this as $d_1 \geqslant d$ for all $d \in D$, where the relation $\geqslant$ is to be read as 'is not worse than'. Then it is customary to suppose that $d_1 \geqslant d_2$ and $d_2 \geqslant d_3$ together imply $d_1 \geqslant d_3$: in other words, the relation is transitive. A second axiom can be expressed most easily by remarking that preferences between decisions imply preferences between consequences. For consider a decision d^* which, whatever θ obtains, gives the same consequence $c^* = (d^*, \theta)$. Then if d^{**} similarly always

Harper and Hooker (eds.), Foundations of Probability Theory, Statistical Inference, and Statistical Theories of Science, Vol. II, 353–363. All Rights Reserved. Copyright © 1976 by D. Reidel Publishing Company, Dordrecht-Holland.

yields c^{**}, we can write $c^* \geqslant c^{**}$ iff $d^* \geqslant d^{**}$. Having preferences between consequences one axiom can be expressed by saying that if $(d_1, \theta) \geqslant (d_2, \theta)$ for all θ, then $d_1 \geqslant d_2$. This is usually strengthened by supposing that if, in addition, $(d_1, \theta) > (d_2, \theta)$ for some θ, then $d_1 > d_2$; where $>$ is to be read as 'strictly better than'. (Compare the distinction between $\geqslant$ and $>$ for real numbers.) This axiom is called the 'sure-thing' principle.

These are only two of the axioms that we commonly use, but they are the two that seem to me to be the most important. Notice that they express essentially *simple* requirements on the behaviour of the decision-maker, in the sense that any decision-maker who violated them would feel that he had made such an elementary error that he could hardly be forgiven. A comparison with the most famous of all axiom sets, namely those of plane Euclidean geometry, is not inappropriate. A person who drew two straight lines through the same pair of points would be committing a rather stupid error. The simplicity of the ideas is an important ingredient of the work of Popper, though expressed somewhat differently. Notice, again like Euclidean geometry, that there is considerable choice in the selection of axioms. Those of Savage, for example, are different from those of de Groot. It doesn't matter too much where you begin, though simple axioms of general appeal, that are easy to verify, are to be preferred to others not having all of these properties.

Having set up an axiom system one can begin to prove theorems. Most axiom systems that have been discussed in the literature lead to three important results. Firstly, that over Θ there exists a probability measure. Secondly, that each consequence (d, θ) has associated with it a real number $u(d, \theta)$ called its utility. Thirdly, that preferences over D can be expressed in terms of the expected utilities of the decisions. In an informal, but hopefully clear, notation, $d_1 > d_2$ iff $\int u(d_1, \theta) p(\theta) d\theta$ is larger than the corresponding expression with d_2 for d_1. In other words, decisions should be selected on the basis of maximizing expected utility.

There are several things to notice about this result but perhaps the most important is to recognize that the existence of probabilities and utilities is *proved*: they are not part of the axiom system. The recognition of this feature of utilities is due to von Neumann, and it distinguishes sharply between those who use utility theory because it seems sensible, and those who recognize it as inevitable, granted the axioms. The effect of the similar recognition about probabilities, due to Savage, will be discussed later in connexion with statistics. The basic idea of proceeding in the way described is due to Ramsey.

In summary, we have a generally satisfactory theory for a single decision-maker that leads to the concept of maximizing expected utility as the only sensible decision procedure. There are still some unsettled points, for example, whether the utility function is bounded, or, perhaps, more importantly, exactly what we mean by a probability measure, is it σ-additive or only finitely additive? Must it be proper, that is, integrate to 1? These are important, both theoretically and practically, but do not affect the general spirit of the argument which seems well-established.

2. FOUNDATIONS OF STATISTICS

What has this theory of decision-making to do with statistics? After all, the greater part of statistics has no overt mention of decisions, being concerned with the collection and handling of data. But this handling of data is with a view to making inferences that go *beyond* the data. A primary task of a statistician is to proceed from one data set to say what another data set is expected to look like[1]. But what is the purpose of the inference? The answer was given by Ramsey. The purpose is to enable decisions to be made. When the inference is made no decisions need be contemplated, but the inference should be put in such a way that any decision-maker whose action depends on the unknown quantity about which the inference has been made can use the inference to help him reach a decision. Philosophers will be interested in Ramsey's reply to certain criticisms by Russell of this point of view. Part of the reply is worth quoting: analogously, Ramsey says "a lump of arsenic is called poisonous not because it actually has killed or will kill anyone, but because it would kill anyone if he ate it".

Once we accept Ramsey's argument that inferences are to be made for decision purposes, we immediately see the form that the inference must take. Namely, we have to state the probability distribution of θ given the data; for it is that ingredient, and that ingredient only, which is needed for any decision problem concerning θ. When a specific decision problem arises we can construct D, consider $u(d, \theta)$ and then calculate the expected utilities using the probability obtained from the inference. Consequently, if we accept the axioms of decision theory and agree that inferences are made with a view to decision-making (and if they are not, what are they for?) we must express the inference in the form of a probability statement about θ; the measure referred to above.

To appreciate the significance of the conclusion reached in the last paragraph we must return to the first of the three theorems already mentioned, namely the existence of a probability measure over Θ, and emphasize that the statement that it is a probability means that it satisfies all the rules of the probability calculus. In particular, the rule for the manipulation of conditional probabilities

$$p(A \cap B \,|\, C) = p(A \,|\, C)\, p(B \,|\, A \cap C).$$

(This rests on an extra axiom, often called the axiom of 'called-off' bets.) This leads immediately to Bayes' theorem and hence to the form of that theorem most useful for inference, namely

$$p(\theta \,|\, x) \propto p(\theta)\, p(x \,|\, \theta),$$

where x refers to the data and all probabilities are conditional on the knowledge available before x was obtained – this conditioning is suppressed in the notation. The three terms in this result are respectively the posterior (to x) distribution of θ, the prior (to x) distribution of θ, and the likelihood of θ (for given x); the omitted constant of proportionality depending only on x, and not θ.

We now see that the inference needs to be expressed in terms of $p(\theta \,|\, x)$, and that this can be found in terms of the likelihood and the original distribution, $p(\theta)$, prior to x, by using Bayes' theorem. (This central role played by Bayes' theorem is why the subject is referred to as Bayesian statistics.) It is the occurrence here of the prior distribution, $p(\theta)$, that is most commonly cited as the reason for not using Bayesian statistics. It certainly does mean that inferences cannot be made without it, but, on the other hand, any methods that do not use it are liable to produce unsatisfactory decision methods. One school of thought maintains that the statistician should confine himself to stating the likelihood function. There is, however, a real difficulty here because in most cases the parameter θ is an ordered pair, (θ_1, θ_2) say, and the inference (or decision) involves only θ_1 and not θ_2. (The latter is often referred to as a nuisance parameter.) We require, therefore, a statement about θ_1 and the likelihood approach is, in general, incapable of providing this. The same difficulty does not arise in the full Bayesian viewpoint since $p(\theta_1 \mid x)$ can easily be found by the usual marginalization device as $\int p(\theta_1, \theta_2 \mid x) \, d\theta_2$.

The difficulty over the prior distribution seems to me grossly exaggerrated. As emphasized earlier, the distribution does exist for any person who subscribes to the axioms and to describe it as unknown, as many statisticians do, is only to say he has not tried to measure it. A major research effort in statistics should be devoted to determining the prior distribution, but that is another topic. A second reason for thinking that objections to the prior are exaggerrated is that we all do have such views and that the conclusions agree well with practical requirements. It is this operational aspect that is discussed in the third section.

3. Operational Aspects of Bayesian Statistics

The considerations just advanced may not impress a scientist or an engineer who feels the need to use statistical ideas in his work. He will be much more interested in whether he can use the concepts in studying what to him is the real problem, namely the scientific or engineering one. Orthodox statistical methods have provided him with a tool that he can use and which is clearly helpful; why should he desert these for different ones? My argument at this conference is much simplified by the presence of Professor Jaynes and the paper that he has given. He shows, rather clearly, that the new methods answer the real questions that the scientist or engineer poses and do not answer a different, and only tangentially relevant, one.

Consider his first example concerning the difference of means. This is formulated as a decision problem but we can extract its inferential element which concerns a single quantity, namely the difference in mean lifetimes of the components supplied by the two manufacturers. Denoting this by θ_1, as above, the parameter of interest, we require a statement about what values of θ_1 are reasonable, and what unreasonable. Surely $p(\theta_1 \mid x)$ x being the data, provides just this for it gives the probability (density) for each value of the difference. From it we can calculate $p(\theta_1 > 0 \mid x)$, the chance that one supplier is better than another (analogous to a significance test) or provide an interval within which θ_1 most probably lies (analogous to a confidence interval). $p(\theta_1 \mid x)$ is, in principle, obtainable from the full posterior probability $p(\mu_1, \mu_2, \sigma_1^2, \sigma_2^2 \mid x)$ by integration, where μ_i, σ_i^2 are the means and variances for the two manufacturers. Here one of the three integrations is tedious and is perhaps best done numerically in any particular case.

My experience tells me that Bayesian statistics is better able to answer a practical problem than is the orthodox theory, because it can, as in the example just considered, go direct to the quantity of interest and make probability statements about it. Furthermore, since the statements are *direct* probability ones (and not indirect ones such as are involved in significance tests or confidence intervals) the procedure for calculating them is rather clear since the rules of the probability calculus can, and indeed must, be invoked. Let me give an example. As mentioned above, many statistical problems are of the form, given one data set what can one say about a second, similar data set. (Knowledgeable readers will know that orthodox theory has cumbersome techniques like tolerance intervals for dealing with this problem: they are rarely used.) Jaynes' example may really be of this form because the customer will experience a single component, constituting the second data set, and will want to know how this is likely to behave. Denoting the second data set by y, Bayesian ideas say that we need $p(y \mid x)$. By a standard result in the probability calculus we have

$$p(y \mid x) = \int p(y \mid \theta, x)\, p(\theta \mid x)\, d\theta$$

where θ is, as before, the unknown parameter. (In the example $\theta = (\mu_1, \mu_2, \sigma_1^2, \sigma_2^2)$.) The model usually supposes that, given θ, y and x are in-independent. If so we can simplify and write

$$p(y \mid x) = \int p(y \mid \theta)\, p(\theta \mid x)\, d\theta$$

which is simply a product of the likelihood given y and the probability given x, integrated.

I have yet to meet a statistical problem that cannot be put into Bayesian terms and, more importantly, where the method of solution cannot be found using the rules of the probability calculus. The solution itself may be difficult for technical reasons – for example, it may involve laborious integrations – but what has to be done is clear. It is like a problem in mechanics where one can write down the equations, it may nevertheless be hard to solve them. A great strength of the Bayesian argument is that, for the first time in statistics, it provides a formal system into which the subject fits. Given a problem, the mode of solution is reasonably clear, and there is no place for *ad hoc* procedures. It is this adhockery that makes standard statistics so difficult to teach. Should we condition on this ancillary? Should we take an unbiased estimate or the maximum likelihood one? Such questions disappear within the Bayesian framework. Or, if you prefer, the only element of arbitrariness comes at one place; the choice of the prior.

I am often asked if the method gives the *right* answer: or, more particularly, how do you know if you have got the *right* prior. My reply is that I don't know what is meant by 'right' in this context. The Bayesian theory is about *coherence*, not about right or wrong. To understand what is meant by coherence let us go back to the idea of an axiomatic structure as discussed in the first part of this paper. The axioms are all concerned with the relationship between one decision and another. For example if you think $d_1 \geqslant d_2$ and $d_2 \geqslant d_3$, then you think $d_1 \geqslant d_3$. It is not concerned with whether $d_1 \geqslant d_2$ is right or not, but merely with whether three judgements agree or not. We speak of the three judgements as *cohering* and a person who has $d_1 \geqslant d_2$ and $d_2 \geqslant d_3$ but nevertheless, in violation of the axioms, has $d_3 \geqslant d_1$, as *incoherent*. Consequently the theory that springs from it is only concerned with coherence; with how different views fit together, not with judgements of right or wrong. Perhaps the most obvious example of coherence is the way in which views prior to the data must cohere with those posterior to it: namely by Bayes' theorem.

It may be objected that coherence is not enough, and that the question of rightness or wrongness remains. I am not aware of any definition here, nor do I see how there could be one. Several people have tried to produce one, usually based on some invariance concept, the best-known being the attempt of Jeffreys. So far as I am aware, all have failed. (The attempt of Fraser will be discussed elsewhere in this conference.) The idea of being right is usually confused with another notion. Suppose, in the general framework with D and Θ, you were informed of the true value, θ_0, of θ. (Notice that this is a piece of data, often called perfect information.) Consequently provided there is agreement on the utility structure – and since utility is not involved in the inference aspect, this proviso is statistically irrelevant – there is a unique, and presumably, right decision. But usually we do not know θ, that is the essence of the problem, and such judgements cannot be made. When we say "he made the wrong decision" we usually mean "he made a decision which, in the light of subsequent information which he did not have at the time of making the decision, proved to be incorrect". Looked at in probability terms, someone who has a distribution $p(\theta)$ which is concentrated about what ultimately turns out to be the true value, might be better than someone who had a more diffuse distribution or one concentrated about another value, but this seems dangerously like hindsight judgement which could express luck (in assessing the prior) rather than skill.

It should be noted that the concept of coherence is entirely lacking from the orthodox theory of statistics. Neyman and Pearson never asked themselves whether what they were advocating for a sample of size 10 cohered with their recipe for 20 – had they done so they would have found it didn't. More importantly, statisticians have never asked themselves whether two judgements about the same quantity cohere. For instance, suppose, in Jaynes' first example, that an alternative method of comparing the lifetimes had involved Poisson distributions rather than normal; would the two sets of inferences cohere? The coherence is most easily achieved by using a probability distribution for the unknown quantities. Here μ_1 and μ_2 would be common to the normal and Poisson formulations and a single choice of $p(\mu_1, \mu_2)$ would ensure that the axioms would not be violated. As I have said elsewhere; statisticians should stop thinking about their parameters as Greek letters but remember they refer to real things. The idea of assigning a probability distribution to a normal mean is quite different from assigning one to the mean lifetime of these components; for the latter refers to real things, the former to abstractions. Bayesian statistics is about the physical world, and not about Greek letters. This is another aspect of how it is more suited to practical applications than the usual theory.

I have mentioned that the concept of coherence is absent from orthodox statistics. This absence means that Bayesian statistics is quite different from standard teaching. Although there is a close similarity between some results in the two theories, this is essentially because the standard ideas are only capable of dealing with very special cases, and the peculiarities of these lead to accidental agreements. To do good Bayesian statistics [2] it is best to forget the orthodox ideas: just follow the precepts of the probability calculus and the answer will appear. Many so-called Bayesian papers that I am asked to referee fail in this respect: they bowdlerize the argument with irrelevant references to, say, linear estimates. If a linear estimate is appropriate the Bayesian argument will tell you so – there is no need to invoke a new principle.

The differences between the two schools of thought are most easily understood by considering the likelihood principle. Returning to Bayes' theorem in its usual form

$$p(\theta \mid x) \propto p(x \mid \theta) \, p(\theta),$$

with data x and parameter θ, we see that in calculating $p(\theta \mid x)$, our inference about θ, the only contribution of the data is through the likelihood function, $p(x \mid \theta)$. In particular, if we have two pieces of data x_1 and x_2 with the same likelihood function, $p(x_1 \mid \theta) = p(x_2 \mid \theta)$, the inferences about θ from the two data sets should be the same. This is not usually true in the orthodox theory, and its falsity in that theory is an example of its incoherence. The standard example is that of a binomial sequence of trials with successes, S, and failures, F, and θ the chance of success at any trial. Suppose the data consists of the sequence

$$S\,S\,F\,S\,F\,S\,F\,F\,S\,S\,F$$

then the likelihood is $\theta^6(1-\theta)^5$ and the inference is through $\theta^6(1-\theta)^5 p(\theta)$. In orthodox theory this is not so. If the above sequence had been obtained as a result of performing 11 trials, the usual estimate of θ is 6/11: had it been obtained by carrying out trials until the fifth failure had been observed, the usual estimate is 5/10, despite the fact that the likelihood is the same in both situations.

The reason for the difference is that the orthodox theory typically considers results that *might* have occurred in the experiment but did not. For example, with the sequence above, if 11 trials had been performed, all sequences of 11 S's and F's would have been considered: with continuation until the fifth failure sequences like $F\,F\,S\,F\,F\,F$ with 6 elements would have been employed. As Jeffreys has said, what has what might have happened, but did not, got to do with inferences from the experiment? The most obvious violation of the likelihood principle occurs with the idea of a confidence interval, with its concept of repetition of the experiment. Enough counter-examples to show the absurdity of confidence intervals are now available to discredit the idea. In Bayesian statistics every quantity is a random variable until its numerical value is observed, when it becomes a number and its random nature is irrelevant. Once the data is to hand, the distribution of X (or $\tilde{x}$), the random variable that generated the data, is irrelevant.

I have tried, in this brief paper, to state what seem to me to be the main reasons for adhering to the Bayesian position. They are both foundational and operational. I believe it is the statistics of the future because it does work in many practical situations and because, having sound foundations, it will work in others yet to be explored. A fuller account has been provided in my monograph *Bayesian Statistics: A Review* published by SIAM, Philadelphia (1971). It contains a fairly complete bibliography. Since that was compiled the important book by George E. P. Box and George C. Tiao *Bayesian Inference in Statistical Analysis*, Addison-Wesley (1973) has appeared.

University College, London

NOTES

[1] He often passes from the given data set to a statement about parameters, θ. But this introduction of parameters is essentially a technical device and the basic problem is as stated.

[2] And therefore good statistics. Melvin Novick has suggested to me that we ought not to use the term 'Bayesian statistics' implying that it is some special form of statistics. It is not: it is only special in that it is coherent statistics. The rest is incoherent.

DISCUSSION

Commentator Good: Professor Lindley's paper gives an excellently succinct account of the Bayesian position *with sharp probabilities*. I wish to emphasize that this is by no means the only kind of Bayesian position, although I made this clear in my paper at this Workshop.

On his own ground, I have one criticism of his summary. He says that the probability distribution of θ given the data is the only ingredient needed for a decision problem, since, when a decision problem arises, we can take the utility structure into account. I agree with this for the situation *after* the data are obtained, but in the *design of an experiment*, decisions must be made in advance of a knowledge of the data, and then the idea of quasi-utilities often becomes relevant, as discussed in my paper. One reference in that paper is indeed to some work of Lindley himself, on the use of Shannon information, so he has not overlooked the point. But without this comment his brief summary of the 'sharp probability' Bayesian thesis might be misleading because the design of experiments makes up a substantial portion of statistics.

Lindley: Professor Good mentions a difficulty in the design of experiments, particularly in connexion with Shannon information. A research student of mine, J-M. Bernardo, has recently pointed out to me how the difficulty can be overcome within the framework outlined in my paper. The device is to suppose the *decision* space in experimental design to consist of posterior probability distributions of θ. We then have to assign a utility $U(p(\cdot), \theta)$ to each posterior $p(\cdot)$ when θ is the true value. If this depends on $p(\cdot)$ only through $p(\theta)$, the only utility function that is sensible is Shannon's. If the possible values of θ are finite in number the result is well explained by L. J. Savage (*J. Amer. Statist. Assoc.* **66** (1971) 783–801) and the extension to more general situations is straightforward.

GLENN SHAFER

A THEORY OF STATISTICAL EVIDENCE

TABLE OF CONTENTS

Harper and Hooker (eds.), Foundations of Probability Theory, Statistical Inference, and Statistical Theories of Science, Vol. II, 365–436. All Rights Reserved. Copyright © 1976 by D. Reidel Publishing Company, Dordrecht-Holland.

1. Evidence

There are at least two ways in which the impact of evidence on a proposition may vary. On the one hand, there are various possible degrees to which the evidence may support the proposition: taken as a whole, it may support it strongly, just a little, or not at all. On the other hand, there are various possible degrees to which the evidence may cast doubt on the proposition: taken as a whole, it may cast serious doubt on it, thus rendering it extremely doubtful or implausible; it may cast only moderate doubt on it, thus leaving it moderately plausible; or it may cast hardly any doubt on it, thus leaving it entirely plausible.

1.1. Degrees of support and plausibility

In this essay, I formally distinguish these two aspects of the evidence's impact. I say that the evidence supports a proposition to a certain extent, thus endowing it with a certain *degree of support*, and that it casts doubt on it to a certain extent, thus endowing it with a certain *degree of plausibility*.

I approach these degrees of support and plausibility with two ambitions. First, I hope that in some situations they can actually be represented by numbers. And secondly, I hope that in these situations such numerical degrees of support and plausibility for relevant propositions will be sufficient to *completely summarize* the evidence's impact on our knowledge and opinion.

A proposition's degree of support and its degree of plausibility are obviously related, and it might seem that they are so strongly related that the one should determine the other. But they are not. The fact is that while a high degree of support does imply a high degree of plausibility, a low degree of support is compatible both with a low degree of plausibility and with a high degree of plausibility. If the evidence supports the proposition not at all and casts a great deal of doubt on it, then it endows it with a low degree of support and a low degree of plausibility. But if the evidence fails to provide much support for the proposition and also fails to cast much doubt on it, then it endows it with a low degree of support and yet leaves it with a high degree of plausibility. Actually, this latter situation is all too common, for it arises whenever the evidence is scanty. When there is little evidence bearing on a proposition, that proposition cannot be said to be supported by the evidence, but it is plausible even in light of the evidence.

I want these degrees of support and plausibility to be numbers. What numbers?

Consider first the range of numbers that we might want for degrees of support.

At one extreme, when there is no support for a proposition, we will want to say that its degree of support is zero. At the other extreme, we will want a maximum degree of support, corresponding to the case where the evidence establishes the proposition for certain. A convenient convention is to set this maximum degree of support equal to one. So when we measure the degree of support for a proposition we will assign the proposition a number between zero and one.

This same scale from zero to one also seems appropriate for degrees of plausibility.

A proposition will have degree of plausibility zero when the evidence is conclusively against it, and degree of plausibility one when there is no evidence against it.

1.1.1. *The Support and Plausibility Functions*

We are usually interested in degrees of support and plausibility for more than one proposition at a time. For example, when we are concerned with the true value of some quantity θ, we are interested in any proposition that asserts that the true value is included in a given subset of the set of possible values.

Denoting the set of possible values by Θ, the propositions of interest are precisely those of the form 'The true value of θ is in A', where A is a subset of Θ.

Hence the propositions of interest are in a one-to-one correspondence with the subsets of Θ, and for the sake for convenience we can 'identify' them with these subsets.

So denoting the set of all subsets of Θ, or the *power set* of Θ, by the symbol 2^{Θ}, we can describe our problem as that of specifying two functions on 2^{Θ}. First we want a function

$$S: 2^{\Theta} \rightarrow [0, 1]$$

such that $S(A)$ is the degree of support for the subset A. And secondly, we want a function

$$PL: 2^{\theta} \rightarrow [0, 1]$$

such that $PL(A)$ is the degree of plausibility of A.

This formalism may seem fairly special. For we are often interested in propositions that do not deal with the value of a numerical quantity. But it becomes quite general if we allow θ to be a 'parameter' that takes possibly non-numerical values. For example, we might let θ be 'the date and place of origin of the relic in my hand'. In this case, the possible values of θ would be pairs, each pair consisting of a date and a place.

Whatever θ is, it should be noted that whenever a subset $A \subset \Theta$ is thought of as a proposition, its complement $\bar{A}$, the set of all elements of Θ not in A, must be thought of as the negation of that proposition. Notice also that the empty set $\emptyset$ is in 2^{θ}; it corresponds to the proposition that is necessarily false, for the true value of θ cannot be in $\emptyset$. And the entire set Θ is also in 2^{θ}; it corresponds to the proposition that is necessarily true, for by assumption the true value of θ is in Θ.

1.1.2. *The Relation Between S and PL*

I have already pointed out the relation between a proposition's degree of support and its degree of plausibility: high support implies high plausibility, but low support is compatible with both low and high plausibility. We can formalize this relation by requiring that the degree of plausibility be at least as great, but possibly greater than the degree of support. In symbols:

(1) $S(A) \leqslant PL(A)$

for each $A \in 2^{\theta}$. In words: plausibility is easier to come by than support.

A fundamental relation between support and plausibility can be discerned when one compares the degree of plausibility of a proposition $A \in 2^{\theta}$ with the degree of support for its negation $\bar{A}$. Recall that a proposition A is plausible to the extent that the evidence fails to cast doubt on it. But casting doubt on A is really the same thing as supporting $\bar{A}$. Hence A is plausible to the extent that $\bar{A}$ fails to be supported; $PL(A)$ is large to the extent that $S(\bar{A})$ is small.

Since both $PL(A)$ and $S(\bar{A})$ are measured on a scale from zero to one, the most natural way to make this relation precise is to set

(2) $\qquad PL(A) = 1 - S(\bar{A})$

for all $A \in 2^{\Theta}$.

This relation implies not only that we can obtain the function PL from knowledge of S, but also that we can obtain S from knowledge of PL. For (2) implies that

(3) $\qquad S(A) = 1 - PL(\bar{A})$

for all $A \in 2^{\Theta}$. Hence the functions PL and S convey exactly the same information.

1.1.3. *Elementary Rules for S and PL*

It is worth noting that the relations (1) and (2) imply that

$$S(A) + S(\bar{A}) \leqslant 1$$

for all $A \in 2^{\Theta}$. Verbally: it is impossible for both a proposition and its negation to be well supported. Similarly, (1) and (3) imply that

$$PL(A) + PL(\bar{A}) \geqslant 1$$

for all $A \in 2^{\Theta}$. Verbally: for every proposition, either it, its negation or both must be fairly plausible.

There are several other rules that S and PL should obey. For one thing, the elements $\emptyset$ and Θ of 2^{Θ} are rather special. No matter what the evidence is, $\emptyset$ is impossible and hence $S(\emptyset) = PL(\emptyset) = 0$. Similarly, Θ is always taken to be certain, and even in the absence of any evidence we would set $S(\Theta) = PL(\Theta) = 1$. The function S ought also to obey the rule of monotonicity:

$$\text{If} \quad A \subset B, \quad \text{then} \quad S(A) \leqslant S(B).$$

This rule is unavoidable, for when $A \subset B$, any support for the value of θ being in A is also support for the value of θ being in B. Finally, the rule of monotonicity for S implies exactly the same rule for PL:

$$\text{If} \quad A \subset B, \quad \text{then} \quad PL(A) \leqslant PL(B).$$

The rules for S and PL are summarized in Table I.

TABLE I

Rules for S and PL. To the right of each rule for PL is the corresponding rule for S, based on the relation $S(A)=1-PL(A)$

Rules for plausibility	Rules for support
$PL(\emptyset)=0$	$S(\Theta)=1$
$PL(\Theta)=1$	$S(\emptyset)=0$
If $A\subset B$, then $PL(A)\leqslant PL(B)$	If $A\subset B$, then $S(A)\leqslant S(B)$
$PL(A)+PL(\bar{A})\geqslant 1$	$S(A)+S(\bar{A})\leqslant 1$

1.2. The case of two alternatives

The simplest support and plausibility functions occur when θ consists of only two alternatives; say $\Theta=\{\theta_1, \theta_2\}$. In this case the support function $S: 2^\Theta \to [0, 1]$ will be completely determined by two numbers: $S(\{\theta_1\})=$ $=s_1$ and $S(\{\theta_2\})=s_2$. And since $\{\theta_2\}=\overline{\{\theta_1\}}$, these two numbers must obey $s_1+s_2\leqslant 1$.

TABLE II

The general form for S and PL when
$\Theta=\{\theta_1, \theta_2\}$ $(s_1+s_2\leqslant 1)$

A	$S(A)$	$PL(A)$
$\emptyset$	0	0
$\{\theta_1\}$	s_1	$1-s_2$
$\{\theta_2\}$	s_2	$1-s_1$
Θ	1	1

1.2.1. *No Evidence*

It is generally difficult if not impossible to actually assess the evidence and arrive at numerical degrees of support and plausibility. But in one case it is easy – the case where there is no evidence. When there is no evidence none of the propositions in 2^Θ can be supported, and hence all must have degree of support zero. Dually, none can have doubt cast on them and hence all must have degree of plausibility one. Of course, we must make two exceptions: $\emptyset$ being logically impossible, we must have $PL(\emptyset)=0$; and Θ being logically certain, we must have $S(\Theta)=1$. So when there is no evidence we obtain the *vacuous* support function, which assigns every proposition except Θ support zero, and the *vacuous* plausibility function, which assigns every proposition except $\emptyset$ plausibility one.

In terms of Table II, such a vacuous support function would be represented by setting $s_1 = s_2 = 0$. Let me give a concrete example. My friend the art collector shows me a vase and tells me that it has been represented as a product of the Ming dynasty. He asks me what I think – is it genuine or is it counterfeit? Now the only thing I know about vases is that flowers can be kept in them, and the only thing I know about the Mings is that they were not Frenchmen. Surely you will agree that I have no evidence and hence should adopt the vacuous support function, shown in Table III.

TABLE III

The vacuous support function
when $\Theta = \{$genuine, counterfeit$\}$

A	$S(A)$	$PL(A)$
$\emptyset$	0	0
$\{$genuine$\}$	0	1
$\{$counterfeit$\}$	0	1
Θ	1	1

1.2.2. *Conflicting Evidence*

The case of no evidence is easy. But we usually have some evidence. Indeed, there is often so much evidence that we can find some against each possibility.

Consider again the question of whether the vase is genuine or counterfeit: $\Theta = \{$genuine, counterfeit$\}$. By definition it is one or the other, but I will not be surprised if I find some evidence pointing to its being genuine and some evidence pointing to its being counterfeit. Suppose I do. For the sake of concreteness, suppose my evidence comes from a study of the painting on the vase. My evidence for the vase's being genuine may be the particular skill exhibited in the intricate design, a skill that is not believed to have survived the Ming period. And my evidence for its being counterfeit may be the presence of a certain pigment previously believed to have been unavailable during the Ming period. Suppose both of these items of evidence seem to be fairly weighty, but equally weighty. Then the total body of evidence can be aptly described as internally conflicting: there may be quite a bit of it, but is points in both directions at once.

What does the plausibility function look like? Both alternatives have evidence against them and thus doubt cast on them; hence neither remains completely plausible. Let us suppose they both retain plausibility 3/4. Then we obtain the support and plausibility functions shown in Table IV. Notice that the set $\Theta = \{$genuine, counterfeit$\}$ retains plausibility one even though neither of its individual elements are that plausible any more.

Table IV might strike you as a complicated way of saying nothing. "Sure you have all that evidence", you might argue. "But it cancels itself

TABLE IV

A support function that indicates internally
conflicting evidence

A	$S(A)$	$PL(A)$
$\emptyset$	0	0
{genuine}	1/4	3/4
{counterfeit}	1/4	3/4
Θ	1	1

out. Why should you say that the evidence supports both alternatives to a degree of 1/4? It would be simpler to say that the tendencies to provide support in the two opposite directions cancel each other completely, leaving support zero and plausibility one for both alternatives". In other words, you might argue that this precisely balanced conflicting evidence really comes down to the same thing as no evidence at all.

Admittedly, not all the vocabulary we use nowadays to discuss evidence is adapted to distinguishing between a lack of evidence and the presence of conflicting evidence. For example, conflicting evidence does no better than no evidence in providing us with 'information' or helping us make a 'decision'. But the difference between no evidence and conflicting evidence is both real and practical and should be basic to any theory of evidence.

1.2.3. *The Combination of Evidence*

The importance of the difference between no evidence and conflicting evidence emerges clearly when we undertake to combine the evidence we already have with new evidence. Suppose, for example, that we have the conflicting evidence summarized in Table IV, and that we then obtain new evidence strongly in favor of the vase's being genuine. Indeed, suppose the new evidence supports that alternative fairly strongly and does not cast any doubt on it at all. Then this new evidence, taken by itself or combined with 'no evidence', will result in $S(\{\text{genuine}\})$ being fairly high and in $PL(\{\text{genuine}\})$ being equal to one. But when this new evidence is combined with the previous conflicting evidence, the result will be different. Certainly the new evidence will shift the balance in favor of the vase's being genuine, and perhaps it will raise the degree of support for that alternative. But the original evidence against the vase's being genuine (the suspicious pigment) will remain part of the combined evidence, and hence $PL(\{\text{genuine}\})$ will not be equal to one.

Let me make the example more concrete. Say the new evidence is the testimony of an expert who has analyzed the chemical composition of the clay in the vase and has concluded that it could only have come from the Ming period. Now we may put some faith in his expertise and his honesty, so his testimony will provide positive support for the vase's being genuine. But experts can be wrong, so we will not regard his testimony as conclusive evidence. Suppose we think it provides a degree of support of 1/2 for the vase's being genuine. Then when considered alone or combined with no evidence it will produce the support and plausibility functions given in Table V.

TABLE V

The support function based on the new evidence

A	$S(A)$	$PL(A)$
$\emptyset$	0	0
{genuine}	1/2	1
{counterfeit}	0	1/2
Θ	1	1

Let us consider the degrees of support that might be expected to result from the combination of the old evidence represented by Table IV with the new evidence represented by Table V. It seems reasonable that the combination of the two bodies of evidence should produce a fairly high degree of support for the vase's being genuine, perhaps higher than the degree of support in either table. On the other hand, the degree of support for the vase's being counterfeit ought to fall between the 0 in Table V and the 1/4 in Table IV. There will be positive support, deriving from the old evidence, but the one-sidedness of the new evidence will erode it considerably. Numbers in Table VI seem to meet these general requirements.

TABLE VI

The support function resulting
from combination

A	$S(A)$	$PL(A)$
$\emptyset$	0	0
genuine	4/7	6/7
counterfeit	1/7	3/7
Θ	1	1

1.2.4. Lambert's Rule

Actually, Table VI can be obtained from Tables IV and V by the application of a simple rule that was first proposed by J. H. Lambert in 1764 and was rediscovered by A. P. Dempster in 1966.

In order to describe the application of this rule, we need to learn how to represent a support function over two alternatives by a mass that is uniformly distributed over a line segment. This is done in Figure 1 for the support function in Table IV.

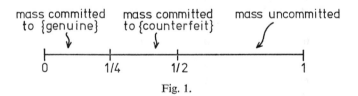

Fig. 1.

In that figure, 1/4 of the mass is committed to {genuine}, corresponding to the degree of support of 1/4 for that alternative, and 1/4 is committed to {counterfeit}, corresponding to the degree of support of 1/4 for that alternative. The support function in Table V is similarly represented in Figure 2. Since the degrees of support for two alternatives must always

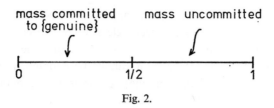

Fig. 2.

add to some number less than or equal to one, any support function over two alternatives can be represented in this way.

In order to combine these two support functions, we combine the two line segments orthogonally, obtaining the square shown in Figure 3. The division of the line segment of Figure 1 into three pieces then induces a division of the square into three vertical strips, while the division of the line segment of Figure 2 into two pieces induces a division of the square into two horizontal strips. Altogether the square is thus partitioned into $3 \times 2 = 6$ rectangles which I have labeled with the letters A through F.

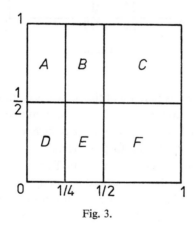

Fig. 3.

Let us consider how each of these six rectangles is affected by the two support functions. Rectangle A is committed to {genuine} by the first support function – a commitment that is not challenged by the second support function; hence it may be counted as committed to {genuine}. Rectangle B is similarly committed to {counterfeit}. Rectangle C, not being committed by either of the two, must be counted as uncommitted.

Rectangle D is committed to {genuine} by the first support function, and the second concurs. But for rectangle E there is a conflict: the first support function would commit it to {genuine} and the second would commit it to {counterfeit}; hence it cannot be counted at all. Finally, rectangle F is committed to {genuine} by the second support function, and this is not challenged by the first one (Figure 4).

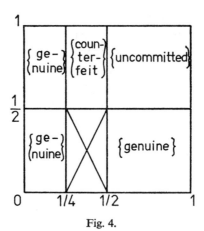

Fig. 4.

The net result, then, is that 4/8 of the square is committed to {genuine}, 1/8 is committed to {counterfeit}, 2/8 is uncommitted, and 1/8 cannot be counted. So of the 7/8 of the square that can be counted, 4/7 is committed to {genuine}, 1/7 is committed to {counterfeit}, and 2/7 is uncommitted. This result corresponds to the numbers in Table VI.

It should be obvious that the procedure used here can be used to combine any pair, or any larger number, of support functions over the same two alternatives. The rule is so simple as to seem almost silly, but its results are always intuitively reasonable.

1.2.5. *Caveat*

Lambert formulated his rule only for the case of two alternatives, but as we will see in §3, Dempster's more general rule of combination can be applied when Θ has any number of elements, provided that the support functions satisfy certain further conditions. This greater generality is important, for the use of the rule for only two alternatives has severe limitations.

The fact is that our evidence often refers to rather large sets of alternatives and loses much of its precision when we narrow our attention to a single dichotomy. In the case of the vase, for example, we might obtain new evidence not bearing directly on the vase's origin, but merely tending to impeach one of the sources of our old evidence. Such new evidence certainly ought to affect the degrees of support and plausibility based on our old evidence, but it can hardly do so by means of Lambert's rule. For it will produce only a vacuous support function when applied directly to the dichotomy $\Theta = \{$genuine, counterfeit$\}$. In order to combine the old and new evidence properly, we would have to apply Dempster's rule to support functions over a set Θ large enough to distinguish not only whether the vase is genuine but also whether the various sources of evidence are trustworthy.

In general, then, Dempster's rule ought always to be applied to support functions over sets Θ that are large enough to make all relevant distinctions. If Θ represents too coarse a division of the possibilities, then the result of the combination may be inaccurate. We can achieve our ambition of completely summarizing the evidence by a support function only if we make Θ sufficiently large.

1.3. CONSONANCE AND DISSONANCE

When Θ consists of two alternatives, a support function that awards positive degrees of support to both alternatives betrays internal conflict or *dissonance* in the evidence. Dissonance can be revealed in a similar way when Θ is larger. Indeed, whenever $A \in 2^\Theta$ and both $S(A)$ and $S(\bar{A})$ are positive, the evidence and the support function S are clearly internally conflicting and hence dissonant. Is this the only way in which a support function can betray dissonance in the evidence?

It is not the only way. As we will see in the following example, a support function can betray dissonance even when it avoids the outright conflict involved in supporting both sides of a dichotomy.

1.3.1. *Another Vase*

Suppose $\boldsymbol{\theta}$ is the exact place of origin of a Chinese vase. We may take Θ to be the set of all places in China, or we may abstract a bit and take Θ to be the set of all points on a map of China; this will give us an infinite number of 'places' in China, many of them very close together. And suppose the evidence is dissonant.

What might it mean for the evidence to be dissonant in this example? Consider a subset A of the map and two subsets A_1 and A_2 of A such that $A_1 \cap A_2 = \emptyset$ and $A_1 \cup A_2 = A$; say A is the province of Honan and A_1 and A_2 are western and eastern Honan, respectively. For the evidence to be dissonant with respect to A_1 and A_2 would mean that some evidence points in one direction, say towards A_1 and away from A_2, while other evidence points in the other direction, towards A_2 and away from A_1. It is not hard to imagine how such conflicting evidence might arise; some aspects of the vase's design might resemble other pottery from sites in eastern Honan, while other aspects might seem more likely to have come from the west.

But we must draw some distinctions. Since Honan does not make up the whole map, evidence pointing towards western Honan is not exactly the same thing as evidence pointing away from eastern Honan. Hence, we can distinguish two different kinds of dissonance. First, there might be some evidence pointing towards, and hence supporting western Honan, and other evidence pointing towards, and hence supporting eastern Honan. Since eastern and western Honan are disjoint, this would certainly be a conflict. Secondly, there might merely be some evidence pointing away from western Honan without pointing away from eastern Honan and other evidence pointing away from eastern Honan without pointing away from western Honan. In other words, there might be some evidence casting doubt on western Honan but not on eastern Honan and other evidence casting doubt on eastern Honan but not on western Honan. In this case, the evidence taken as a whole would have to be considered dissonant, even though we might not want to say that it is in clear conflict with itself. Let us consider both kinds of dissonance in turn.

In the first case, we have positive support both for $A_1 =$ western Honan and for $A_2 =$ eastern Honan. These two subsets are disjoint: $A_1 \cap A_2 = \emptyset$. So dissonance is revealed in this case by the fact that

$$(4) \qquad A_1 \cap A_2 = \emptyset, \qquad S(A_1) > 0 \quad \text{and} \quad S(A_2) > 0.$$

Notice that the disjointness of A_1 and A_2 implies that $A_2 \subset \bar{A}_1$; hence

$S(\bar{A}_1) \geqslant S(A_2) > 0$. So whenever (4) occurs, we have

(5) $S(A_1) > 0$ and $S(\bar{A}_1) > 0$.

Hence (4) is merely another way of saying that there is positive support for both sides of a dichotomy.

Fig. 5.

The second kind of dissonance, which is weaker but more interesting, is best described in terms of the plausibilities $PL(A)$, $PL(A_1)$ and $PL(A_2)$. There may be some doubt cast on Honan as a whole and that doubt will also apply to eastern Honan and to western Honan, thus pushing down all three values $PL(A)$, $PL(A_1)$ and $PL(A_2)$. But in the case we are describing, there is evidence casting additional doubt on western Honan but not on eastern Honan and hence not on Honan as a whole, and also evidence casting additional doubt on eastern Honan, but not on western Honan and hence not on Honan as a whole. Hence, both $PL(A_1)$ and $PL(A_2)$ will be pushed down farther than $PL(A)$; we will have both $PL(A_1) < PL(A)$ and $PL(A_2) < PL(A)$. So the occurrence of the relations

(6) $PL(A_1) < PL(A_1 \cup A_2)$ and $PL(A_2) < PL(A_1 \cup A_2)$

marks our second kind of dissonance.

The relations (6) can occur even without the occurrence of positive degrees of support for both sides of any dichotomy. In order to verify this, let us make our example numerical. Let the value of $PL(B)$ for nonempty B be given by (i) $PL(B) = 1$ if B is not contained in Honan, (ii) $PL(B) = 1/2$ if B is contained in Honan but includes points from both western and eastern Honan, and (iii) $PL(B) = 1/4$ if B is completely contained in either western or eastern Honan.

These values assure that (6) does occur, and they result in the following quantities $S(B)$ for proper subsets B of Θ: (i) $S(B) = 0$ if B does not include the complement of Honan, (ii) $S(B)=1/2$ if B contains the complement of Honan but excludes points from both western and eastern Honan, and (iii) $S(B)=3/4$ if B contains both the complement of Honan, and all of either western Honan or eastern Honan. It is evident that S does not obey both $S(B)>0$ and $S(\bar{B})>0$ for any $B \in 2^{\Theta}$, for it is impossible for both B and $\bar{B}$ to contain the complement of Honan.

We should pause to remark that (6) does not require A_1 and A_2 to be disjoint. This is appropriate, for (6) is a symptom of dissonance even when A_1 and A_2 overlap. Suppose, for example, that A_1 is the western two-thirds of Honan, while A_2 is the eastern two-thirds. Then (6) would still betray the presence of some evidence pointing in both directions. In fact, it would imply the same relation for the case where A_1 and A_2 are the western and eastern halves of Honan.

We have isolated (5) and (6) as different symptoms of dissonance. In fact, however, (5) is merely a special case of (6). For if we set A_1 in (6) equal to A and A_2 equal to $\bar{A}$, we obtain

$$PL(A) < PL(A \cup \bar{A}) = 1 \quad \text{and} \quad PL(\bar{A}) < PL(A \cup \bar{A}) = 1;$$

and when this is translated by (3), it becomes (5). So (6) is the most general symptom of dissonance we have discerned so far.

1.3.2. *The Definition of Consonance*

Let us consider now a plausibility function that is completely non-dissonant. Such a function will fail to satisfy (6) and hence will satisfy

$$(7) \qquad PL(A_1 \cup A_2) = \max_{i=1,2} PL(A_i)$$

for all pairs $A_1, A_2 \in 2^{\Theta}$. In fact, it will satisfy

$$PL(A_1 \cup \cdots \cup A_n) = \max_{i=1,\ldots,n} PL(A_i)$$

for all finite collections $A_1, \ldots, A_n$ of elements of 2^{Θ}, for (7) implies (8).

Now a relation like (8) naturally causes one to ask whether the analogous relation for infinite collections should hold. Does (8) imply that

$$(9) \qquad PL\left(\bigcup_{\gamma} A_{\gamma}\right) = \sup_{\gamma} PL(A_{\gamma})$$

for all non-empty collections $\{A_{\gamma}\}$ of elements of 2^{Θ}? Unfortunately, it does not. If Θ is infinite, then (9) is a stronger condition on PL than (8).

Nonetheless, if *PL* were based on strictly non-dissonant evidence, then it would be reasonable to expect it to obey (9) as well as (7) and (8). Hence I will take (9) to be the criterion for completely non-dissonant evidence. I will say that a plausibility function *PL* is *consonant* if it obeys (9).

One virtue of this definition is that it reduces to a much simpler form. It is easily verified that a plausibility function $PL: 2^\theta \rightarrow [0, 1]$ is consonant if and only if

(10) $$PL(A) = \sup_{\theta \in A} PL(\{\theta\})$$

for all non-empty $A \in 2^\theta$. This implies that a consonant plausibility function is completely determined by its values on singletons.

Conflicting evidence is the rule rather than the exception in this life, and we can expect most of the plausibility functions we meet to be dissonant. But the definition of consonance is interesting because it shows what this dissonance costs in terms of complexity. Consonant plausibility functions have a very simple structure, but dissonant ones do not, and the more dissonant they are the more complex their structure is.

1.4. THE LIMITS OF DISSONANCE

Though most plausibility functions exhibit some dissonance, the nature of empirical evidence seems to set limits on the degree of possible dissonance. Let us return to the example involving the map of China to see how such limits arise.

In that example, I raised the possibility that both western Honan and eastern Honan might be less plausible as places of origin of the vase than the province as a whole. If we partition the province more finely, we can adduce examples of more severe dissonance. Consider, for example, a partition into three regions as shown in Figure 6. It is possible for Honan as a whole to be more plausible than any of the regions $A_1 \cup A_2$,

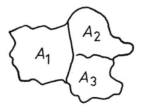

Fig. 6.   Honan partitioned into western, northern and southern regions.

$A_1 \cup A_3$ or $A_2 \cup A_3$ taken singly. This would happen if for each of the three regions $A_1 \cup A_2$, $A_1 \cup A_3$ and $A_2 \cup A_3$ there were some evidence casting doubt on it but not on the remaining third of Honan.

We need not stop with thirds: given any integer n, we can partition Honan into n regions and raise the possibility that the union of any $(n-1)$ regions is subject to doubt not applying to the remainder of the providence and hence that all such unions are less plausible than the province as a whole. Following this line of thought to its furthest extreme, we might postulate that every proper subset of Honan is subject to some doubt that does not apply to all of the remainder, and hence is less plausible than the province as a whole.

Clearly, though, such extreme dissonance could never be attained on the basis of empirical evidence. Consider, for example, a subset B of Honan that falls short of including the whole province only by the exclusion of a single point. Are we to suppose that empirical evidence could cast doubt on every point of B without casting doubt on the single remaining point? One could hardly expect such precision. And it is even more farfetched that the evidence could be sufficiently voluminous and discordant for this to happen for every possible choice of the excluded point.

Surely any subset of Honan that excludes only a single point has to be accounted fully as plausible as Honan as a whole. And the same should hold true for many much smaller subsets of Honan. Consider, for example, a proper subset B that is so dense that it includes some point within an inch of every point of Honan. Such a subset B might contain only a finite number of points, but it seems hard to imagine any evidence casting doubt on it without casting doubt on all of Honan.

This example reveals two limitations of empirical evidence and two corresponding restrictions on the dissonance of plausibility functions based on empirical evidence. First, our ability to distinguish points on the map is limited, so much of the doubt cast on any point must also apply to many neighboring points. And secondly, the total amount of evidence we can acquire is limited, so our evidence cannot point in too many directions at once. Let us consider each of these limitations in turn.

1.4.1. *Topological Rules*

The mathematician will recognize the first restriction as essentially topological, and an attempt to express it precisely would lead him to adduce several rules framed in terms of the topology of Θ. Several of these rules would correspond to the regularity conditions used by Gustave Choquet to define his 'capacities'. Another rule, more easily stated but less familiar to mathematicians, would require the plausibility of any subset of Θ to equal the plausibility of its 'closure'. These topological rules are interesting, but I will not investigate them in this essay.

1.4.2. Condensability

The second restriction on dissonance derives not from the topology of Θ but rather from the fact that the evidence itself can have only limited complexity. There can only be so many distinct aspects of the evidence, and hence only so many distinct subsets of Θ that are touched by distinct bases of doubt. In fact, one might expect only a finite number of such fundamentally distinct aspects, or at most a countably infinite number with a finite number predominating in importance.

Fixing our attention on a given subset A of Θ, we can adduce similar considerations relative to the bases of doubt that apply to proper subsets of A but not to all of A. If there were only a finite number of these, then by choosing a point of A immune to each we could form a finite subset B of A which, as a whole, was immune to all the bases of doubt. This finite subset $B \subset C$ would be fully as plausible as A itself. If, on the other hand, there were an infinite number of bases of doubt, a finite number of which carried most of the weight, then we could expect to find a finite subset $B \subset A$ which was nearly as plausible as A.

This second restriction leads us, then, to the rule that

$$(11) \qquad PL(A) = \sup \{ PL(B) \mid B \subset A : B \text{ is finite} \}$$

for all $A \in 2^{\Theta}$. A plausibility function that obeys this rule is called *condensable* – a name based on the intuitive idea that at least most of the plausibility of A can be 'condensed' onto a finite subset.

The considerations just adduced are meant to explain the intuitive significance of condensability; they hardly constitute a demonstration that plausibility functions ought always to be condensable. I have found, however, that condensability is characteristic of empirical evidence, and it plays an important role in the theory developed in this essay. As will see in §3, condensability is one of the conditions that must be met if Dempster's rule of combination is to be fully applicable.

It is also interesting to note that the criterion for consonance, (9), is equivalent to (7) once condensability is assumed.

2. STATISTICAL EVIDENCE

I have explained at length what degrees of support would look like if we could calculate them, but it remains painfully obvious that we usually cannot. In most cases where we want to assess the evidence for a proposition there is simply no quantitative structure that can be exploited to produce numbers. But we find an exception in *statistical evidence* – a type of evidence that has so rich a quantitative structure that the calculation of numerical degrees of support is quite conceivable.

2.1. THE PROBLEM OF STATISTICAL SUPPORT

The notion of statistical evidence depends on the notion of an aleatory law, or an objective probability law. An aleatory law is a law that tells an experiment's propensity for producing each of its various possible outcomes. The first step in specifying an aleatory law is to specify the set X of all possible outcomes of the experiment. If X is conceived of as a topological space, then the further description of the aleatory law is somewhat complicated. But in the case where X is 'discrete' it is quite simple. One simply specifies, for each $x \in X$, a quantity $P(x)$ which is the experiment's propensity for producing the outcome x. The quantity $P(x)$ is also called x's objective probability, and it is the frequency with which x will occur in a long sequence of physically independent trials of the experiment. Each of the quantities $P(x)$, $x \in X$, is non-negative, and they add to one. In the following discussion I will concern myself mainly with the case where X is discrete.

Let us return to the problem of support for a parameter θ whose possible values constitute a set Θ. Suppose that θ is related to a given experiment by a *statistical specification* $\{P_\theta\}_{\theta \in \Theta}$. This means that to each element $\theta \in \Theta$ there corresponds an aleatory law P_θ on X, and that the experiment is actually governed by the aleatory law corresponding to the true value of θ. Since the different aleatory laws may attribute different propensities or objective probabilities to the various possible outcomes in X, the outcomes that are observed in a sequence of physically independent trials will constitute evidence as to which aleatory law actually governs the process and hence as to which element of Θ is the true value of θ. It is reasonable to call this type of evidence *statistical evidence*. If it is the only kind of evidence we have about θ. then the problem of measuring degrees of support for θ is a *problem of statistical support*.

It is useful to distinguish between statistical specifications that are complete and those that are restricted. A *complete* specification $\{P_\theta\}_{\theta \in \Theta}$ on X is one that includes every possible aleatory law on X; a *restricted* one is one that is not complete. (This terminology is not standard.) Obviously, every restricted specification on X can be thought of as a subset of the essentially unique complete specification on X.

The idea of using the outcomes of an experiment as evidence about which of a class $\{P_\theta\}_{\theta \in \Theta}$ of aleatory laws governs it is a familiar one. I should enlarge, though, on what is meant when an aleatory law P is said to govern an experiment. This is taken to mean not only that the experiment's propensity to produce the outcome x in a single trial is $P(x)$, but also that its propensity to produce the sequence $(x_1, ..., x_n)$ of outcomes in a sequence of physically independent trials is $P(x_1)...P(x_n)$.

2.1.1. *Dissonance*

I just asserted that the outcomes observed in a sequence of trials of an experiment constitute a body of evidence about which aleatory law governs the experiment. The thought behind this assertion is that the observation of an outcome $x \in X$ is evidence in favor of those laws that attribute to the experiment the greatest propensity for producing x. More generally, it tends to favor any law that attributes a given probability to x over a law that attributes a smaller probability to x.

The straightforward appearance of the evidence provided by a single observation x might lead us to think of it as highly consonant evidence. After all, it points in just one direction – towards those aleatory laws that attribute a high probability to x. In some cases, however, there will be several quite different aleatory laws that attribute a high probability to x, and since the evidence points towards each of these, it might be said to point in many directions at once. Hence it is not clear that the evidence provided by a single observation will always be consonant; in some cases it might be better to think of it as dissonant.

While there may often be some question about the consonance or dissonance of the evidence in the case of a single observation, there will usually
be little question in the case of several observations. In that case, the
evidence will almost always be highly dissonant. For even if each single
observation points in a single direction, the different observations will
most likely point in different directions.

So we must consider two possible types of dissonance: dissonance
arising from the combination of observations and dissonance arising from
a single observation.

2.1.2. *Dissonance from the Combination of Observations*

Let us consider the simplest of statistical specifications: the *binomial speci-
cation*. Suppose $X = \{\text{Heads, Tails}\}$ and $\Theta = [0, 1]$, with $P_\theta(\text{Heads}) = \theta$
$P_\theta(\text{Tails}) = 1 - \theta$. In other words, θ is a coin-tossing experiment's propensity or probability for producing heads, and θ might have any value between zero and one. Obviously, a flip resulting in tails will be evidence
for a low value. Any single flip by itself will be quite straightforward
evidence, but a flip resulting in heads will point in the direction opposite
to a flip resulting in tails. So while we may expect a single observation x
to produce a consonant support function, we must expect a sequence
$\mathbf{x} = (x_1, ..., x_n)$ of observations to produce a dissonant support function,
at least if $\mathbf{x}$ includes both some heads and some tails. In other words, we
might expect the support function S_{x_i} arising from the ith trial to be consonant, but we would expect the support function $S_{\mathbf{x}}$ resulting from the
combination of the support functions $S_{x_1}, ..., S_{x_n}$ to be dissonant.

2.1.3. *Dissonance from a Single Observation*

Suppose we are confronted with a closed box of recently hatched chicks.
We know that the chicks include both Rhode Island Reds and White
Leghorns, and both males and females, but we are uncertain of the proportions in the different categories. There may be equal numbers of each
breed, but either more males or fewer males than females. Or there may
be equal numbers of each sex, but either more Rhode Island Reds or
fewer Rhode Island Reds than White Leghorns. For the sake of concreteness, let us suppose that there are only four possibilities: (i) equal
numbers of each breed but four times as many males as females, (ii) equal
numbers of each breed but four times as many females as males, (iii)

equal numbers of each sex but four times as many Reds as Whites, or
(iv) equal numbers of each sex but four times as many Whites as Reds.
Denote these four possibilities by θ_M, θ_F, θ_R and θ_W, respectively, and
set $\Theta = \{\theta_M, \theta_F, \theta_R, \theta_W\}$.

Now suppose we draw a chick from the box 'at random'. Then the
element of Θ that correctly represents the contents of the box will determine an aleatory law that will govern the result of the draw. That aleatory
law will be defined, of course, on the set $X = \{$Red Female, Red Male,
White Female, White Male$\}$. The four possible aleatory laws are shown
in Table VII.

TABLE VII

The four aleatory laws

	Red Female	Red Male	White Female	White Male
θ_F	0.4	0.1	0.4	0.1
θ_M	0.1	0.4	0.1	0.4
θ_R	0.4	0.4	0.1	0.1
θ_W	0.1	0.1	0.4	0.4

Finally, suppose our single draw results in a Red female chick. What
sort of evidence does this provide as to the correct aleatory law?

Clearly, the evidence points both towards θ_F and towards θ_R. The fact
that a female chick was drawn points towards θ_F; and the fact that a Red
check was drawn points towards θ_R. So the result of the single draw
points in two different directions at once; it provides dissonant evidence.

2.2. THE FIRST POSTULATE OF PLAUSIBILITY

In a problem of statistical support, the propositions about θ are sometimes
called hypotheses, and one distinguishes between simple hypotheses and
composite hypotheses. A subset $\{\theta\} \subset \Theta$ that contains a single element θ
is a simple hypothesis; it asserts that the experiment is governed by the
aleatory law P_θ. A subset of Θ that has more than one element, on the
other hand, is a composite hypothesis; it asserts only that the experiment
is governed by one of several aleatory laws. As it turns out, it is easier to
investigate the plausibilities of simple hypotheses than the plausibilities
of composite hypotheses.

Suppose, indeed, that one has observed the outcome x in a trial of the experiment and wants to compare the plausibilities of the two simple hypotheses $\{\theta_1\}$ and $\{\theta_2\}$. Is it more plausible that the experiment is governed by P_{θ_1} or that it is governed by P_{θ_2}? We answer this question, of course, by comparing $P_{\theta_1}(x)$ and $P_{\theta_2}(x)$: the simple hypothesis that attributes the greater objective probability to the actual observation x will be the more plausible one.

It seems reasonable to go even farther and postulate that the degree of plausibility of a simple hypothesis $\{\theta\}$ should be proportional to the quantity $P_\theta(x)$. Denoting by PL_x the plausibility function on 2^Θ resulting from the observation x, this postulate can be written in symbols as

$$PL_x(\{\theta\}) = c(x)\,P_\theta(x),$$

where $c(x)$ depends on x but not on θ.

For the sake of economy, I will state this postulate formally for the case where $\{P_\theta\}_{\theta \in \Theta}$ is complete:

(I)   Suppose $\{P_\theta\}_{\theta \in \Theta}$ is a complete statistical specification on the discrete space X, and suppose $PL_x: 2^\Theta \to [0, 1]$ is the plausisibility function based on the single observation $x \in X$. Then

$$PL_x(\{\theta\}) = c(x)\,P_\theta(x),$$

where $c(x)$ depends on x but not on θ.

This is the *first postulate of plausibility.*

In the presence of other postulates that I will adopt in §4, this postulate implies the following more general statement, which applies to both complete and restricted specifications and to any number of observations:

(I′)   Suppose $\{P_\theta\}_{\theta \in \Theta}$ is a statistical specification on the discrete space X, and suppose $PL_x: 2^\Theta \to [0, 1]$ is the plausibility function based on the observations $\mathbf{x} = (x_1, \ldots, x_n)$. Then

$$PL_\mathbf{x}(\{\theta\}) = c(\mathbf{x})\,P_\theta(x_1)\ldots P_\theta(x_n),$$

where $c(\mathbf{x})$ depends on $\mathbf{x}$ but not on θ.

The first postulate of plausibility is hardly a novel idea. Every statistician will agree that the objective probability that a simple hypothesis atrributes to the actual observations is a measure of the simple hypothesis' 'relative plausibility', 'likelihood', 'probability', or some such thing. The idea can be traced back at least to Johann Heinrich Lambert's *Photometria*, published in 1760. Daniel Bernoulli toyed with the idea about the same time, and shortly later it was incorporated into the 'Bayesian' framework that was firmly imposed on statistics by the consummate politician named Pierre Simon Laplace. In this century it was forcefully re-extracted from that framework by R. A. Fisher, who called the quantity $P_\theta(x)$, considered as a function of θ, the 'likelihood' of θ.

The aspect of the present formulation that is novel is the explicit recognition that it is only the plausibilities of simple hypotheses that are proportional to the objective probabilities. The first postulate tells us only about the quantities $PL_x(\{\theta\})$ for $\theta \in \Theta$, and when the evidence is dissonant these will not determine the quantities $PL_x(A)$ for composite hypotheses $A \in 2^\Theta$.

I will now consider the first postulate of plausibility in the light of two examples, one involving a complete specification, and the other involving a restricted specification.

2.2.1. *The Binomial Specification*

Consider first the example mentioned earlier, where $X = \{\text{Head, Tails}\}$, $\Theta = [0, 1]$, and P_θ (Heads) $= \theta$, P_θ (Tails) $= 1 - \theta$. This specification is complete, and since X has two elements it is called a *binomial* specification.

Suppose we have six observations, $\mathbf{x} = (x_1, ..., x_6)$ from this specification, and three of them are heads while the other three are tails. Then by the first postulate of plausibility,

$$PL_\mathbf{x}(\{\theta\}) = c(\mathbf{x}) P_\theta(x_1) ... P_\theta(x_6)$$
$$= c(\mathbf{x}) [P_\theta(\text{Heads})]^3 [P_\theta(\text{Tails})]^3$$
$$= c(\mathbf{x}) \theta^3 (1 - \theta)^3.$$

Since $\theta^3(1 - \theta)^3$ takes its maximum value when $\theta = 1/2$, $1/2$ will be the most plausible single value for $\mathbf{\theta}$. Any other value θ will have a degree of plausibility that is only $64\theta^3(1 - \theta)^3$ as great as the degree of plausibility for $1/2$.

But this is all the first postulate of plausibility tells us. It does not tell us the absolute value of $PL_\mathbf{x}\{1/2\}$ or of $PL_\mathbf{x}(\{\theta\})$ for any other $\theta \in \Theta$. And it tells us nothing about the degrees of plausibility for composite hypotheses.

2.2.2. *A Restricted Binomial Specification*

Now let us consider the restricted statistical specification that is obtained from the preceding example by taking Θ to be the pair $\{1/3, 2/3\}$ instead of the whole interval $[0, 1]$. In other words, let us suppose that θ, the coin's propensity for coming up heads, is known to have either the value $1/3$ or the value $2/3$. Still assuming that we have obtained three heads and three tails, the plausibilities for the simple hypotheses $\{1/3\}$ and $\{2/3\}$ will be

$$PL_{\mathbf{x}}(1/3) = c(\mathbf{x}) (1/3)^3 (2/3)^3$$

and

$$PL_{\mathbf{x}}(2/3) = c(\mathbf{x}) (2/3)^3 (1/3)^3.$$

In other words, they will be the same. But how great will their common value be?

Both alternatives are equally plausible, and they are the only alternatives, so one might be inclined to award them both plausibility one. But this would correspond to saying that the six tosses have really produced no evidence at all. In fact, they have produced conflicting evidence; each of the two values has had some doubt cast on it as a result of the six tosses. Doubt was cast on the value $1/3$ every time heads came up, and doubt was cast on the value $2/3$ every time tails came up.

Another way to see that our evidence is internally conflicting rather than null is to observe that it has an effect when it is combined with further evidence. Suppose, for example, that another six tosses of the coin result in four heads and two tails. Now this new evidence, considered on its own, should produce a mild degree of support and a higher degree of plausibility for $\{2/3\}$. But when we combine it with the old evidence, we end up with seven heads out of twelve tosses. This overall result still lends more support to $\{2/3\}$ than to $\{1/3\}$, but it surely casts more doubt on $\{2/3\}$ than did the observation of four heads out of six tosses. So the three heads and three tails do differ from no evidence when combined with more tosses.

We can describe this phenomenon more generally. The effect of the three heads and three tails will be to counter any more one-sided evidence arising from further tosses. And the same countering effect will result from any equal number of heads and tails – the greater the number, the stronger the effect. If, for example, our initial evidence consists of fifty heads out of a hundred tosses, then it will practically nullify the more one-sided evidence provided by four heads and six tosses.

So the common degree of plausibility resulting from three heads and three tails must be less than one. On the other hand, it must be greater than one-half, for the two plausibilities must obey

$$PL_x(\{1/3\}) + PL_x(\{2/3\}) = PL_x(\{1/3\}) + PL_x(\overline{\{1/3\}}) \geqslant 1.$$

The theory of §4 below gives the value 0.6, resulting in the dissonant plausibility function in Table VIII.

TABLE VIII

S_x and PL_x when x consists of three heads and three tails.

A	$S_x(A)$	$PL_x(A)$
$\emptyset$	0	0
$\{1/3\}$	0.4	0.6
$\{2/3\}$	0.4	0.6
Θ	1	1

2.3. THE SECOND POSTULATE OF PLAUSIBILITY

The first postulate of plausibility concerns the plausibilities of subsets of Θ that consist of single elements. As a first step beyond the first postulate, it is natural to consider subsets of Θ that consist of two elements.

Consider a doubleton $\{\theta_1, \theta_2\} \in 2^\Theta$, and suppose we have a single observation $x \in X$. If PL_x were consonant, we would have

$$PL_x(\{\theta_1, \theta_2\}) = \max_{i=1, 2} PL_x(\{\theta_i\}).$$

But if PL_x were dissonant with respect to the pair θ_1, θ_2, we would have

$$(12) \qquad PL_x(\{\theta_1, \theta_2\}) > \max_{i=1, 2} PL_x(\{\theta_i\}).$$

When ought (12) to occur? In other words, when is the evidence x dissonant with respect to θ_1, θ_2?

We may assume that $P_{\theta_1}(x) \geqslant P_{\theta_2}(x)$, so that $PL_x(\{\theta_1\}) \geqslant PL_x(\{\theta_2\})$ by the first postulate. Our question then becomes whether we ought to have

$$(13) \qquad PL_x(\{\theta_1, \theta_2\}) > PL_x(\{\theta_1\})$$

even though $PL_x(\{\theta_1\}) \geqslant PL_x(\{\theta_2\})$. In other words, when ought the addition of θ_2 to the hypothesis $\{\theta_1\}$ cause an increase in plausibility even though P_{θ_2} attributes no greater objective probability to the actual observation than P_{θ_1}?

This is a difficult question. On the whole, of course, the evidence x points towards θ_1 more strongly than towards θ_2. (Or equally strongly if $P_{\theta_1}(x) = P_{\theta_2}(x)$.) But if some aspect of the evidence x points more towards θ_2 than towards θ_1, we will have an example of dissonance, and (13) should hold.

Without answering definitely the question of when (13) should hold, we can specify one situation where nearly everyone would agree that it should not hold. Suppose that for every $y \in X$,

$$(14) \qquad \frac{P_{\theta_1}(x)}{P_{\theta_1}(y)} \geqslant \frac{P_{\theta_2}(x)}{P_{\theta_2}(y)}.$$

This is stronger than saying that P_{θ_1} attributes a greater probability to x than P_{θ_2} does. For it says that for every other possible outcome y, P_{θ_1} attributes a greater probability to x relative to y than P_{θ_2} does. If (14) holds, then it would seem that every aspect of the evidence points towards θ_1 more strongly than towards θ_2. Certainly, the comparison of the actual outcome with any other possible outcome favors θ_1 over θ_2. Hence we may conclude that (13) should not hold if (14) holds.

We have arrived at the second postulate. As in the case of the first postulate, I will state it formally for the case where the specification is complete:

(II) Suppose $\{P_\theta\}_{\theta \in \Theta}$ is a complete statistical specification on the discrete space X, and suppose $PL_x : 2^\Theta \to [0, 1]$ is the plausibility function based on the single observation $x \in X$. Then

$$PL_x(\{\theta_1, \theta_2\}) = PL_x(\{\theta_1\})$$

whenever θ_1 and θ_2 are elements of Θ and

$$\frac{P_{\theta_1}(x)}{P_{\theta_1}(y)} \geqslant \frac{P_{\theta_2}(x)}{P_{\theta_2}(y)}$$

for all $y \in X$.

This is the *second postulate of plausibility*.

In the presence of the further postulates adopted in §4, this postulate implies the following more general statement:

(II′) Suppose $\{P_\theta\}_{\theta \in \Theta}$ is a statistical specification on a discrete space X, and suppose $PL_x: 2^\Theta \to [0, 1]$ is the plausibility function based on the observations $x = (x_1, ..., x_n)$. Then

$$PL_x(\{\theta_1, \theta_2\}) = PL_x(\{\theta_1\})$$

whenever θ_1 and θ_2 are elements of Θ and

$$\frac{P_{\theta_1}(x_1) \dots P_{\theta_1}(x_n)}{P_{\theta_1}(y_1) \dots P_{\theta_1}(y_n)} \geq \frac{P_{\theta_2}(x_1) \dots P_{\theta_2}(x_n)}{P_{\theta_2}(y_1) \dots P_{\theta_2}(y_n)}$$

for all sequences $y = (y_1, ..., y_n)$ of elements of X.

2.3.1. *The Binomial: One Observation*

Consider the complete binomial specification: $X = \{\text{Heads, Tails}\}$, $\Theta = [0, 1]$, and $P_\theta(\text{Heads}) = \theta$, $P_\theta(\text{Tails}) = 1 - \theta$. And suppose we have a single observation $x = \text{Heads}$. Then (14) becomes

$$\frac{P_{\theta_1}(\text{Heads})}{P_{\theta_1}(y)} \geq \frac{P_{\theta_2}(\text{Heads})}{P_{\theta_2}(y)}$$

for all $y \in X$. But this means that

$$\frac{\theta_1}{\theta_1} \geq \frac{\theta_2}{\theta_2} \quad \text{and} \quad \frac{\theta_1}{1 - \theta_1} \geq \frac{\theta_2}{1 - \theta_2},$$

and this is equivalent to $\theta_1 \geq \theta_2$. Hence the second postulate says in this case that

$$PL_{\text{Heads}}(\{\theta_1, \theta_2\}) = PL_{\text{Heads}}(\{\theta_1\})$$

whenever $\theta_1 \geq \theta_2$.

Hence

$$PL_{\text{Heads}}(\{\theta_1, \theta_2\}) = \max_{i=1, 2} PL_{\text{Heads}}(\{\theta_i\})$$

for all pairs θ_1, θ_2; there is no pair θ_1, θ_2 for which dissonance is exhibited. This reflects the consonant nature of the evidence consisting of a single outcome of heads. Such evidence points unambiguously towards the aleatory laws that attribute greater probability to heads.

Notice in particular that

$$PL_{\text{Heads}}(\{1, \theta\}) = PL_{\text{Heads}}(\{1\})$$

for all $\theta \in [0, 1]$; there is no aleatory law whose addition will increase the plausibility of the simple hypothesis that attributes probability one to heads.

2.3.2. *The Binomial: Many Observations*

While discussing the first postulate, I considered an example of six observations $\mathbf{x} = (x_1, \ldots, x_6)$ from the restricted binomial specification $\Theta = \{1/3, 2/3\}$. Assuming that $\mathbf{x}$ consisted of three heads and three tails, I obtained the dissonant plausibility function in Table VIII. Is the dissonance in that plausibility function permitted by (II′)?

The plausibility function $PL_{\mathbf{x}}$ in Table VIII is dissonant because

$$PL_{\mathbf{x}}(\{1/3, 2/3\}) > PL_{\mathbf{x}}(\{1/3\}) = PL_{\mathbf{x}}(\{2/3\}).$$

This is permitted by our second postulate only if

$$(15) \qquad \frac{P_{1/3}(x_1) \ldots P_{1/3}(x_6)}{P_{1/3}(y_1) \ldots P_{1/3}(y_6)} < \frac{P_{2/3}(x_1) \ldots P_{2/3}(x_6)}{P_{2/3}(y_1) \ldots P_{2/3}(y_6)}$$

holds for some $\mathbf{y}$. But (15) will indeed hold if we choose a $\mathbf{y}$ consisting, say, of six tails. For then (15) will become

$$\frac{(1/3)^3 (2/3)^3}{(2/3)^6} < \frac{(2/3)^3 (1/3)^3}{(1/3)^6},$$

or $(1/2)^3 < 2^3$. Hence the second postulate does allow the dissonance in this example. This can be explained by pointing out that while $\{2/3\}$ attributes no greater likelihood to the overall observations $\mathbf{x}$ than $\{1/3\}$ does, it does attribute a greater likelihood to $\mathbf{x}$ relative to a sequence $\mathbf{y}$ containing even more tails.

2.3.3. *A Trinomial Example*

Set $X = \{\text{Azure, Brown, Crimson}\}$, and let $\{P_\theta\}_{\theta \in \Theta}$ be the complete statistical specification on X. This specification is most easily described by setting.

$$\Theta = \{(a, b, c) \mid a \geqslant 0, b \geqslant 0, c \geqslant 0; a + b + c = 1\}$$

and setting $P_\theta(\text{Azure})=a$, P_θ (Brown)$=b$, and $P_\theta(\text{Crimson})=c$ when $\theta=(a, b, c)$.

In order to think about the second postulate, consider the following elements of Θ:

$$\theta_1 = (3/4, 1/3, 1/8) \qquad \theta_3 = (1/2, 1/6, 1/3)$$
$$\theta_2 = (1/2, 1/4, 1/4) \qquad \theta_4 = (1/4, 0, 3/4)$$
$$\theta_5 = (1, 0, 0)$$

Now suppose we have a single observation $x=$ Azure. Then what does the second postulate tell us about the plausibilities of the various doubletons that can be formed from these five elements of Θ?

Well, the second postulate will require that

$$PL_{\text{Azure}}(\{\theta_i, \theta_j\}) = PL_{\text{Azure}}(\{\theta_i\})$$

whenever

$$\frac{P_{\theta_i}(\text{Azure})}{P_{\theta_i}(y)} \geq \frac{P_{\theta_j}(\text{Azure})}{P_{\theta_j}(y)}$$

for all $y \in X$: i.e.; whenever

$$\frac{P_{\theta_i}(\text{Azure})}{P_{\theta_i}(\text{Brown})} \geq \frac{P_{\theta_j}(\text{Azure})}{P_{\theta_j}(\text{Brown})}$$

$$\text{and} \quad \frac{P_{\theta_i}(\text{Azure})}{P_{\theta_i}(\text{Crimson})} \geq \frac{P_{\theta_j}(\text{Azure})}{P_{\theta_j}(\text{Crimson})}.$$

This means that the second postulate will require

$$PL_{\text{Azure}}(\{\theta_1, \theta_2\}) = PL_{\text{Azure}}(\{\theta_1\}),$$
$$PL_{\text{Azure}}(\{\theta_1, \theta_3\}) = PL_{\text{Azure}}(\{\theta_1\}),$$

and

$$PL_{\text{Azure}}(\{\theta_5, \theta_i\}) = PL_{\text{Azure}}(\{\theta_5\})$$

for $i=1,\ldots, 4$. But it will *not* require

$$PL_{\text{Azure}}(\{\theta_2, \theta_3\}) = PL_{\text{Azure}}(\{\theta_2\}),$$

nor

$$PL_{\text{Azure}}(\{\theta_1, \theta_4\}) = PL_{\text{Azure}}(\{\theta_1\}).$$

In other words, it will allow

$$(16) \qquad PL_{\text{Azure}}(\{\theta_2, \theta_3\}) > PL_{\text{Azure}}(\{\theta_2\})$$

and

$$(17) \qquad PL_{\text{Azure}}(\{\theta_1, \theta_4\}) > PL_{\text{Azure}}(\{\theta_1\}),$$

even though $PL_{\text{Azure}}(\{\theta_3\}) = PL_{\text{Azure}}(\{\theta_2\})$ and $PL_{\text{Azure}}(\{\theta_4\}) <$
$< PL_{\text{Azure}}(\{\theta_1\})$.

Notice that if (16) and (17) actually hold then this will be an example in which dissonance is displayed even though there is but a single observation. This may seem strange, for the observation $x =$ Azure might be thought to point unambiguously towards those aleatory laws which attribute the greater probability to Azure. Actually, it does point towards the aleatory laws attributing a greater probability to Azure, but this does not define an unambiguous 'direction'. In comparing θ_2 and θ_3, for example, we see that θ_2 attributes the greater likelihood to Azure relative to Brown, while θ_3 attributes the greater likelihood to Azure relative to Crimson.

2.3.4. *The Geometry of the Trinomial Specification*

The application of the second postulate to the trinomial specification can be understood more easily if the set Θ is represented by an equilateral triangle.

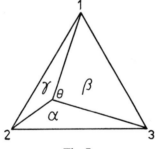

Fig. 7.

As we see in Figure 7, any point θ of an equilateral triangle divides the triangle into three smaller triangles α, β, and γ, each based on one of the three sides of the equilateral triangle. Let us suppose that the total area of the triangle is equal to one, and denote the areas of the triangles α, β and γ by a, b and c, respectively. Then $a + b + c = 1$. The triplet (a, b, c) is called the *barycentric coordinates* of the point θ. As the position of

θ varies, the areas a, b and c will vary; and it is evident that by placing θ in the right place they can be made to assume any triplet of non-negative values adding to one. Hence the points θ of the triangle are in a one-to-one correspondence with the elements θ of Θ.

Now consider a ray emanating from vertex 3, as in Figure 8, and consider any two points $\theta_1 = (a_1, b_1, c_1)$ and $\theta_2 = (a_2, b_2, c_2)$ on that ray. It is easily seen that $a_1/b_1 = a_2/b_2$. In other words, the ratio a/b is constant for points on a given ray from vertex 3. Furthermore, that ratio increases as the ray is raised; in Figure 9, for example, the ratio is higher for θ_1 than for θ_2.

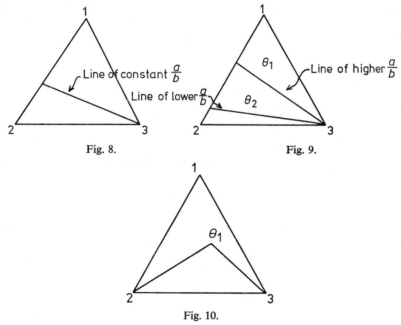

Fig. 8.   Fig. 9.

Fig. 10.

Similarly, rays from vertex 2 are lines of constant a/c; the higher the ray the higher that constant.

Now as we saw above, the plausibility function PL_{Azure} is required by the second postulate to obey $PL_{\text{Azure}}(\{\theta_1, \theta_2\}) = PL_{\text{Azure}}(\{\theta_1\})$ whenever both

$$(18) \qquad \frac{a_1}{b_1} \geqslant \frac{a_2}{b_2} \quad \text{and} \quad \frac{a_1}{c_1} \geqslant \frac{a_2}{c_2},$$

where $\theta_1 = (a_1, b_1, c_1)$ and $\theta_2 = (a_2, b_2, c_2)$. And it is easy to see that (18) holds only when θ_2 is inside the triangle $\theta_1 23$. If θ_2 is outside that triangle, then the second postulate will permit $PL_{\text{Azure}}(\{\theta_1, \theta_2\}) >$ $> PL_{\text{Azure}}(\{\theta_1\})$.

3. BELIEF

Our theory of statistical evidence seems promising, but it has reached an impasse. I have adduced several general rules for support functions, and I have added two postulates that are specific to statistical evidence. Yet all of these conditions together fail to determine a unique real function $S_x: 2^\theta \to [0, 1]$ that can be regarded as measuring the degrees to which the statistical evidence x supports the different statistical hypotheses in 2^θ. We evidently need yet further conditions on the function S_x.

I propose to search for some of these conditions in a theory of partial belief. I am going to adduce general rules governing a function Bel: $2^\theta \to [0, 1]$ that purports to measure a person's degrees of belief in the various propositions in 2^θ, and then I am going to require that the support functions S_x also obey these rules. As we will see in §4, this requirement will still fail to determine the support functions S_x uniquely, but it will considerably narrow the range of possibilities.

Why do I want to cross our theory of evidence with a theory of partial belief? Since the degree of support for a proposition ought to determine one's degree of belief in it, I might excuse myself by claiming that degrees of support should obey rules applicable to degrees of belief. But this would be at most a partial justification; for the fact that the degree of support ought to determine one's degree of belief hardly implies that the first should obey any rule applicable to the second. The real reason for turning to a theory of partial belief lies in the differences rather than in the similarities between the notions of support and belief. The fact is that the subjective notion of partial belief has a stronger structure than the logical notion of partial support, so that it is possible to adduce rules for degrees of belief that cannot be adduced so naturally for degrees of support. Indeed, it is natural to think not only that partial belief is related to complete belief as a part is to a whole, but also that the partial beliefs accorded to different propositions correspond to parts of the same whole. This amounts to reifying 'our total belief' – to thinking of it as a fixed substance that can be divided up in various ways. It is not nearly so natural to reify 'the total support' in this way.

This section outlines a theory based on the notion that we have a certain degree of belief in a proposition when we have committed to it that proportion of our total belief – or if you prefer, that proportion of our total 'probability'. A fuller account of this theory can be found in my *Allocations of Probability: A Theory of Partial Belief.*

3.1. BELIEF FUNCTIONS

A function $\text{Bel}: 2^{\theta} \rightarrow [0, 1]$ purporting to give a person's degrees of belief is called a *belief function* if it obeys these three axioms:

(I) $\text{Bel}(\emptyset) = 0$.

(II) $\text{Bel}(\Theta) = 1$.

(III) $\text{Bel}(A_1 \cup \ldots \cup A_n) \geqslant \sum_i \text{Bel}(A_i) - \sum_{i<j} \text{Bel}(A_i \cap A_j) +$
$$+ - \ldots + (-1)^{n+1} \text{Bel}(A_1 \cap \ldots \cap A_n)$$

whenever
$$A_1, \ldots, A_n \in 2^{\Theta}.$$

These axioms imply all the rules for support functions listed in Table I. Hence any belief function will formally qualify as a support function. On the other hand, these axioms are more restrictive than the rules for support functions, and hence not every function satisfying those rules necessarily qualifies as a belief function. Nevertheless, all the support functions for the case of two alternatives – i.e., all the support functions described in Table V – do qualify as belief functions. Other support functions that qualify as belief functions include the vacuous ones, all consonant ones, and many dissonant ones.

3.1.1. *Deriving the Axioms*

The axioms for belief functions follow naturally from the idea that $\text{Bel}(A)$, as our degree of belief in A, is the measure of that portion of our belief that is committed to A.

Axioms (I) and (II) are unexceptionable; we already adduced them for support functions. But here they acquire a stronger intuitive meaning. The statement that $\text{Bel}(\emptyset)=0$ reflects the fact that none of our belief ought to be committed to the impossible proposition $\emptyset$. And the statement that $\text{Bel}(\Theta)=1$ reflects the fact that all of our belief ought to be committed to the sure proposition Θ and the convention that the measure of our total belief is equal to one.

Axiom (III) looks a little more formidable. To begin with, it is really an infinite number of axioms, one for each natural number n. Before asking you to swallow it whole, I will ask you to consider some of its simpler consequences. For $n=2$, the axiom becomes

(19) $\text{Bel}(A_1 \cup A_2) \geqslant \text{Bel}(A_1) + \text{Bel}(A_2) - \text{Bel}(A_1 \cap A_2)$

for all pairs A_1, A_2 of subsets of Θ. Now when A_1 and A_2 are disjoint, or $A_1 \cap A_2 = \emptyset$, $\text{Bel}(A_1 \cap A_2) = 0$. So one consequence of (19) is:

(20)   If $A_1 \cap A_2 = \emptyset$, then $\text{Bel}(A_1 \cup A_2) \geqslant \text{Bel}(A_1) + \text{Bel}(A_2)$.

Now suppose $A \subset B$ and set $A_2 = B - A$. Then $B = A \cup A_2$ and $A \cap A_2 = \emptyset$. Hence (20) will give $\text{Bel}(B) = B(A \cup A_2) \geqslant \text{Bel}(A) + \text{Bel}(A_2)$. And hence $\text{Bel}(B) \geqslant \text{Bel}(A)$. So one consequence of (20) is our familiar rule of monotonicity:

(21)   If $A \subset B$, then $\text{Bel}(A) \leqslant \text{Bel}(B)$.

Let us develop the intuitive arguments for (19), (20), and (21), beginning with (21) and working backwards.

I would argue for (21) as follows: Since $A \subset B$, the proposition A implies the proposition B. Hence any belief I commit to A I must also commit to B; and the total portion I commit to B will therefore include the total portion I commit to A. And hence $\text{Bel}(B)$, the measure of the total portion of belief committed to B, will be at least as great as $\text{Bel}(A)$, the measure of the total portion of belief committed to A.

The defense of (20) is similar. First we note that any belief committed to A_1 must also be committed to $A_1 \cup A_2$, since $A_1 \subset A_1 \cup A_2$. Similarly, any belief committed to A_2 must also be committed to $A_1 \cup A_2$. And there can be no overlap between the belief committed to A_1 and the belief committed to A_2; the relation $A_1 \cap A_2 = \emptyset$ means that as propositions A_1 and A_2 are incompatible, and a single portion of belief can hardly be committed to both of two incompatible propositions. Hence the total belief committed to $A_1 \cup A_2$ will include the two disjoint portions that are committed to A_1 and A_2, respectively; and its measure will be at least as great as the sum of the measures of these two disjoint portions.

When the two propositions A_1 and A_2 are compatible (i.e., $A_1 \cap A_2 \neq \emptyset$), there may be some overlap between the portion of belief committed to A_1 and the portion of belief committed to A_2. In fact, the overlap – or the belief that is committed both to A_1 and to A_2 – will consist precisely of the belief that is committed to $A_1 \cap A_2$. This fact provides the basis for (19). For it means that the quantity $\mathrm{Bel}(A_1) + \mathrm{Bel}(A_2) - \mathrm{Bel}(A_1 \cap A_2)$ measures the total belief that is committed either to A_1, to A_2 or to both; and all that belief must be included in the total belief committed to $A_1 \cup A_2$, which is measured by $\mathrm{Bel}(A_1 \cup A_2)$.

The versions to Axiom (III) for other values of n can be justified by similar, but progressively more convoluted arguments.

3.1.2. *Allocations of Probability*

The axioms for belief functions are based on an intuitive picture wherein various portions of our belief, or various of our probability masses, are committed to various propositions. This intuitive picture can be made more precise by the notion of an *allocation of probability*.

The first step in formalizing the intuitive picture is the assumption that the set M of probability masses has the mathematical structure of a *complete Boolean algebra*. Let me outline roughly what this means. First we suppose that for every pair M_1, M_2 of probability masses in M there is another probability mass $M_1 \vee M_2$ in M, which is their 'union', and yet another, $M_1 \wedge M_2$, which is their overlap or 'intersection'. Of course, M_1 and M_2 may be disjoint, in which case the probability mass $M_1 \wedge M_2$ will be null – there will be no probability in it. We can use the symbol Λ to represent the null probability mass and the symbol V to represent the probability mass consisting of all our probability. Besides unions and intersections for pairs we also require unions and intersections for larger collections of probability masses. For any collection $\{M_\gamma\}$ of elements of M we require the existence of a union $\vee\, M_\gamma$ and an intersection $\wedge\, M_\gamma$. And for each $M \in \mathsf{M}$ we require the existence of a probability mass in M that consists precisely of all the probability not in M. This probability mass is called the complement of M and is denoted $\bar{M}$; M and $\bar{M}$ always obey $M \wedge \bar{M} = \Lambda$ and $M \vee \bar{M} = V$. Finally, we write $M_1 \leqslant M_2$ to indicate that all the probability in M_1 is also in M_2.

The second step is to assume there is a *measure* on $\mathbf{M}$ – a function $\mu: \mathbf{M} \to [0, 1]$ such that $\mu(M)$ is the measure of the probability mass M. This function must obey the following rules:

(i) $\mu(\Lambda) = 0$.

(ii)   If $M \neq \Lambda$, then $\mu(M) > 0$.

(iii) $\mu(V) = 1$.

(iv)   If $\{M_\gamma\}$ is disjoint, then $\mu(\vee M_\gamma) = \sum \mu(M_\gamma)$.

Notice that (iv) applies to both finite and infinite collections; the measures of disjoint probability masses always add.

Finally, we specify a mapping $\rho: 2^\Theta \to \mathbf{M}$ which satisfies three rules:

(i) $\rho(\emptyset) = \Lambda$.

(ii) $\rho(\Theta) = V$.

(iii) $\rho(A_1 \cap A_2) = \rho(A_1) \wedge \rho(A_2)$ for all $A_1, A_2 \in 2^\Theta$.

This function is called the *allocation of probability*: for each $A \in 2^\Theta$, $\rho(A)$ is the total probability mass committed to A.

The allocation $\rho: 2^\Theta \to \mathbf{M}$ does two things. It tells which probability masses are committed to which propositions, and it tells the degree of belief in each proposition. In order to tell whether a probability mass M is committed to a proposition A, we need only check whether M is included in the total probability committed to A; i.e., whether $M \leqslant \rho(A)$. In order to find the degree of belief in a proposition A, we need only find the measure of $\rho(A)$. In other words, $\mathrm{Bel}(A) = \mu(\rho(A))$ for all $A \in 2^\Theta$, or $\mathrm{Bel} = \mu \circ \rho$.

As it turns out, this formalization corresponds exactly to the structure of belief functions. In other words, whenever $\rho: 2^\Theta \to \mathbf{M}$ is an allocation of probability, the function $\mu \circ \rho$ is a belief function on 2^Θ. And any belief function on 2^Θ can be represented in this way by some complete Boolean algebra $\mathbf{M}$, some measure μ on $\mathbf{M}$, and some allocation $\rho: 2^\Theta \to \mathbf{M}$.

3.1.3. *Upper Probabilities*

Since the degree of belief $\mathrm{Bel}(A)$ corresponds to the degree of support for A, it is natural to think of the quantity $1 - \mathrm{Bel}(\bar{A})$ as corresponding to the degree of plausibility of A. In other words, $1 - \mathrm{Bel}(\bar{A})$ should measure the degree to which one finds A to be plausible, or the extent to which one regards A as plausible. This interpretation fits the intuitive picture of probability masses, for since $\mathrm{Bel}(\bar{A})$ measures the probability mass committed to $\bar{A}$, $1 - \mathrm{Bel}(\bar{A})$ measures the portion of our probability that is not committed to $\bar{A}$, i.e., is not committed against A.

Following A. P. Dempster, I will call the quantity $1 - \mathrm{Bel}(\bar{A})$ the upper probability of A, and denote it by $P^*(A)$. And I will call a function $P^*: 2^\theta \rightarrow [0, 1]$ an *upper probability function* if and only if the function $\mathrm{Bel}: 2^\theta \rightarrow [0, 1]$ defined by $\mathrm{Bel}(A) = 1 - P^*(\bar{A})$ is a belief function.

Notice that alongside the logical vocabulary of support, we now have a subjective vocabulary. The full correspondence between the two vocabularies is shown in Table IX.

TABLE IX

The two vocabularies

Subjective		Logical	
Degree of Belief in A.	$\mathrm{Bel}(A)$	Degree of Support for A.	$S(A)$
Degree of Doubt for A.	$\mathrm{Bel}(\bar{A})$	Degree of Dubiety of A.	$S(\bar{A})$
Upper Probability of A.	$1 - \mathrm{Bel}(\bar{A})$	Degree of Plausibility of A.	$1 - S(\bar{A})$

3.2. CONDENSABILITY

In §1.4, I argued that a plausibility function based on empirical evidence should be condensable. The same should evidently apply to an upper probability function $P^*: 2^\theta \rightarrow [0, 1]$ based on empirical evidence; it ought to obey

$$P^*(A) = \sup \{ P^*(B) \mid B \subset A; \ B \text{ is finite} \}$$

for all $A \in 2^\theta$.

Remarkably enough, an upper probability function P^* will obey this rule if and only if the corresponding allocation $\rho: 2^\theta \rightarrow \mathbf{M}$ obeys

$$\rho(\bigcap A_\gamma) = \bigwedge \rho(A_\gamma)$$

for all collections $\{A_\gamma\}$ of elements of 2^θ. In other words, the requirement of condensability corresponds to the requirement that ρ should preserve all intersections – infinite as well as finite. (Any allocation preserves all finite intersections.)

The requirement that ρ should preserve all intersections seems to be natural and necessary if ρ is to faithfully represent the intuitive picture underlying the axioms for belief functions. For it corresponds to the intuition that a probability mass committed to each of a collection $\{A_\gamma\}$ of propositions should also be committed to their logical conjunction $\bigcap A_\gamma$. Hence condensability emerges as a condition both appropriate to empirical evidence and natural to our notion of partial belief.

3.2.1. *A Geometric Intuition*

A condensable allocation lends itself to a very vivid geometric intuition. Think of Θ as a geometric set of points, and think of our probability as being spread over the set Θ. But instead of requiring that it be distributed in a fixed way, as in the picture associated with a 'distribution of probability', let us permit it a limited freedom of movement. Indeed, each time that a probability mass is 'committed' to a set A, let us say that it is 'constrained' to A, meaning that even though it may enjoy some freedom of movement, none of it can manage to escape from A.

This picture fits our rules for the commitment of probability masses to propositions perfectly. For example, a probability mass that is constrained to stay within a subset A is obviously constrained to stay within any subset B such that $A \subset B$. And any probability mass that is constrained to stay within each of a collection $\{A_\gamma\}_{\gamma \in \Gamma}$ of subsets must stay within $\bigcap A_\gamma$.

Furthermore, the belief function and the upper probability function can be interpreted very simply in terms of this picture: the degree of belief in A is the measure of all the probability that cannot get out of A, while A's upper probability is the measure of all the probability that can get into A.

The word 'condensability' itself acquires a vivid meaning in this picture. The fact is that no matter how diffusely the probability that can get into a set A might spread itself out over A, it is always possible to 'condense' it into a (possibly countably infinite) number of discrete pieces, each of which can get into some singleton $\{\theta\}$, where $\theta \in A$.

A simple example will show how this picture can help develop our intuition for belief functions. Consider Figure 11, where a subset $A \subset \Theta$

is shown with disjoint subsets A_1 and A_2 of A such that $A_1 \cup A_2 = A$. When we think of our probability mass as being semi-mobile over Θ, we can easily see how it might happen that $\text{Bel}(A_1) = 0$, $\text{Bel}(A_2) = 0$ and yet $\text{Bel}(A_1 \cup A_2) > 0$. And more generally, we can see how there might often be some probability that is constrained neither to A_1 nor to A_2 yet is constrained to $A_1 \cup A_2$. Indeed, this will happen whenever a probability mass M is free to move back and forth from A_1 to A_2 but is not free to move outside of $A_1 \cup A_2 = A$.

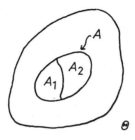

Fig. 11.

3.2.2. Commonality Numbers

Consider a finite collection $M_1, \ldots, M_n$ of probability masses. If one knows the measure of each of the M_i, and also how much overlap there is between each pair, among each triplet, etc., then one can deduce the measure of $M_1 \vee \ldots \vee M_n$ by the formula

$$\mu(M_1 \vee \ldots \vee M_n) = \sum_i \mu(M_i) - \sum_{i<j} \mu(M_i \wedge M_j) +$$
$$(22) \qquad\qquad + - \ldots + (-1)^{n+1} \mu(M_1 \wedge \ldots \wedge M_n).$$

A similar formula enables one to obtain the measure of an intersection from the measures of unions:

$$(23) \qquad \mu(M_1 \wedge \ldots \wedge M_n) = \sum_i \mu(M_i) - \sum_{i<j} \mu(M_i \vee M_j) +$$
$$+ - \ldots + (-1)^{n+1} \mu(M_1 \vee \ldots \vee M_n).$$

These two relations play an important role in the theory of condensable allocations.

Returning to our geometric picture, let $\zeta(\theta)$ denote the total probability mass that can get into the singleton $\{\theta\}$, and consider the intuitive significance of unions and intersections of the probability masses $\zeta(\theta)$. The intuitive significance of a union is obvious; $\zeta(\theta_1) \vee \ldots \vee \zeta(\theta_n)$ is the total probability mass that can get into the finite set $\{\theta_1, \ldots, \theta_n\}$. The intuitive significance of an intersection is a bit more subtle, though; $\zeta(\theta_1) \wedge \ldots \wedge \zeta(\theta_n)$ is the total probability mass that can get to each and every one of the points $\theta_1, \ldots, \theta_n$ – i.e., the total probability mass that has complete freedom of movement within the set $\{\theta_1, \ldots, \theta_n\}$. It is convenient to speak of $\zeta(\theta_1) \wedge \ldots \wedge \zeta(\theta_n)$ as the probability mass that is 'common' to the points $\theta_1, \ldots, \theta_n$.

Now $\mu(\zeta(\theta))$, being the measure of the total probability that can get into $\{\theta\}$, is simply $P^*(\{\theta\})$. Similarly,

$$P^*(\{\theta_1, \ldots, \theta_n\}) = \mu(\zeta(\theta_1) \vee \ldots \vee \zeta(\theta_n)).$$

The quantities $\mu(\zeta(\theta_1) \wedge \ldots \wedge \zeta(\theta_n))$, on the other hand, are new to us. I will denote

$$Q(\{\theta_1, \ldots, \theta_n\}) = \mu(\zeta(\theta_1) \wedge \ldots \wedge \zeta(\theta_n)),$$

and I will call $Q(\{\theta_1, \ldots, \theta_n\})$ the *commonality number* for $\{\theta_1, \ldots, \theta_n\}$.

The relations (22) and (23) provide us, of course, with the connections between upper probabilities and commonality numbers:

$$P^*(\{\theta_1, \ldots, \theta_n\}) = \sum_i Q(\{\theta_i\}) - \sum_{i<j} Q(\{\theta_i, \theta_j\}) + - \ldots +$$
$$+ (-1)^{n+1} Q(\{\theta_1, \ldots, \theta_n\}),$$

and

$$Q(\{\theta_1, \ldots, \theta_n\}) = \sum_i P^*(\{\theta_i\}) - \sum_{i<j} P^*(\{\theta_i, \theta_j\}) + - \ldots +$$
$$+ (-1)^{n+1} P^*(\{\theta_1, \ldots, \theta_n\}).$$

Or, in a notation that is sometimes more convenient:

$$(24) \qquad P^*(A) = \sum_{\substack{T \subset A \\ T \neq \emptyset}} (-1)^{1 + \operatorname{card} T} Q(T)$$

and

$$Q(A) = \sum_{\substack{T \subset A \\ T \neq \emptyset}} (-1)^{1 + \operatorname{card} T} P^*(T)$$

for all finite subsets A of Θ.

Formula (24) means in particular that the upper probabilities of finite subsets are determined by the commonality numbers. In the condensable case this implies that the entire upper probability function is determined by the commonality numbers. In fact a condensable upper probability function will obey

$$P^*(A) = \sup_{\substack{B \subset A \\ B \text{ finite}}} P^*(B),$$

or

$$P^*(A) = \sup_{\substack{B \subset A \\ B \text{ finite}}} \sum_{\substack{T \subset B \\ T \neq \emptyset}} (-1)^{1 + \text{card } T} Q(T)$$

for all $A \in 2^\Theta$.

So commonality numbers provide us with yet another way of specifying a condensable belief function. As it turns out, they provide the easiest way to specify many important condensable belief functions. And they also provide the simplest way of expressing Dempster's rule of combination.

3.3. DEMPSTER'S RULE OF COMBINATION

In §1.2, I illustrated Lambert's rule for combining support functions over two alternatives, and I mentioned a more general rule, due to A. P. Dempster, which applies when Θ is larger provided the support functions satisfy certain auxiliary conditions. As it turns out, these auxiliary conditions are precisely the rules for condensable belief functions.

The felicitousness of these rules is hardly surprising. For the crucial step in applying Lambert's rule is the representation of each support function by an imaginary mass, some of which is committed to each alternative. And as we have seen, the rules for condensable belief functions are precisely the rules that make an effective representation by imaginary 'probability masses' possible.

The actual derivation of Dempster's rule is rather complicated. After establishing the existence of a 'Boolean algebra of probability masses' representing each belief function, one must 'orthogonally combine' the two Boolean algebras of probability masses in some way analogous to the orthogonal combination of the two line segments in §1.2. One must then determine which of the resulting probability masses are committed to which elements of 2^Θ, and which are committed contradictorily and hence cannot be counted.

Let me describe the result that finally emerges from this process.

Suppose one combines the two condensable belief functions Bel_1: $2^{\Theta} \to [0, 1]$ and Bel_2: $2^{\Theta} \to [0, 1]$. Then the result of the combination is a condensable belief function Bel: $2^{\Theta} \to [0, 1]$ given by

$$(25) \qquad \mathrm{Bel}(A) = \frac{c(A) - c(\emptyset)}{1 - c(\emptyset)},$$

where

$$c(A) = \sup \left\{ \sum_i \mathrm{Bel}_1(A_i)\,\mathrm{Bel}_2(B_i) - \right.$$
$$- \sum_{i<j} \mathrm{Bel}_1(A_i \cap A_j)\,\mathrm{Bel}_2(B_i \cap B_j) + - \ldots +$$
$$\left. + (-1)^{n+1}\,\mathrm{Bel}_1(A_1 \cap \ldots \cap A_n)\,\mathrm{Bel}_2(B_1 \cap \ldots \cap B_n) \right\},$$

the supremum being taken over all collections $A_1, \ldots, A_n$ and $B_1, \ldots, B_n$ of elements of 2^{Θ} such that $A_i \cap B_i \subset A$ for each i.

The only case in which two condensable belief functions Bel_1 and Bel_2 cannot be combined is when they flatly contradict each other – i.e., when there exists $A \in 2^{\Theta}$ such that $\mathrm{Bel}_1(A) = 1$ and $\mathrm{Bel}_2(\bar{A}) = 1$. When such a contradiction occurs, $c(\emptyset) = 1$, and (25) cannot be applied. As long as such a contradiction does not occur, however, $c(\emptyset) < 1$ and (25) can be applied.

The derivation of the commonality numbers for Bel from the commonality numbers for Bel_1 and Bel_2 is quite simple; except for a constant of renormalization, one simply multiplies. More precisely, if the commonality numbers for Bel_1, Bel_2 and Bel are denoted by $Q_1(A)$, $Q_2(A)$ and $Q(A)$, respectively, then

$$Q(A) = k\,Q_1(A)\,Q_2(A)$$

for all $A \in 2^{\Theta}$, where

$$k = \frac{1}{1 - c(\emptyset)}.$$

The constant k is also determined by the requirement that

$$P^*(\Theta) = \sup_{\substack{A \subset \Theta \\ A\,\text{finite}}} \sum_{\substack{T \subset A \\ T \neq \emptyset}} (-1)^{1+\operatorname{card} T}\, Q(T) =$$
$$= k \sup_{\substack{A \subset \Theta \\ A\,\text{finite}}} \sum_{\substack{T \subset A \\ T \neq \emptyset}} (-1)^{1+\operatorname{card} T}\, Q_1(T)\,Q_2(T) = 1.$$

The actual computation of k is often quite difficult.

Although I have presented (25) as a rule for combining condensable belief functions, it can sometimes be applied in the noncondensable case. Unfortunately, its general usefulness in the noncondensable case is questionable. It cannot always be applied, for the lack of condensability will sometimes allow $c(\emptyset) = 1$ even when there is no apparent contradiction. And even when it can be applied, the interpretation of the results may be problematic.

3.3.1. *The Murder of Mr Green*

A simple example will do more to convey the nature of Dempster's rule than the preceding formulae.

Mr Green has been murdered, and we are certain that the murder was committed by either Dr Black, Dr Gray, or Mr White. Besides the evidence that allows us to narrow the field to these three suspects, we have evidence based on the mode of the murder, and evidence based on motives for the murder.

Mr Green was poisoned by a rare chemical. This fact provides Mr White's strongest defense, for there seems to have been little way he could have obtained the chemical. But as physicians, both Dr Black and Dr Gray had access to the chemical. Of these two, Dr Black is particularly implicated, for chemicals seem to be tightly controlled at the hospital where Dr Gray works.

When we consider motives, we obtain quite a different picture. There is no apparent motive that can be ascribed to Dr Black. On the other hand, both Dr Gray and Mr White are alleged to have been Mrs Green's lovers, and hence might have plotted with Mrs Green to inherit Mr Green's fortune. The evidence is weaker in the case of Dr Gray, but Mrs Green's involvement with Mr White was practically public knowledge.

In order to apply our theory to this example, we must set $\Theta = \{$Black, Gray, White$\}$ and postulate support functions S_1 and S_2 based on the two separate sources of evidence.

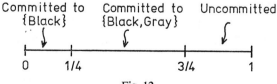

Fig. 12.

The support function S_1, derived from the mode of the murder, is represented in Figure 12. According to this figure, 3/4 of our probability is committed to the guilt of one of the doctors, and part of that is committed more specifically to the guilt of Dr Black. Explicitly,

$$S_1(\{Black\}) = 1/4 \qquad S_1(\{Black, Gray\}) = 3/4$$
$$S_1(\{Gray\}) = 0 \qquad S_1(\{Black, White\}) = 1/4$$
$$S_1(\{White\}) = 0 \qquad S_1(\{Gray, White\}) = 0.$$

And hence

$$PL_1(\{Gray, White\}) = 3/4 \qquad PL_1(\{White\}) = 1/4$$
$$PL_1(\{Black, White\}) = 1 \qquad PL_1(\{Gray\}) = 3/4$$
$$PL_1(\{Black, Gray\}) = 1 \qquad PL_1(\{Black\}) = 1$$

The support function S_2, derived from the consideration of motives, is similarly represented in Figure 13. In view of their possible motives, we have commited 2/3 of our probability to the proposition that either

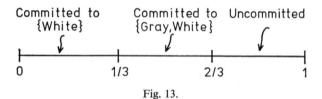

Fig. 13.

Dr Gray or Mr White is the murderer. One-half of this, or 1/3 of our probability, is committed more specifically to Mr White, whose involvement with Mrs Green is certain. Explicitly,

$$S_2(\{Black\}) = 0 \qquad S_2(\{Black, Gray\}) = 0$$
$$S_2(\{Gray\}) = 0 \qquad S_2(\{Black, White\}) = 1/3$$
$$S_2(\{White\}) = 1/3 \qquad S_2(\{Gray, White\}) = 2/3.$$

And hence

$$PL_2(\{Gray, White\}) = 1 \qquad PL_2(\{White\}) = 1$$
$$PL_2(\{Black, White\}) = 1 \qquad PL_2(\{Gray\}) = 2/3$$
$$PL_2(\{Black, Gray\}) = 2/3 \qquad PL_2(\{Black\}) = 1/3.$$

The combination of S_1 and S_2, as illustrated in Figure 14, is quite analogous to the application of Lambert's rule in §1.2. The only novelty

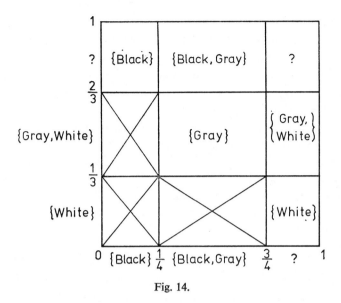

Fig. 14.

occurs in assigning the central rectangle in Figure 14. S_1 commits that rectangle to the proposition that either Black or Gray is the murderer, while S_2 commits it to the proposition that either Gray or White is the murderer; so together they commit it to the conjunction of these two propositions – the proposition that Gray is the murderer.

The support function S based on this combined evidence is thus as follows

$$S(\text{Black}) = 1/8 \qquad S(\text{Black, Gray}) = 5/8$$
$$S(\text{Gray}) = 1/4 \qquad S(\text{Black, White}) = 1/4$$
$$S(\text{White}) = 1/8 \qquad S(\text{Gray, White}) = 1/2.$$

And the corresponding plausibility function is given by

$$PL(\text{Gray, White}) = 7/8 \qquad PL(\text{White}) = 3/8$$
$$PL(\text{Black, White}) = 3/4 \qquad PL(\text{Gray}) = 3/4$$
$$PL(\text{Black, Gray}) = 7/8 \qquad PL(\text{Black}) = 1/2.$$

Notice that Dr Gray is the most seriously implicated suspect on the basis of the combined evidence, even though neither of the separate sources of evidence gave positive support to his guilt. It is the combination of means and motive that points strongly in his direction.

Notice that both PL_1 and PL_2 are consonant, while PL is dissonant.

3.3.2. *The Combination of Evidence*

Dempster's rule of combination is best understood as a rule for combining separate bodies of evidence. For reasons that are now evident, I have presented it as a rule for combining 'belief functions', but I do not want to suggest that the rule provides a method for 'pooling beliefs'. Evidence can be pooled; beliefs cannot be.

If two individuals' belief functions are based on separate sources of evidence, then those individuals can pool their evidence by using Dempster's rule to combine their belief functions. But if the two belief functions are based even partially on the same evidence, then Dempster's rule is inappropriate and will give misleading results. For it always treats the separate belief functions it combines as if they were based on separate sources of evidence.

Suppose, for example, that $\Theta = \{\theta_1, \theta_2\}$ and that the two individuals have exactly the same belief function, namely $\mathrm{Bel}_0: 2^\Theta \to [0, 1]$, where $\mathrm{Bel}_0(\{\theta_1\}) = 2/10$ and $\mathrm{Bel}_0(\{\theta_2\}) = 1/10$. Now if we combine Bel_0 with itself we will obtain a belief function that indicates, roughly speaking, twice as much evidence in both directions as Bel_0 indicates. Indeed, the belief function Bel obtained by combining Bel_0 with itself is given by $\mathrm{Bel}(\{\theta_1\}) = \frac{32}{96}$ and $\mathrm{Bel}(\{\theta_2\}) = \frac{15}{96}$. Clearly, the two individuals should retain the belief function Bel_0 rather than adopt Bel.

3.4. CONDITIONING BELIEF FUNCTIONS

Suppose we begin with a belief function $\mathrm{Bel}: 2^\Theta \to [0, 1]$ and then obtain new evidence showing that the true value of θ is not only in Θ but also in some proper subset Θ_0 of Θ. Can we use Dempster's rule of combination to combine this new knowledge with the evidence underlying Bel?

We can, provided that we can represent the new evidence by a belief function. But that can be done quite simply; the knowledge that the true value of θ is in Θ_0 is conveyed by the belief function $\mathrm{Bel}_0: 2^\Theta \to [0, 1]$, where

$$\mathrm{Bel}_0(A) = \begin{cases} 1 & \text{if} \quad \Theta_0 \subset A \\ 0 & \text{otherwise} \end{cases}.$$

So we should use Dempster's rule to combine Bel_0 and Bel.

This combination results in the belief function $\mathrm{Bel}(\cdot \mid \Theta)$, given by

$$(26) \qquad \mathrm{Bel}(A \mid \Theta_0) = \frac{\mathrm{Bel}(A \cup \bar{\Theta}_0) - \mathrm{Bel}(\bar{\Theta}_0)}{1 - \mathrm{Bel}(\bar{\Theta}_0)}$$

for all $A \in 2^\Theta$. A more succinct formula can be given in terms of the upper probability functions. If we denote the upper probability functions corresponding to Bel and $\mathrm{Bel}(\cdot \mid \Theta_0)$ by P^* and $P^*(\cdot \mid \Theta_0)$, respectively, then (26) becomes

$$P^*(A \mid \Theta_0) = \frac{P^*(A \cap \Theta_0)}{P^*(\Theta_0)}$$

for all $A \in 2^\Theta$.

This rule for obtaining $\mathrm{Bel}(\cdot \mid \Theta_0)$ from Bel is called *Dempster's rule of conditioning*, and the symbols '$\mathrm{Bel}(A \mid \Theta_0)$' may be read 'the degree of belief in A, given Θ_0'.

Since this rule is a special case of Dempster's rule of combination, it can also be expressed in terms of the commonality members. Indeed, if the commonality numbers for Bel are denoted by $Q(A)$ and those for $\mathrm{Bel}(\cdot \mid \Theta_0)$ are denoted by $Q^0(A)$, then $Q(A)$ and $Q^0(A)$ will be related by

$$Q^0(A) = \begin{cases} \dfrac{Q(A)}{P^*(\Theta_0)} & \text{if } A \subset \Theta_0 \\ 0 & \text{otherwise} \end{cases}.$$

3.4.1. *The Analogy with Bayes' Theory*

The general notion of conditioning, and formula (27) in particular, is strongly reminiscent of Bayes' theory. That theory explores the possibility of expressing degrees of belief or support by functions $P: 2^\Theta \to [0, 1]$ that obey the rules

(i) $P(\emptyset) = 0$,

(ii) $P(\Theta) = 1$,

(iii) $P(A \cup B) = P(A) + P(B)$ whenever $A \cap B = \emptyset$;

and it has at its core a rule of conditioning closely analogous to (27). By that rule, conditioning P on Θ_0 will produce another function obeying (i)–(iii) and given by the formula

$$(28) \qquad P(A \mid \Theta_0) = \frac{P(A \cap \Theta_0)}{P(\Theta_0)}$$

for all $A \in 2^\Theta$.

Actually, Bayes' theory can be regarded as a special case of our theory of belief, and (28) can be regarded as the corresponding special case of (26). For any function satisfying (i)–(iii) will qualify as a belief function, and in the case of a belief function P obeying (iii), (26) reduces to (28). Actually, a belief function obeying (iii) is identical with its upper probability function, so both (26) and (27) reduce to (28).

Students of Bayes' theory have explained (28) in various ways, but one of the most appealing explanations is in terms of probability masses. In these terms, rule (iii) corresponds to the situation where our probability, instead of being allowed some freedom of movement, is distributed in a fixed way over Θ. In such a situation, conditioning on Θ_0 can be thought of as discarding the probability that is distributed over $\bar{\Theta}_0$ and 'renormalizing' the measure of the rest.

We can verify that such a procedure produces (28) by studying Figure 15. Referring to that figure, we see that when the probability distributed over $\bar{\Theta}_0$ is discarded, part of that which was distributed over A will be

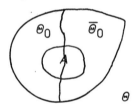

Fig. 15.

included in what is discarded, so that only the probability that was distributed over $A \cap \Theta_0$ will remain. This accounts for the numerator in (28). The denominator results from the 'renormalization'. The difficulty is that when the probability mass over $\bar{\Theta}_0$ is discarded, the total remainder will have a measure of only $1 - P(\bar{\Theta}_0) = P(\Theta_0)$. Since we want our probability to have total measure 1, we must multiply the measures of all our probability masses by $(P(\Theta_0))^{-1}$.

3.4.2. The Geometric Argument for (26)

This intuitive argument can be generalized to explain our rule of conditioning. Indeed, when we condition a belief function Bel: $2^{\theta} \rightarrow [0, 1]$

on Θ_0, how must we treat our semi-mobile probability masses? Clearly, we should eliminate the probability that is committed to $\bar{\Theta}_0$ and renormalize the measure of the remainder. The measure of the probability thus eliminated will be $\text{Bel}(\bar{\Theta}_0)$, so the constant of renormalization will be $(1 - \text{Bel}(\bar{\Theta}_0))^{-1}$. There is only one further idea that must be introduced: since our probability is allocated in a semimobile way over Θ rather than being distributed in a fixed way, we must recognize that the restriction to Θ_0 may further restrict the mobility of some of our probability without eliminating it entirely. This means that some of our probability that was not committed to A before may become committed to A by the restriction to Θ_0.

In fact, any probability that was committed to $A \cup \bar{\Theta}_0$ before will now be committed to A, unless it is eliminated. In general, then, the amount of probability committed to A after conditioning will be the measure of the probability previously committed to $A \cup \bar{\Theta}_0$ less the measure of the probability eliminated, or

$$\text{Bel}(A \cup \bar{\Theta}_0) - \text{Bel}(\bar{\Theta}_0).$$

But of course this must be renormalized, so we obtain

$$\text{Bel}(A \mid \Theta_0) = \frac{\text{Bel}(A \cup \bar{\Theta}_0) - \text{Bel}(\bar{\Theta}_0)}{1 - \text{Bel}(\bar{\Theta}_0)}$$

for all $A \in 2^{\Theta}$.

4. STATISTICAL SUPPORT

Our present goal is to define statistical support functions that obey both the postulates of plausibility and the rules for condensable belief functions. In this section, I will present two different methods for doing so. These two methods are based on quite different rationales, and produce different systems of support functions.

The first method is based on the assumption that support functions from single observations should be consonant. This assumption is contrary to the opinions I expressed in §2, but it does not contradict the postulates of plausibility. It leads to the *linear support functions*.

The second method is based on a theory of partial belief that embraces aleatory laws. It leads to the *simplicial support functions*, which are both more interesting and more difficult to study than the linear ones.

In §2, I distinguished between elements of the set Θ and the aleatory laws $\{P_\theta\}_{\theta \in \Theta}$ corresponding to them. Such a distinction serves to emphasize that the true value of θ may have a broader substantive significance than the aleatory law P_θ that it associates with the particular experiment at hand. In particular, it serves to remind us that two distinct values θ_1 and θ_2 might have $P_{\theta_1} \equiv P_{\theta_2}$. Unfortunately, though, the distinction complicates our notation. In this section, I will dispense with the distinction and think of each element θ of Θ as an aleatory law $\theta \colon X \to [0, 1]$.

In this notation, the postulates of plausibility read as follows:

Suppose Θ is the complete statistical specification on the discrete space X, and suppose $PL_X \colon 2^\Theta \to [0, 1]$ is the plausibility function based on the single observation $x \in X$. Then

(I) $PL_x(\{\theta\}) = c(x)\,\theta(x)$, where $c(x)$ depends on x but not on θ. And

(II) $PL_x(\{\theta_1, \theta_2\}) = PL_x(\{\theta_1\})$ whenever
$$\frac{\theta_1(x)}{\theta_1(y)} \geq \frac{\theta_2(x)}{\theta_2(y)} \quad \text{for all } y.$$

And the stronger versions of these postulates become:

Suppose Θ is a statistical specification on the discrete space X, and suppose $PL_x \colon 2^\Theta \to [0, 1]$ is the plausibility function based on the observations $\mathbf{x} = (x_1, \ldots, x_n)$. Then

(I′) $PL_{\mathbf{x}}(\{\theta\}) = c(\mathbf{x})\,\theta(x_1) \ldots \theta(x_n)$, where $c(\mathbf{x})$ depends on $\mathbf{x}$ but not on θ. And

(II′) $PL_{\mathbf{x}}(\{\theta_1, \theta_2\}) = PL_{\mathbf{x}}(\{\theta_1\})$ whenever
$$\frac{\theta_1(x_1) \ldots \theta_1(x_n)}{\theta_1(y_1) \ldots \theta_1(y_n)} \geq \frac{\theta_2(x_1) \ldots \theta_2(x_n)}{\theta_2(y_1) \ldots \theta_2(y_n)} \quad \text{for all } y_i \text{ from } X.$$

4.1. THE THREE POSTULATES OF SUPPORT

When I say that our statistical support functions should obey the rules for condensable belief functions, I mean both that they should qualify as condensable belief functions and that they should obey Dempster's rules of combination and conditioning.

The relevance of Dempster's rule of combination is obvious. If $\mathbf{x} = (x_1, \ldots, x_n)$, then the support function $S_\mathbf{x}$ resulting from the n observations ought to be the same as the support function obtained by combining the n support functions $S_{x_1}, \ldots, S_{x_n}$ by Dempster's rule of combination.

Dempster's rule of conditioning is relevant because some statistical specifications are subsets of others. Suppose, indeed, that Θ is the complete statistical specification on X and we obtain a support function $S_\mathbf{x} : 2^\Theta \rightarrow [0, 1]$ based on the observations $\mathbf{x}$. And suppose that we then change our minds and want to consider the restricted specification $\Theta_0 \subset \Theta$. Then according to our theory of partial belief, we should obtain our support function on 2^{Θ_0} by conditioning $S_\mathbf{x}$ on Θ_0.

Notice that Dempster's rules reduce our problem to that of finding support functions in the special case of a single observation from a complete specification. It is for this reason that I have stated our two postulates of plausibility, (I) and (II), in terms of this special case.

So in addition to (I) and (II), we now have three postulates of support:

(III)   *The First Postulate of Support.* Suppose Θ is the complete statistical specification on a discrete space X. Then the support function $S_X : 2^\Theta \rightarrow [0, 1]$ based on an observation $x \in X$ must be a condensable belief function.

(IV)   *The Second Postulate of Support.* Suppose Θ is the complete statistical specification on a discrete space X, and $S_{x_1}, \ldots, S_{x_n}$ are the support functions on 2^Θ based on the observations $x_1, \ldots, x_n$, respectively. Then the support function $S_X : 2^\Theta \rightarrow [0, 1]$ based on all the observations $\mathbf{x} = (x_1, \ldots, x_n)$ must be the belief function obtained by combining $S_{x_1}, \ldots, S_{x_n}$ by Dempster's rule of combination.

(V)   *The Third Postulate of Support.* Suppose Θ is the complete statistical specification on a discrete space X, and $S_\mathbf{x} : 2^\Theta \rightarrow [0, 1]$ is the support function based on the observations $\mathbf{x} = (x_1, \ldots, x_n)$. Then the support function $S_\mathbf{x}^0 : 2^{\Theta_0} \rightarrow [0, 1]$ based on the observations $\mathbf{x}$ and the restricted specification $\Theta_0 \subset \Theta$ must be the belief function obtained by conditioning $S_\mathbf{x}$ on Θ_0 by Dempster's rule of conditioning.

These are the further postulates that I promised in §2; they allow one to deduce (I') from (I) and (II') from (II).

They also allow one to strengthen (II) so as to apply to larger subsets than doubletons:

(II″).   Suppose $A \subset B \subset \Theta$, and suppose that for every $\theta_2 \in B$ there exists $\theta_1 \in A$ such that

$$\frac{\theta_1(x)}{\theta_1(y)} \geq \frac{\theta_2(x)}{\theta_2(y)}$$

for all $y \in X$. Then $PL_x(B) = PL_x(A)$.

In other words, the enlargement of a hypothesis cannot increase its plausibility as long as for each now simple hypothesis added there is already a simple hypothesis present under which the actual observation has greater likelihood relative to every other possible observation.

4.1.1. *The Dissonance of Statistical Evidence*

In general, our five postulates do not suffice to completely determine the support functions S_x and S_x. They are obeyed, for example, by both the linear and the simplicial support functions, and as we will see below, these differ in important respects. But any method of computing support functions that obeys the five postulates will emphasize the dissonant nature of statistical evidence.

It is Dempster's rule of combination that is responsible for this prominence of dissonance. As more and more observations are accumulated, this rule will operate to produce more and more dissonance. And eventually it will lead to a support function indicating strong evidence against each possibility – i.e., against each possible aleatory law. In other words, the plausibility of every simple hypothesis will decline; typically, the plausibility of the most plausible simple hypothesis will tend to zero as the number of observations grows.

In this respect, the present theory contrasts sharply with some other methods of assessing statistical evidence, which seem to treat it as consonant. The method of nested confidence regions provides a case in point. The theory of confidence regions is, of course, an operational rather than an epistemic theory. But nested confidence regions corresponding to different confidence coefficients are often used informally to summarize statistical evidence. (Cf. p. 62 of Erich Lehman's book.) Such a nested family can sometimes be interpreted as defining a support function: the degree of support for a given subset A of Θ will be the largest confidence coefficient such that A contains the confidence region with that coefficient. A support function so defined will be consonant, no matter how conflicting the evidence may appear to be.

4.1.2. *The Binomial Specification*

While our five postulates do not uniquely determine the support functions S_x and S_x in general, they do uniquely determine them in the case where X has only two elements.

Suppose, indeed, that X has two elements and that Θ is the complete specification on X. Then it may be deduced from (I) and (II″) that PL_x must be given by

$$PL_x(A) = \sup_{\theta \in A} \theta(x)$$

for all $A \in 2^\Theta$. The plausibility functions for restricted specifications and/or many observations can be obtained, of course, by combination and conditioning; they are the upper probability functions for which A. P. Dempster gave detailed formulae in his 1966 article.

4.2. THE LINEAR PLAUSIBILITY FUNCTIONS

In §2.1, I argued that statistical evidence can be dissonant even when it consists of only a single observation. It is possible, however, to take the opposite view and insist that plausibility functions based on a single observation should always be consonant. It turns out that this assumption of consonance suffices to completely determine the plausibility function based on a single observation $x \in X$ and a complete specification Θ on X. Indeed, that plausibility function will be given by

$$(29) \qquad PL_x(A) = \sup_{\theta \in A} \theta(x)$$

for all $A \in 2^\Theta$, just as in the special case of the binomial specification.

In the case of a restricted specification, Θ_0, we must condition (29), obtaining

$$(30) \qquad PL_x(A) = \frac{\sup\limits_{\theta \in A} \theta(x)}{\sup\limits_{\theta \in \Theta_0} \theta(x)}$$

for all $A \in 2^{\Theta_0}$. This too will be a consonant plausibility function.

Using Dempster's rule of combination, we can also obtain the plausibility functions based on many observations. Hence the assumption of
consonance for single observations, together with our five postulates,
determines a complete system of plausibility functions. I call them the
linear plausibility functions.

The simplicity of (29) and (30) is a strong point in favor of the linear
plausibility functions. The linear plausibility functions for multiple
observations are somewhat more complicated, but still manageable. This
simplicity contrasts with the complexity of the simplicial plausibility
functions, described in §4.3 below. The simplicial plausibility functions
are theoretically more attractive than the linear ones, but they present
formidable computational difficulties.

4.2.1. *The Continuous Case*

Thus far I have dealt exclusively with aleatory laws on discrete spaces.
It is possible, however, to define the linear plausibility functions for the
continuous case as well.

A continuous statistical specification on a space X is usually described
with reference to a fixed measure on X; each aleatory law is given by a
density with respect to that measure. In the simplest case, X is the real
line and the reference measure is Lebesque measure. In that case, an
aleatory law given by a non-negative function θ such that

$$\int_{-\infty}^{\infty} \theta(x)\, dx = 1 \;;$$

θ is called a *probability density*, and it is understood to assign probability

$$\int_{A} \theta(x)\, dx$$

to any (measurable) subset A of the real line.

This description makes no reference to any topology on X. On the
other hand, it is commonly and quite correctly argued that continuous
statistical specifications are extremely idealized and should be understood
as limiting cases of discrete specifications. And the notion of approx-
imating a continuous specification by a discrete one can be made in-

telligible only in the context of a topology on X and some requirements of continuity on the densities in the specification.

It is not clear just what requirements of continuity should be imposed, but in the case of the real line, we might require each density to be continuous, with the moduli of continuity for the different densities bounded at each point. More precisely, we might say that a set Θ_0 of functions on the real line X is a statistical specification on X provided that

(i) $\quad \theta(x) \geqslant 0 \quad$ for all $\quad x \in X \quad$ and $\quad \theta \in \Theta_0$.

(ii) $\quad \int_{-\infty}^{\infty} \theta(x) = 1 \quad$ for all $\quad \theta \in \Theta_0$.

(iii) $\quad$ For each $x \in X$ there exists $K < \infty$ such that
$$|\theta(x') - \theta(x)| < K|x - x'| \quad \text{for all} \quad x' \in X \quad \text{and} \quad \theta \in \Theta_0.$$

One consequence of these requirements is that $\sup_{\theta \in \Theta_0} \theta(x) < \infty$ for each $x \in X$. So whenever these requirements are met, (30) can be used to define the plausibility function $PL_x: 2^{\Theta_0} \to [0, 1]$ based on the observation $x \in X$. And plausibility functions based one many observations can then be obtained by combination.

4.3. The Simplicial Plausibility Functions

As I mentioned above, the simplicial support functions can be justified by a theory of belief that embraces aleatory laws. In this section, I will sketch this theory of belief and then briefly describe the simplicial plausibility functions.

4.3.1. *Aleatory Laws and Degrees of Belief*

The idea of an aleatory law has been intertwined with the idea of degrees of belief throughout the history of both ideas, but the connection can be perplexing. For though an aleatory law can always supply us with degrees of belief, it is appropriate to adopt those degrees of belief only when we know the law to hold. And such knowledge is rarely available.

Let me be more precise. If an experiment has outcomes in X, and we know it is governed by the aleatory law $\theta: X \to [0, 1]$, then we will naturally adopt $\theta(x)$ as our degree of belief that the experiment will result in $x \in X$. And more generally, we will adopt $\sum_{x \in A} \theta(x)$ as our degree of belief that the experiment will result in one of possible outcomes in a subset A of X. In other words, we will adopt the belief function $\mathrm{Bel}_\theta \colon 2^X \to [0, 1]$ given by $\mathrm{Bel}_\theta(A) = \sum_{x \in A} \theta(x)$. In fact, though, we can never really be certain that the experiment is governed by the aleatory law θ. So the belief function Bel_θ is appropriate only conditionally upon knowledge that we do not and cannot have.

This perplexing situation could be made intelligible within our theory of belief if we could somehow represent our knowledge by another belief function which produces Bel_θ only when conditioned on the fact that θ is the true aleatory law governing the experiment. Is this possible?

The belief function Bel_θ applies to propositions about the outcome of a forthcoming trial, and we wish to obtain it by conditioning another belief function, say Bel, on the proposition that the aleatory law θ governs the experiment. Hence Bel must apply both to propositions about the outcome of the trial and to propositions about what aleatory law governs the experiment. And it must also apply to propositions that simultaneously assert something about the outcome of the trial and something about which aleatory law governs the experiment.

We are led, then, to postulate the existence of a belief function Bel: $2^{\Theta \times X} \to [0, 1]$, where Θ is the collection of all possible aleatory laws on X, $\Theta \times X$ is the Cartesian product of Θ and X, and $A \in 2^{\Theta \times X}$ is taken to be the proposition that the pair (θ, x) is in $A \subset \Theta \times X$, where θ is the true aleatory law governing the experiment and x is the outcome of the forthcoming trial. Such a belief function Bel can indeed be conditioned on the proposition that a given aleatory law $\theta \in \Theta$ is the true one; we simply condition Bel on the subset $\{\theta\} \times X$ of $\Theta \times X$, obtaining a belief function on $2^{\{\theta\} \times X}$. A belief function on $2^{\{\theta\} \times X}$ amounts to the same thing as a belief function on 2^X, so we can require this belief function to be Bel_θ, as given above.

The notion of such an overall belief function Bel is very fruitful, for we can express many of our other ideas in terms of Bel. Lack of any prior opinion about the identity of the true aleatory law can be expressed, for example, by saying that $P^*(\{\theta\} \times X) = 1$ for all $\theta \in \Theta$. And most importantly, it becomes natural to obtain our support function S_x based on the observation x by conditioning Bel on x – i.e., on $\Theta \times \{x\}$. Hence conditions on the support functions S_x become conditions on Bel.

So for every discrete space X, we find ourselves demanding a function Bel: $2^{\Theta \times X} \to [0, 1]$, where Θ is the set of all aleatory laws on X, and Bel satisfies at least the following four conditions:

(1)   Bel is a condensable belief function.

(2) $P^*(\{\theta\} \times X) = 1$ for all $\theta \in \Theta$, where P^* is the upper probability function corresponding to Bel.

(3)   For every $\theta \in \Theta$, conditioning Bel on θ results in the belief function $\text{Bel}_\theta: 2^X \to [0, 1]$ given by $\text{Bel}_\theta = \sum_{x \in A} \theta(x)$ for all $A \in 2^X$.

(4)   For every $x \in X$, conditioning Bel on x results in an upper probability function on 2^Θ that satisfies the second postulate of plausibility for the observation x. (More precisely, if $PL_x: 2^\Theta \to [0, 1]$ is the upper probability function obtained by conditioning Bel on $\{x\} \times \Theta$, then PL_x should satisfy $PL_x(\{\theta_1, \theta_2\}) = PL_x(\{\theta_1\})$ for all doubletons $\{\theta_1, \theta_2\} \in 2^\Theta$ such that $\theta_1(x)/\theta_1(y) \geqslant \theta_2(x)/\theta_2(y)$ for all $y \in X$.)

It turns out that for every discrete space X there is one and only one belief function Bel: $2^{\theta \times X} \to [0, 1]$ satisfying these four requirements. I will not prove this fact here, but I will describe the upper probability functions $\{PL_x\}_{x \in X}$ that result from conditioning this unique belief function Bel on the various possible observations $x \in X$. These upper probability functions PL_x are, of course, the *simplicial plausibility functions*.

4.3.2. *Some Formulae*

Let me begin my description of the functions PL_x by supplying some formulae. I will then turn to a more illuminating geometric description.

I continue to denote by Θ the set of all aleatory laws on the discrete space X. Fix $x \in X$, and for each finite non-empty subset A of Θ, set

$$(31) \quad Q_x(A) = \begin{cases} \dfrac{1}{\displaystyle\sum_{y \in X} \max_{\theta \in A} \dfrac{\theta(y)}{\theta(x)}} & \text{if } \theta(x) > 0 \text{ for all } \theta \in A, \\ 0 & \text{otherwise}. \end{cases}$$

These are the commonality numbers for the simplicial plausibility function PL_x. In other words,

$$PL_x(A) = \sum_{\substack{T \subset A \\ T \neq \theta}} (-1)^{1 + \mathrm{card}\, T} Q_x(T)$$

for all finite non-empty subsets A of Θ. And, of course, the values of $PL_x(A)$ for infinite subsets A are given by condensability:

$$PL_x(A) = \sup_{\substack{A' \subset A \\ A' \subset \mathrm{finite}}} PL_x(A').$$

The commonality numbers $Q_x^0(A)$ and the plausibility function PL_x^0 for a restricted specification $\Theta_0 \subset \Theta$ can be obtained from these formulae by the rule of conditioning. The commonality numbers $Q_x^0(A)$ for finite subsets A of Θ_0 will be given, for example, by

$$(32) \qquad Q_x^0(A) = \begin{cases} \displaystyle k \overline{\sum_{y \in X} \max_{\theta \in A} \frac{\theta(y)}{\theta(x)}} & \text{if } \theta(x) > 0 \quad \text{for all} \quad \theta \in A, \\ 0 & \text{otherwise}, \end{cases}$$

where the constant k is equal to $(PL_x(\Theta_0))^{-1}$.

4.3.3. The Geometric Representation

In the case where X has three elements, the function PL_x that we have just defined can be described in terms of the geometric picture developed in §2.3.

Suppose, indeed, that $X = \{1, 2, 3\}$. Then as we saw in §2.3, the aleatory laws on X are in a one-to-one correspondence with the points of an equilateral triangle of unit area. Let me review that one-to-one correspondence, using a slightly different notation.

First, denote by $\mathbf{M}$ the set of measurable subsets of the triangle, and by $\mu(M)$ the measure of a subset $M \in \mathbf{M}$. Then note that the point θ of the triangle determines three smaller triangles α_θ, β_θ and γ_θ, as in Figure 16. The barycentric coordinates of θ are then

$$(\mu(\alpha_\theta), \mu(\beta_\theta), \mu(\gamma_\theta));$$

these three numbers are always non-negative and add to one. Finally, the aleatory law $\theta: X \to [0, 1]$ corresponding to the point θ of the triangle

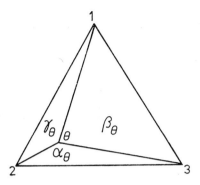

Fig. 16.

is given by

$$\theta(1) = \mu(\alpha_\theta),$$
$$\theta(2) = \mu(\beta_\theta),$$
$$\theta(3) = \mu(\gamma_\theta).$$

Now suppose our single observation x is equal to 1, and let us use this picture to describe the allocation of probability corresponding to PL_1. The key is to think of M as the collection of our probability masses, and to think of α_θ as the probability mass than can get into the singleton $\{\theta\}$. Hence

$$\bigvee_{\theta \in A} \alpha_\theta$$

is the total probability mass that can get into $A \in 2^\Theta$, and

$$PL_1(A) = \mu\left(\bigvee_{\theta \in A} \alpha_\theta\right).$$

The probability mass $\bigvee_{\theta \in A} \alpha_\theta$ is shown for several different finite subsets in Figure 17.

Now consider a finite subset A of Θ, and consider the probability mass

$$\bigwedge_{\theta \in A} \alpha_\theta.$$

This will be the total probability mass that can get into each and every point of A. Its measure will be the commonality number for A:

$$(33) \qquad Q_1(A) = \mu\left(\bigwedge_{\theta \in A} \alpha_\theta\right).$$

Can we compute $Q_1(A)$ from our geometric picture?

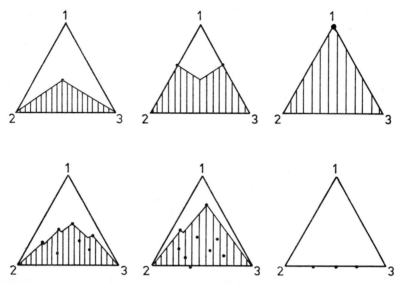

Fig. 17. In each example, the elements of A are represented by dots, and the probability mass $\bigvee_{\theta \in A} \alpha_\theta$ is shaded. In the last example each of the three elements θ of A has $\theta(1)=0$ and hence lies on the base of the triangle; the probability mass $\bigvee_{\theta \in A} \alpha_\theta$ is therefore empty.

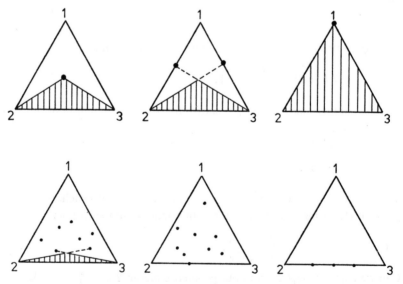

Fig. 18. In each example, the elements of A are represented by dots, and the probability mass $\bigwedge_{\theta \in A} \alpha_\theta$ is shaded. In the last two examples, there is an element θ of A with $\theta(1)=0$; hence the probability mass $\bigwedge_{\theta \in A} \alpha_\theta$ is empty.

The probability mass $\bigwedge_{\theta \in A} \alpha_\theta$ is shown for a few finite sets A in Figure 18. From these pictures, it is evident that we need to consider two cases: the case where $\theta(1)=0$ for some $\theta \in A$, and the case where $\theta(1)>0$ for all $\theta \in A$. In the first case,

$$(34) \qquad Q_1(A) = \mu\left(\bigwedge_{\theta \in A} \alpha_\theta\right) = 0.$$

In the second case, there will be a unique point θ_A of the triangle such that

$$(35) \qquad \alpha_{\theta_A} = \bigwedge_{\theta \in A} \alpha_\theta,$$

$$(36) \qquad \frac{\theta_A(1)}{\theta_A(2)} = \min_{\theta \in A} \frac{\theta(1)}{\theta(2)},$$

and

$$(37) \qquad \frac{\theta_A(1)}{\theta_A(3)} = \min_{\theta \in A} \frac{\theta(1)}{\theta(3)}.$$

(θ_A may be one of the points of A, as in the first and third examples of Figure 18, or it may not be, as in the second and fourth examples there.) From (33) and (34), we find that

$$Q_1(A) = \mu\left(\bigwedge_{\theta \in A} \alpha_\theta\right) = \mu(\alpha_{\theta_A}) = \theta_A(1).$$

Can we compute $\theta_A(1)$ from (36) and (37)?

We can. Inverting these two relations gives

$$\max_{\theta \in A} \frac{\theta(2)}{\theta(1)} = \frac{\theta_A(2)}{\theta_A(1)}$$

and

$$\max_{\theta \in A} \frac{\theta(3)}{\theta(1)} = \frac{\theta_A(3)}{\theta_A(1)}.$$

Of course,

$$\max_{\theta \in A} \frac{\theta(1)}{\theta(1)} = \frac{\theta_A(1)}{\theta_A(1)} = 1;$$

so

$$\sum_{y \in X} \max_{\theta \in A} \frac{\theta(y)}{\theta(1)} = \sum_{y \in X} \frac{\theta_A(y)}{\theta_A(1)} = \frac{1}{\theta_A(1)},$$

and

(38) $$Q_1(A) = \theta_A(1) = \frac{1}{\displaystyle\sum_{y \in X} \max_{\theta \in A} \frac{\theta(y)}{\theta(1)}}.$$

Notice that (34) and (38) do indeed agree with (31).

Several of the properties of PL_1 should be evident from this geometric representation. Condensability, for example, means that the area of $\bigvee_{\theta \in A} \alpha_\theta$ can always be approximated arbitrarily closely by the area of $\bigvee_{\theta \in A'} \alpha_\theta$ for some finite subset subset A' of A. The second postulate of plausibility is satisfied, for whenever

$$\frac{\theta_1(1)}{\theta_1(2)} \geqslant \frac{\theta_2(1)}{\theta_2(2)} \quad \text{and} \quad \frac{\theta_1(1)}{\theta_1(3)} \geqslant \frac{\theta_2(1)}{\theta_2(3)},$$

we have $\alpha_{\theta_2} \leqslant \alpha_{\theta_1}$, whence

$$PL_1(\{\theta_1, \theta_2\}) = \mu(\alpha_{\theta_1} \vee \alpha_{\theta_2}) = \mu(\alpha_{\theta_1}) = PL_1(\{\theta_1\}).$$

And the first postulate of plausibility is also satisfied, for

$$PL_1(\{\theta\}) = \mu(\alpha_\theta) = \theta(1).$$

I have developed the geometric representation for $x=1$, but the analogous development for $x=2$ and $x=3$ should be obvious. We would have, for example,

$$PL_2(A) = \mu\left(\bigvee_{\theta \in A} \beta_\theta\right)$$

and

$$PL_3(A) = \mu\left(\bigvee_{\theta \in A} \gamma_\theta\right).$$

Finally, I should remark that an analogous geometric representation is possible when X has any number of elements. In this case, we have used a triangle, but in the general case we would use a $(k-1)$-dimensional simplex, where k is the number of elements in X. Hence the name 'simplicial'.

4.3.4. *The Continuous Case*

The simplicial method of defining plausibility functions can also be extended to the continuous case, though the extension is more problematic than in the linear case. The key to the extension is formula (32).

Suppose, indeed, that Θ_0 is a specification on a continuous space X. Then the division of X into discrete categories will result in commonality numbers

$$(39) \qquad Q_x^\gamma(A) = \frac{k_\gamma}{\displaystyle\sum_{g \in \gamma} \max_{\theta \in A} \frac{\int_g \theta(y)\, dy}{\int_{g_0} \theta(y)\, dy}},$$

where γ consists of all the categories and g_0 is the category containing x. As the categories are made finer and finer, we might expect (39) to tend to

$$(40) \qquad \frac{k}{\displaystyle\int \max_{\theta \in A} \frac{\theta(y)}{\theta(x)}\, dy}$$

for some constant k. Such a convergence does occur in many cases, and if the constant k is not zero, then the quantities defined by (40) can be taken as the commonality numbers for a plausibility function $PL_x: 2^{\Theta_0} \to [0, 1]$.

5. THE HISTORICAL BACKGROUND

I have exposited the ideas in this essay as directly as possible, avoiding extensive historical references lest they evoke controversies or preconceptions. But I will now briefly review a few of the more prominent features in the historical background of these ideas.

5.1. DEGREES OF BELIEF

The notion of degree of belief has roots in the seventeenth and eighteenth centuries. The writers of that period usually used the terms 'degree of certainty' or 'degree of probability' rather than 'degree of support' or 'degree of belief', but they were concerned with the same questions and reached some of the same answers as I have discussed here. Notable examples are James Bernoulli, in his *Ars Conjectandi* (published post-humously in 1713) and J. H. Lambert in his *Neues Organon* (1764). Both of these writers considered degrees of probability for the case of two alternatives that took the general form exhibited in Table II, but Lambert seems to have had the clearer view of the matter, for it was in the process of correcting Bernoulli's faulty and less general rule that he arrived at the rule described in §1.1.

The modern student of probability, whether statistician or philosopher, may find it strange to hear numbers like those in Table II called probabilities. For we have been indoctrinated in the view that probabilities must be additive. In other words, the numbers s_1 and s_2 in Table II are nowadays required to add to one before they can be called probabilities. But no such doctrine appears to have been known to Bernoulli or Lambert, for they discuss non-additive examples of probabilities without apology.

How then did the rule of additivity come to be applied to probabilities? The answer to this question is surely to be found in the fact that there are two kinds of probability. On the one hand there is the aleatory kind – the objective probabilities that are given by aleatory laws. And on the other hand there is the epistemic kind – the degrees of certainty, of support, of belief, or what-have-you. Bernoulli seems to have regarded the first kind as a special case of the second, and I took a similar view in §4.3. But in any case, the first kind incontestably must obey the rule of additivity; whereas, contrary to some contemporary opinion, it is highly doubtful that the second kind ought always to obey the rule of additivity. But it is easy to confuse and plausible to identify the two kinds of probability and thus apply the rules for aleatory probabilities to epistemic probabilities. Unfortunately, Laplace rather deliberately made precisely such an identification. Perhaps he wanted to deal with aleatory probabilities but found it suited his determinism to call them epistemic probabilities. Whatever his motivation, one effect of his great synthesis was the suppression, for a century and a half, of non-additive probabilities.

Indeed, it is remarkable how thoroughly Laplace's successors, and students of the history of probability down to this very day, have ignored Bernoulli's and Lambert's work on non-additive probabilities. Lambert's ideas were exposited by Prevost and Lhuilier in 1797, but since then they seem to have plunged into almost total obscurity. Issac Todhunter did discuss Prevost and Lhuilier's article in his *History* in 1865 (pp. 461–463), but apparently with little comprehension; and I know of no further reference to these ideas by any student or historian of probability during the past century.

5.2. STATISTICAL INFERENCE

I have spoken in quite general terms in the preceding exposition, and I may have left the impression that my subject has been the whole field of statistical inference. But my subject has been much smaller than that. It has been the problem of statistical support – the problem of measuring the support for various subsets of Θ when a statistical specification $\{P_\theta\}_{\theta \in \Theta}$ is taken as known. In practice, the statistical specification may be far from known; and the problem of providing a specification – or of assessing the adequacy of a proposed one – seems to be the harder part of the theory of statistical inference.

Fifty years ago, R. A. Fisher distinguished between the problems of specification and the problems of 'estimation', declaring that the problems of specification were entirely 'a matter for the practical statistician'. For Fisher, it was the problem of estimating the correct value of θ on the basis of the specification $\{P_\theta\}_{\theta \in \Theta}$ and the observations x that was central to theoretical, as opposed to practical statistics. Today we would probably modify Fisher's judgment in several respects. First of all, his notion of estimation turned out to be too narrow, for the evidence about θ often cannot be summarized by an estimate and a measure of the estimate's accuracy. Hence many of us would replace the 'problem of estimation' with the 'problem of support' in a general description of the business of theoretical statistics. Secondly, we have come to see the problem of specification as a quite theoretical problem. As we have come to admit, the practical statistician's struggle with data is not only antecedent to the statistical specification – it is also antecedent to the specification of an 'observation space' X and even to the notion that anything is being governed by an aleatory law.

So in a modern view of theoretical statistics we might distinguish two broad problems – the problem of support and the problem of specification. The problem of support, which is the subject of this exposition, is probably the more modest and manageable of the two.

5.3. APOLOGIA

This essay is in part a defense and in part a reformulation of earlier work by A. P. Dempster. The ideas in it were inspired by my study of Dempster's work – a study that began when I attended his seminar on statistical inference at Harvard in the spring of 1971; and it culminates, in §4, with a justification of some of his methods of assessing statistical evidence. But, quite naturally, it differs in important respects from Dempster's work.

Most importantly, my philosophical account differs from Dempster's own. For far from rejecting the Laplacean synthesis, Dempster saw it as the source of 'much of the motivation and fascination of the modern science of probability'. (Research Report S-3, Harvard University, p. 8.) And instead of thinking of his lower probabilities as degrees of belief or degrees of support, he preferred, at least originally, to think of his upper and lower probabilities as bounds for some true but somehow unknowable probabilities, thus retaining the identification of degrees of belief with additive probabilities.

The mathematical account in §3 also differs from Dempster's earlier account. The differences derive mainly from the replacement of multivalued mappings by allocations of probability and from the isolation of the notion of condensability – innovations that permit the role of the commonality numbers to be fully developed.

Finally, the degrees of plausibility adduced in §4 overlap with but are not identical with the upper probabilities adduced in Dempster's papers. Those obtained by the simplicial method in §4.3 are identical with the upper probabilities produced by Dempster's 'structures of the second kind', (see p. 349 of his 1966 article.) Those obtained by the linear method in §4.2 are not discussed in any of Dempster's published work, though he has privately encouraged interest in the method. And the upper probabilities that result from Dempster's 'structures of the first kind' bear no relation to the present essay – the arguments given here provide no justification for them.

It is the new understanding of the meaning of Dempster's upper probabilities that I offer as the primary contribution of this essay. Since Dempster's own interpretation has not proven widely appealing, I hope that the more radical understanding of the present essay will inspire a wider interest.

Whether or not this hope is fulfilled, I must express my gratitude to my wife Terry and my many other friends, teachers and fellow students who have helped me with these ideas.

ACKNOWLEDGEMENT

This essay was written while I was supported by a graduate fellowship from the National Science Foundation, and it was revised while I was supported by contract N00014-67A0151-0017 from the Office of Naval Research.

Dept. of Statistics, Princeton University

REFERENCES

Bernoulli, James: 1713, *Ars Conjectandi*, Basel. Especially pp. 217–223. See also pp. 22–34 of the translation by Bing Sung, issued as Technical Report No. 2 of the Department of Statistics, Harvard University, February 12, 1966, and available in microfiche from the Clearinghouse for Federal Scientific and Technical Information, Washington, D.C.

Bernoulli, Daniel: 1777, 'The Most Probable Choice Between Several Discrepant Observations and the Formation Therefrom of the Most Likely Induction'. *Acta Acad. Petrop.*, pp. 3–33. Reprinted as pp. 157–167 of Pearson and Kendall's *Studies in the History of Statistics and Probability*, Griffin, 1970.

Choquet, Gustave: 1953–4, 'Theory of Capacities', *Annales de l'Institut Fourier, Université de Grenoble*, V, pp. 131–296.

Dempster, A. P.: 1966, 'New Methods for Reasoning Towards Posterior Distributions Based on Sample Data'. *Ann. Math. Statist.* **37**, 355–374.

Dempster, A. P.: 1967, 'Upper and Lower Probabilities Induced by a Multivalued Mapping', *Ann. Math. Statist.* **38**, 325–339.

Dempster, A. P.: 1968, 'A Generalization of Bayesian Inference (with discussion)', *J. Roy. Statist. Soc. Ser. B.* **30**, 205–247.

Dempster, A. P.: 1968, *The Theory of Statistical Inference: A Critical Analysis*, Ch. 2, Research Report S-3, Dept. of Statistics, Harvard University. September 27, 1968.

Fisher, R. A.: 1922, 'On the Mathematical Foundations of Theoretical Statistics', *Philosophical Transactions of the Royal Society of London, Series A*, Vol. 222, pp. 309–368. Reprinted in R. A. Fisher, *Contributions to Mathematical Statistics*, Wiley, New York, 1950.

Lambert, Johann Heinrich: 1760, *Photometria*, Augsburg, pp. 131–147.

Lambert, Johann Heinrich: 1764, *Neues Organon*, Zweiter Band, pp. 318–421. Reprinted in 1965 as Vol. II of *Lambert's Philosophische Schriften* by Georg Olms Verlagsbuchhandlung, Hildesheim.

Lehman, E. L.: 1959, *Testing Statistical Hypotheses*, Wiley, New York.

Prevost and Lhuilier: 1797, 'Mémoire sur l'application du Calcul des probabilités à la valeur du témoignage', *Mémoires de l'Académie Royale de Berlin*, pp. 120–152.

Shafer, Glenn: 1973, *Allocations of Probability: A Theory of Partial Belief*. Doctoral dissertation submitted to the Dept. of Statistics, Princeton University, June 26, 1973. Available from University Microfilms, Ann Arbor, Michigan.

Todhunter, Issac: 1865, *A History of the Mathematical Theory of Probability*. Reprinted by the Chelsea Publishing Co., New York, 1949.

DISCUSSION

Commentator Good: Suppose the evidence arises in two experiments with a thousand heads in one and a thousand tails in the other. Would the order in which the data are obtained effect the results?

Shafer: No, the order does not matter. Dempster's rule of combination is symmetric.

Lindley: This theory owes much to Dempster's work. My own view of that theory is that it is upset by Aitchison's counter-example. (This was presented in the discussion to Dempster's 1968 paper: *J. Roy. Statist. Soc. B,* **30**, 205–247 on page 234.) Essentially Aitchison considers two trinomial distributions with probabilities (0.4, 0.5, 0.1) and (0.4, 0.1, 0.5) – the essential point being the equality of the probabilities for the first class. He then shows that, according to Dempster's theory, a single observation falling in the first class can change our opinions about which trinomial obtains. This is very counter-intuitive since the observation would appear to contribute nothing to this question. My question to tonight's speaker is: does the same criticism apply to his theory?

Shafer: Aitchison's criticism applies to the simplicial method, which is indeed the same as the method in Dempster's 1968 paper. The criticism does not apply to the linear method.

Perhaps I should take a paragraph to cast Aitchison's example in the vocabulary of the preceding essay. The statistical specification consists of two aleatory laws: $\Theta = \{\theta_1, \theta_2\}$. The set of possible outcomes is, say, $X = \{$Azure, Brown, Crimson$\}$. Both aleatory laws assign Azure a chance of 0.4, but they disagree on Brown and Crimson. One begins, presumably, with the vacuous belief function: $\text{Bel}(\{\theta_1\}) = \text{Bel}(\{\theta_2\}) = 0$. But what support function over Θ should one have after a single observation $x = $ Azure? (1) Following the simplicial method, one obtains the support function S over Θ given by $S(\{\theta_1\}) = S(\{\theta_2\}) = \frac{2}{9}$. (2) But following the linear method, one obtains $S(\{\theta_1\}) = S(\{\theta_2\}) = 0$ – i.e., there is no change from the vacuous support function with which one began.

Choosing between the linear and simplicial methods in this example obviously amounts to deciding whether the observation $x =$ Azure should be treated as no evidence (linear solution) or as precisely balanced conflicting evidence (simplicial solution). In defense of the simplicial solution, one might argue that the single observation $x =$ Azure is indeed internally conflicting: the observation of Azure rather than Crimson supports θ_1 while the observation of Azure rather than Brown supports θ_2. But this is not very convincing, and I now agree with Mr Aitchison and Mr Lindley that the linear solution is preferable.

Writing in January of 1975, I should point out that the original essay and talk that inspired Mr Lindley's question argued for the simplicial method and did not mention the linear method. The essay printed above, which gives equal billing to the two methods, was written in the summer of 1973. My present preference for the linear method is based primarily on its adaptation to the notion of 'weight of evidence', a notion which allows a much deeper understanding of Dempster's rule of combination. (See my forthcoming book *A Mathematical Theory of Evidence*.)

PATRICK SUPPES

TESTING THEORIES AND THE
FOUNDATIONS OF STATISTICS

1. Historical perspective

In this paper I examine the extent to which problems in the foundations of probability are relevant to the testing of theories, and what view towards probability, if any, can be inferred from the problems and practices found in the scientific literature. I start with a historical perspective and then consider particular examples in contemporary science. In this latter discussion I confront some of the issues made salient by Bayesians.

Far and away the most serious quantitative scientific treatise in ancient times that uses both mathematics and data in a systematic way is Ptolemy's *Almagest*. What is surprising is that in Ptolemy's *Almagest* and in other astronomical treatises of ancient times there is no evidence of a quantitative theory of error; in fact, there is little evidence of any theory of error at all. This is in marked contrast to ancient astronomy's mathematical and observational sophistication. It might be thought that the lack of such a systematic theory of error is simply a reflection of the absence of any developments of a quantitative sort in probability theory in Hellenistic science, and consequently, an explanation for the absence of such an analysis in Ptolemy is easily found.

The story, however, is much more complicated, because what is true of Ptolemy's *Almagest* is also true of Newton's *Principia.* There is, I believe, not one single computation of a quantitative error term in Newton's *Principia.* It contains a few remarks about errors, but all of them are of a qualitative and incidental character. Again, it might be thought that this is simply a consequence of the fact that the theory of probability was just being developed in quantitative form in the seventeenth century, and as a result, detailed applications could hardly be expected. That more complicated explanations of the absence of such a theory of error are needed is testified to by the absence of such systematic computations of errors in Laplace's *Celestial Mechanics*. Of all the

Harper and Hooker (eds.), Foundations of Probability Theory, Statistical Inference, and Statistical Theories of Science,
Vol. II, 437–455. All Rights Reserved. Copyright © 1976 by D. Reidel Publishing Company, Dordrecht-Holland.

classical treatises in which one would expect to find such systematic computations, Laplace's is the one. In spite of the fact that Laplace more than anyone else contributed to the development of the theory of probability in the eighteenth century and the early part of the nineteenth century, and in spite of the fact that he discusses in his treatise on the theory of probability the analysis of data from a probabilistic standpoint in order to determine evidence for 'constant causes', there is in the systematic treatise on the solar system no detailed analysis of error terms or any application directly of a quantitative theory of error.

This Ptolemaic tradition did not end with Laplace, but also is found in Maxwell's treatise on electricity and magnetism. There is little numerical confrontation between data and theory in Maxwell, and certainly no analysis of problems of errors of measurement. In fact, from the standpoint of the confrontation between data and theory, there would seem to be some downhill sliding from the time of Ptolemy to that of Maxwell.

Within astronomy proper, reporting error terms in the analysis of astronomical data did become common in the nineteenth century. I hope on another occasion to trace that history. At the present time, however, my understanding of it is too poor to enter into the details. It is certainly true that, from the standpoint of physical theory, the more important development of electromagnetic theory does not reflect a corresponding development of a sophisticated theory and practice of data analysis at the level characteristic of astronomy in the last half of the nineteenth century.

The conceptual point of importance for this paper is that the verification of the historically important theories of physical phenomena has practically never used a detailed statistical theory of confirmation to test empirical adequacy. In thinking about the ways in which statistics and probability are and should be used in science, it seems to me that this historical fact is important to keep in mind so as not to create a simplified theory of how theories may be tested. By this statement I do not mean to suggest that I am against the use of statistical methods in the verification of theories. I only enter the cautionary note that the verification of theories is a complex business, and any simple view of how to apply statistical methods is bound to be inaccurate.

Numerous other examples from the seventeenth, eighteenth, and nineteenth centuries can easily be found. It might be thought, however, that these historical examples have been superseded by the statistical sophistication of the twentieth century. After all it may properly be claimed that, in spite of the kind of developments begun by Laplace, much of the development of explicit procedures of statistical inference and estimation dates from the second or third decade of the twentieth century, and that to get a true assessment of the situation, we must examine some twentieth-century theories.

2. Twentieth-century theories

A. *Quantum Mechanics*

The most impressive and most extensively tested theory in this century is surely quantum mechanics. When one turns to the evidence in support of quantum mechanics, and the kind of confrontations between data and theory that are used to support the theory, some surprises are in store. First of all, in the standard treatises, no attempt is made to present empirical data or to point out in what respects discrepancies exist between theoretical predictions and empirical data. To support this statement, I casually looked in my own library at three treatises on quantum mechanics: P. A. M. Dirac, *Quantum Mechanics* (3rd ed., 1947); L. D. Landau and E. M. Lifshitz, *Quantum Mechanics: Non-relativistic Theory* (1958); Albert Messiah, *Quantum Mechanics*, Volumes 1 and 2 (1961). By reporting these negative results, I am not suggesting that there is no place one can find experimental evidence. Rather, unlike the tradition of Newton and Laplace, in contemporary treatments of quantum mechanics, there is often no attempt to present supporting data and to examine discrepancies between theory and data. Although data can be found in the experimental literature, and one can track down the examination of data in the classical experiments that are ordinarily cited in support of quantum mechanics, I believe it is fair to say that there exists no book in which these data are brought together in a systematic way and in which a careful examination from a statistical standpoint of the relation between the data and the theory is considered. One classical text, Leonard Schiff's *Quantum Mechanics* (1949), does list in the discussion on the physical basis of quantum mechanics the relations between the classical experiments, running from Young's experiments at the beginning of the nineteenth century on diffraction to the Stern-Gerlach experiment in 1922. But there is in Schiff's book no detailed discussion of the relation between theory and data, but only a development of the theory. On the other hand, turning to one of the more data-oriented books that do not develop the theory of quantum mechanics, for example, F. K. Richtmyer and E. H. Kennard, *Introduction to Modern Physics* (4th ed., 1947), one does find the experimental data and the comparisons with theoretical predictions. However, I believe that this classical text contains not a single statistical inference, nor even a statistically descriptive statistic. Although I have not systematically surveyed the original experimental literature, from my experience what I have said about Richtmyer and Kennard also holds for this literature in almost all cases.

To some extent, these references are to the older experimental literature in physics. Perusal of current issues of *Physical Review* indicates that the actual use of standard statistical tests can be found in a variety of experimental articles. Yet the main thrust of my remark is, I think, still correct. In the testing of highly structured theories of the kind characteristic of physics, there is little use of the vast apparatus of modern statistics.

B. *Econometrics*

Perhaps the sharpest scientific contrast to quantum mechanics and to other parts of physics can be found in economics. Among social scientists, econometricians are probably the most statistically sophisticated and the most careful in their use of statistical procedures. The analysis of data is superb from a statistical standpoint in most of the major work. Because it is true that economists deal with nonexperimental data, there is even more reason to be statistically explicit about the analysis and inferences made. In this respect, economics compares more directly with astronomy or meteorology than with quantum mechanics, where the data are almost all experimental in character. As an example of the kind of theory used in econometrics, I have selected a recent article by Chiswick and Mincer (1972). What is striking about this and other serious applications of mathematical concepts in economics is that when data are involved the model is usually of a relatively simple character without substantial theoretical deductions from the model itself. I quote from the second and third pages of the article (35–36) in which the mathematical model used for the analysis is stated.

The relation between gross earnings and investment in human capital for the ith person in year j can be written as

$$(1) \qquad E_{ji} = E_{oi} + \sum_{t=1}^{j-1} r_{ti} C_{ti},$$

where the gross earnings (E_{ji}) are a function of the 'original' endowment (E_{oi}) and the sum of the returns on previous investments (C_{ti}), r_{ti} being the average rate of return to the investment in the tth year. In this expression, earnings are a linear function of dollars of investment.

An alternative specification of the relation between gross earnings and investment can be obtained by expressing C_{ti} as a fraction of E_{ti} (that is, $C_{ti} = k_{ti} E_{ti}$). If the original endowment is assumed constant across years and individuals (E_0), we can write

$$(2) \qquad E_{ji} = E_0 + \sum_{t=1}^{j-1} r_{ti} k_{ti} E_{ti} = E_0 \prod_{t=1}^{j-1} (1 + r_{ti} k_{ti}).$$

By taking the natural log of both sides of Equation (2), since $r_t k_t$ is small, we obtain (approximately)

$$(3) \qquad \ln(E_{ji}) = \ln E_0 + \sum_{t=1}^{j-1} r_{ti} k_{ti}.$$

What is proposed is a simple linear model that the effects of returns on previous investments, as well as the function of 'the original' endowment, can satisfactorily be expressed in a linear way. This linear model is not derived from any more fundamental assumptions, nor is it the consequence of elementary qualitative assumptions or of some deeper running formulation of economic theory. This kind of regression model is characteristic of applications in econometrics, and the efforts that have been made to understand thoroughly the statistical pitfalls of making inferences by use of such models are thoroughly explored in the literature, for example, in Malinvaud's classic work (1966). Although Malinvaud's book takes us beyond the kind of linear model defined by Chiswick and Mincer, it does not take us far.

It might be thought that I am pushing a kind of conservation thesis: the more theory the less statistics, and the less theory the more statistics. From an empirical standpoint there is something to be said for this. It is even true of the mathematics, for example. Although the mathematical requirements are rather different, the mathematical level of a treatise like Malinvaud's is, in my judgment, about comparable to the treatises on quantum mechanics I mentioned above, although Malinvaud's treatise would satisfy mathematical standards of rigor more explicitly than would the physical treatises.

Thus far I have picked two extremes of theories – one, the highly structured and developed theories of quantum mechanics, and the other, the very simple regression models characteristic of much of econometrics. It is natural to ask if these two examples, each extreme in its own way, represent the whole story. I do not think this is the case, and I want to turn to still a third class of theories, theories that do not have the depth of structure and development of quantum mechanics, but that have a fundamental theory and consequences that lead to more elaborate structures.

C. Psychological Theories

This third kind of case is drawn from psychology, which also is more like physics than econometrics in that the tests of theories are basically experimental rather than nonexperimental in character. Although psychological theories of learning, as an example, are much shallower than the great physical theories of the twentieth century, they can be given a precise formulation in general terms and can lead to rigorous deductions of particular predictions for particular experiments. Surprisingly, even in relatively simple applications of the theory quite intractable stochastic processes arise for which the application of standard statistical procedures of estimation of parameters is essentially hopeless. Without going into detail, let me mention one or two examples. A typical simple learning experiment that constitutes a test of an underlying theoretical model, itself derived from a more general qualitative theory, involves estimating parameters in a chain of infinite order. In other words, the mathematical model itself is a stochastic process that is a chain of infinite order. In most cases the chain of infinite order is an ergodic process, but explicit maximum-likelihood or Bayesian estimate of the parameters is strictly out of the question, and in practice some relatively rough-and-ready approximation to a maximum-likelihood estimate, or sometimes a minimum chi-squared estimate, is used. In almost all cases, the tests are not actually maximum likelihood nor minimum chi-square, but tests that approximate these, and whose characteristics as tests have not in any sense been thoroughly investigated.

In most experiments that test such psychological models, the number of observations is huge, for example, upwards from two or three thousand to twenty or thirty thousand. Given the large number of observations, the relative statistical crudeness of a pseudo-maximum-likelihood estimate is not disturbing to anyone, for it is clear that refinements of statistical procedure will have little effect on the summary estimate of the goodness of fit of the theory to data. Because of computational difficulties in many applications of the suitable maximum-likelihood function, the complete function is computed rather than seeking a solution to the derivative to find the maximum. The graphing of the complete function has been instructive in a number of cases, because when we look at the complete function we see that a fairly wide variation in the value of the parameter estimated makes little difference in the fit of the theory to the data. An example of this for a simple linear learning model (that is, a chain of infinite order) in the observables is shown in Figure 1 (drawn from Suppes

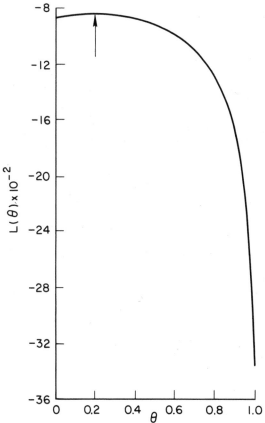

Fig. 1.   The pseudo-likelihood function, $L^*(\theta)$, for the linear model.

and Atkinson, 1960, p. 218). Note that the suitable likelihood function is quite flat, between 0.02 and 0.30, and consequently, any value of the learning parameter θ lying in this interval will give about as good a fit as another. This kind of flatness of suitable likelihood function is another reason that working scientists will not take too seriously refinements of statistical concepts, at least insofar as they are billed as offering something of importance and significance to the scientist in evaluating the relation between theory and experiment.

3. FOUNDATIONS OF STATISTICS

I do not for a moment want to defend the careless statistical practices of physicists; in fact, I think much information in physical experiments is lost because of the lack of care in analysis. A good example is the lack of the analysis of randomness in experiments on radioactive decay of a given substance. Such radioactive decay is often cited by both physicists and philosophers as a prime example of randomness in nature. Personally I tend to accept this view, but I would be much more secure in it if I had at hand the kind of massive analysis of data from radioactive decay that is characteristic of the kind of tests statisticians have given such data in other domains. It is well known that it is extraordinarily difficult to produce data that illustrate the appropriate randomness features in all respects. It would be curious indeed to find deviations from what was expected in the matter of radioactive decay, and only by the application of refined statistical techniques are we at all likely to find such deviations if they do exist, or to confirm the view that randomness all the way is the story.

In spite of this example, and in spite of my willingness to criticize the statistical procedures of physicists, I think we need a way of looking at the foundations of statistics that takes account of the kind of rough-and-ready theoretical tests I have described earlier in this paper. What we need among our concepts of logical inference and statistical inference are appropriate ways of closing the gap between commonsense conclusions that the evidence is indeed decisive and that no refined tests are required.

It may be possible to say on some occasions that it is merely a matter of routine work to produce the statistical distributions of data required to make the inference explicit and firm, but of course this is not the way the matter mainly works. If one examines the discussion of these matters in tests like those of Richtmyer and Kennard, it is evident that the data are not even thought of in a way that would permit a statistical inference to be made. A Bayesian inference could be made only in the crudest and most subjective way, and yet we all admit and agree to the solidity of the evidence in many cases and on the extent to which it supports the theoretical predictions.

There are several ways of expressing skepticism about the realism of objective or Bayesian approaches to this problem. Apart from the sophisticated problem of assuming probability distributions for data structures, there can be and should be proper skepticism about assigning a subjective or prior probability to the agreement between an experimental result and a theoretical prediction. To indicate some of the problems in a quantitative way, let E be the experimental result (which we may think of as a random variable) and let T be the theoretical prediction (which we may also think of as a random variable). We might begin by expressing our confidence in the agreement between the two by the following probabilistic inequality:

(1) $P(E=T)>1-\varepsilon.$

The difficulty with inequality (1) is that we almost never expect the experimental result and the experimental prediction to agree exactly if the random variables are thought of as having an underlying continuous distribution or, put another way, if the empirical quantities in question are assumed to be essentially continuous in nature. A more realistic expression of our confidence in the essential agreement between E and T may be expressed in the following way:

(2) $P(|E-T|<\varepsilon_1)>1-\varepsilon_2.$

In both inequalities (1) and (2) we expect ε, ε_1, and ε_2 to be small numbers, but already inequality (2) has assumed a rather formidable character and seems too elaborate for the purpose at hand.

Inequality (2) has a surface resemblance to a confidence-interval statement, but the lack of an underlying theory of distributions makes development of a theory of confidence intervals for the kind of situation I am talking about even more unrealistic than the use of inequalities like (2).

What seems natural is to analyze these situations of qualitative judgment in terms of a qualitative theory of probability and belief. Thus, in
qualitative terms we replace inequality (2) by the qualitative statement (3).

(3)   The event that $|E - T|$ is small $\approx X$.

In (3), the relation $\approx$ is the relation for indistinguishability and X is the
certain event. (We might think of X as the sample space, but I agree with
the Bayesians in being suspicious of having one definite reference space,
and I prefer simply to let X be a certain event – I am even willing to slide
between saying *a* certain event X and *the* certain event X.)

In the classical theory of qualitative probability, the relation $\approx$ of indistinguishability would be reflexive, symmetric, and transitive. Here I
ask only that it be reflexive and symmetric. For judgments of strictly
greater qualitative probability, I use a semiorder that is a relation $\succ$ that
satisfies the following three axioms:

1.   Not $x \succ x$.
2.   If $x \succ y$ and $y \succ z$ then either $x \succ w$ or $w \succ y$.
3.   If $x \succ y$ and $z \succ w$ then either $x \succ w$ or $z \succ y$.

The indistinguishability relation is just the nontransitive indifference relation that can be defined in terms of strict probability preference by the
following:

(4) $A \approx B$ iff not $A \succ B$ and not $B \succ A$.

Although one can add to the axioms introduced and write axioms that,
in the case of a finite number of events, will lead to the existence of a
probability measure, that is not my purpose here. The matter has been
studied elsewhere, and good results for this particular case have been
given by Domotor and Stelzer (1971). Rather, in the present discussion
I want to pursue the kind of apparatus I have introduced.

As I see it, it is a mistake to make the Bayesian move and to ask the
investigator for a subjective estimate that the agreement, for example,
between experimental and theoretical results is more accurate or less accurate than the accuracy with which the velocity of light is measured, or
the specific heat of sodium. In other words, the natural Bayesian thing
would be to ask the investigator to 'calibrate' his judgment of the quality
of the result by comparing it with other results for corresponding physical
experiments or appropriate experiments in the domain in question. In

practice, this is exactly what we do not do. We expect the investigator to present the numerical results, and in general we expect some discrepancy between the experimental result and the theoretical prediction. A qualitative comment may be made by the investigator, such as, 'the agreement is pretty good', or 'the agreement is not so good as we may ultimately expect to obtain but the results are encouraging'. The reader and colleague is left to draw his own conclusion, and there is a clear restraint from offering a more detailed or more complete analysis.

It is this practical sense of leaving things vague and qualitative that needs to be dealt with and made explicit. In my judgment to insist that we assign sharp probability values to all of our beliefs is a mistake and a kind of Bayesian intellectual imperialism. I do not think this corresponds to our actual ways of thinking, and we have been seduced by the simplicity and beauty of some of the Bayesian models. On the other hand, a strong tendency exists on the part of practicing statisticians to narrow excessively the domain of statistical inference, and to end up with the view that making a sound statistical inference is so difficult that only rarely can we do so, and usually only in the most carefully designed and controlled experiments.

I want to trod a qualitative middle path between these two extremes of optimism and pessimism about the use of probabilistic and statistical concepts. On a previous occasion I have expressed my skepticism about drawing a sharp and fundamental difference between logical inference and statistical inference, if only because this distinction is not present in the ordinary use of language and ordinary thinking (Suppes, 1966). The kind of gap I have attempted to stress in the present discussion is one that lies between the present explicit theory of logical inference and the explicit theory of statistical inference. The appropriate place to look for the theory to close this gap is in the semantics of ordinary language, for although I have been concerned with statements made in the summary about the tests of theories, I think that the character of these statements is by and large consistent with statements of ordinary language about ordinary experience, and that we should not invoke a special language and a special apparatus. What we have left is a residue of common sense and ordinary language, and it is my philosophical belief that not only will this residue remain, but it will also remain robust in the discussion of scientific theories and their verification. To expect these robust uses of commonsense judgments to be eliminated from our judgments of scientific theory is a mistaken search for precision. In fact, as I have attempted to argue, in many ways the stronger the theory and the better the evidence, the less tendency to use any defined statistical apparatus to evaluate the predictions of the theory. I see no reason to think that this broad generalization will not continue to hold.

This means that the intellectual task of closing the gap is almost identical with the task of giving a proper semantics for such ordinary language statements as:

> Almost all observations are in agreement with the experiment.
> Most of the observations are in agreement with the theoretical predictions.
> The agreement between prediction and theory is pretty good.
> See for yourself. The results are not bad.

The tools for providing such a semantical analysis are now being developed by a number of people, and the prospect for having a well-developed theory of these matters in the future looks bright. In the meantime, there are many simpler ways of improving on the situation that have relevance both to the foundations of statistics and to the testing of theories. In the next section I discuss one such approach, which can also be used in the deeper semantical analysis still to be developed in detail.

4. UPPER AND LOWER PROBABILITIES

The first step in escaping some of the misplaced precision of standard statistics is to replace the concept of probability by that of upper and lower probability. The first part of what I have to say is developed in more detail in Suppes (1974) and I shall only sketch the results here. (References to the earlier literature are to be found in this article.)

To begin with, let X be the sample space, $\mathfrak{F}$ an algebra of events on X, and A and B events, i.e., elements of $\mathfrak{F}$. The three essential properties we expect upper and lower measures on the algebra $\mathfrak{F}$ to satisfy are the following:

(I) $\quad P_*(A) \geqslant 0$.

(II) $\quad P_*(X) = P^*(X) = 1$.

(III)   If $A \cap B = \emptyset$ then

$$P_*(A) + P_*(B) \leqslant P_*(A \cup B) \leqslant P_*(A) + P^*(B) \leqslant P^*(A \cup B)$$
$$\leqslant P^*(A) + P^*(B).$$

From these properties we can easily show that

$$P_*(A) + P^*(\neg A) = 1.$$

Surprisingly enough, quite simple axioms on qualitative probability can be given that lead to upper and lower measures that satisfy these properties. The intuitive idea is to introduce standard events that play the role of standard scales in the measurement of weight. Examples of standard events would be the outcomes of flipping a fair coin n number of times for some fixed n.

The formal setup is as follows. The basic structures to which the axioms apply are quadruples $\langle X, \mathfrak{F}, \mathscr{S}, \geqslant \rangle$, where X is a nonempty set, $\mathfrak{F}$ is an algebra of subsets of X, that is, $\mathfrak{F}$ is a nonempty family of subsets of X and is closed under union and complementation, $\mathscr{S}$ is a similar algebra of sets, intuitively the events that are used for standard measurements, and I shall refer to the events in $\mathscr{S}$ as *standard* events S, T, etc. The relation $\geqslant$ is the familiar ordering relation on $\mathfrak{F}$. I use standard abbreviations for equivalence and strict ordering in terms of the weak ordering relation. (A weak ordering is transitive and strongly connected, i.e., for any events A and B, either $A \geqslant B$ or $B \geqslant A$.)

DEFINITION. *A structure* $\mathscr{X} = \langle X, \mathfrak{F}, \mathscr{S}, \geqslant \rangle$ *is a finite approximate measurement structure for beliefs if and only if X is a nonempty set, $\mathfrak{F}$ and $\mathscr{S}$ are algebras of sets on X, and the following axioms are satisfied for every A, B, and C in $\mathfrak{F}$ and every S and T in $\mathscr{S}$:*

AXIOM 1. *The relation* $\geqslant$ *is a weak ordering of* $\mathfrak{F}$;

AXIOM 2. *If* $A \cap C = \emptyset$ *and* $B \cap C = \emptyset$ *then* $A \geqslant B$ *if and only if* $A \cup C \geqslant B \cup C$;

AXIOM 3. $A \geqslant \emptyset$;

AXIOM 4. $X > \emptyset$;

AXIOM 5. $\mathscr{S}$ is a finite subset of $\mathfrak{F}$;

AXIOM 6. If $S \neq \emptyset$ then $S > \emptyset$;

AXIOM 7. If $S \geqslant T$ then there is a V in $\mathscr{S}$ such that $S \approx T \cup V$.
From these axioms the following theorem can be proved.

THEOREM 1. Let $\mathscr{X} = \langle X, \mathfrak{F}, \mathscr{S}, \geqslant \rangle$ be a finite approximate measurement structure for beliefs. Then
 (i) there exists a probability measure P on $\mathscr{S}$ such that for any two standard events S and T

$$S \geqslant T \text{ if and only if } P(S) \geqslant P(T),$$

 (ii) the measure P is unique and assigns the same positive probability to each minimal event of $\mathscr{S}$,
 (iii) if we define P_* and P^* as follows:
   (a) for any event A in $\mathfrak{F}$ equivalent to some standard event S,

$$P_*(A) = P^*(A) = P(S),$$

   (b) for any A in $\mathfrak{F}$ not equivalent to some standard event S, but lying in the minimal open interval (S, S') for standard events S and S'

$$P_*(A) = P(S) \quad and \quad P^*(A) = P(S'),$$

   then P_* and P^* satisfy conditions (I)–(III) for upper and lower probabilities on $\mathfrak{F}$, and
   (c) if n is the number of minimal elements in $\mathscr{S}$ then for every A in $\mathfrak{F}$

$$P^*(A) - P_*(A) \leqslant \frac{1}{n},$$

 (iv) if we define for A and B in $\mathfrak{F}$

$$A *> B \text{ if and only if } \exists S \text{ in } \mathscr{S} \text{ such that } A > S > B,$$

then $*>$ is a semiorder on $\mathfrak{F}$, if $A *> B$ then $P_*(A) \geqslant P^*(B)$, and if $P_*(A) \geqslant P^*(B)$ then $A \geqslant B$.
Following an earlier suggestion of Good (1962), events whose upper probability is 1 can be said to be *almost certain*, and events whose lower probability is 0 can be said to be *almost impossible*.

Moreover, let us consider events that are not exactly equivalent to any of the standard events. This restriction is easy to impose on our measurement procedures if we follow procedures often used in physics by requiring that each nonstandard event measured be assigned to a minimal open interval of standard events. In terms of such properly measured events A and B, as I shall call them, we may define upper and lower conditional probabilities as follows:

$$P_*(A \mid B) = P_*(A \cap B)/P_*(B),$$
$$P^*(A \mid B) = P^*(A \cap B)/P^*(B),$$

provided $P_*(B) > 0$. We can then show that the upper and lower conditional probabilities satisfy properties (I) to (III) except for the condition on $P_*(A) + P^*(A)$. In particular, $P_*(A \mid B) \leqslant P^*(A \mid B)$.

Within this framework we can then develop a reasonable approximation to Bayes' theorem or to the method of maximum likelihood. The point is that we can develop a machinery of statistical inference that is approximate in character and consequently is closer to the ordinary talk of scientists dealing with their summary evaluations of experimental tests of theories.

Moreover, an important feature of the kind of setup I am describing is that it is not meaningful to ask for arbitrary precision in the assignment of upper and lower probabilities to events, but only an assignment in rational numbers to the scale of the finite net of standard events. Further questions about precision do not have a clear meaning.

Using the kind of apparatus outlined, we can then replace, if we so desire, the inequality expressed in (2) in the previous section by the following equation using upper probabilities:

$$P^*(|E - T| < \varepsilon_1) = 1.$$

We could also paraphrase this statement in the manner indicated above by the qualitative statement that almost certainly all deviations between the experimental data and the theoretical predictions are less than ε or, even more qualitatively, are small.

I should also emphasize that the particular upper and lower measures derived from qualitative structures satisfying the axioms given above do not have many of the pathological characteristics of arbitrary upper and lower measures. This may be seen already in the fairly complete theory of conditional upper and lower measures that follows.[1]

I believe that the approximation theory I have sketched forms a natural bridge between the quantitative theory of statistics and the qualitative statements of ordinary scientific language. Further developments of the theory are needed to make clear whether my hopes are realistic or too sanguine.

Stanford University

NOTE

[1] The measurement-theoretic conception of upper and lower probabilities I use is quite different conceptually from the 'uncertainty' approach to such probabilities of Dempster (1967, 1968). Consequently, my approach to the theory of statistical inference is also formally and conceptually different from Dempster's.

BIBLIOGRAPHY

Chiswick, B. R. and Mincer, J., 'Time-Series Changes in Personal Income Inequality in the United States from 1939, with Projections to 1985', *Journal of Political Economy* **80** (1972) 34–66.

Dempster, A. P., 'Upper and Lower Probabilities Induced by a Multivalued Mapping', *Annals of Mathematical Statistics* **38** (1967) 325–340.

Dempster, A. P., 'A Generalization of Bayesian Inference', *Journal of the Royal Statistical Society* **30** (1968) (Series B), 205–247.

Dirac, P. A. M., *Quantum Mechanics* (3rd ed.), Oxford University Press, London, 1947.

Domotor, Z. and Stelzer, J., 'Representation of Finitely Additive Semiordered Qualitative Probability Structures', *Journal of Mathematical Psychology* **8** (1971) 145–158.

Good, I. J., 'Subjective Probability as the Measure of a Non-Measurable Set', in E. Nagel, P. Suppes, and A. Tarski (eds.), *Logic, Methodology and Philosophy of Science: Proceedings of the 1960 International Congress*, Stanford University Press, Stanford, 1962.

Landau, L. D. and Lifshitz, E. M., *Quantum Mechanics: Non-Relativistic Theory*, Pergamon Press, London, 1958.

Malinvaud, E., *Statistical Methods of Econometrics*, Rand McNally, Chicago, 1966.

Messiah, A., *Quantum Mechanics*, Vol. 1, 2, North-Holland, Amsterdam, 1961.

Richtmyer, F. K. and Kennard, E. H., *Introduction to Modern Physics* (4th ed.), McGraw-Hill, New York, 1947.

Schiff, L., *Quantum Mechanics*, McGraw-Hill, New York, 1949.

Suppes, P., 'Probabilistic Inference and the Concept of Total Evidence', in J. Hintikka and P. Suppes (eds.), *Aspects of Inductive Logic*, North-Holland, Amsterdam, 1966.

Suppes, P., 'The Measurement of Belief', *Journal of the Royal Statistical Society* **36** (1974) (Series B), 160–175.

Suppes, P. and Atkinson, R. C., *Markov Learning Models for Multiperson Interactions*, Stanford University Press, Stanford, 1960.

DISCUSSION

Commentator: Lindley: I have a delightful book at home which describes how they made wheels in the middle of the last century. These wheels were masterpieces of design and construction and yet no measuring instrument of any sort was used in their construction. Nowadays we turn out wheels more cheaply and in vastly greater numbers partly because we use precise measuring instruments. My point is that the fact that measurements were not used, does not mean that we will not be better off by using them.

Suppes: A parallel which strikes me here is the situation in which Laplacian determinism now finds itself: Just as we have now discovered that the universe is such that we can no longer carry out the Laplacian deterministic program (determine completely the state of the universe at some given instant of time) so we cannot carry out your suggestions – *we simply cannot get around in our universe, and function at the level at which we wish to function,* without carrying out an extensive program of measurement (among other things). (Indeed, there seems to have been a 'natural line of theological succession' from the early belief that God ran the universe in a definite fashion, to the Laplacian belief that the universe ran itself in a definite fashion to Lindley's belief that we all have access to a unique prior probability.) Just as we cannot carry out Laplace's program, we cannot carry out yours. The fact of the matter is that we all could carry around a suitable gambling apparatus which, upon the assumption of its independence from the rest of the universe, would serve to insure that we always realize Savage's axioms in a conduct of our lives. But the plain fact is that we do not wish to do that kind of thing, we are not constructed so as to operate with such a precise judgment of probabilities and in formulating theories of the sort that I have been discussing probability theorists have committed an error of misplaced precision.

Teller: I'd like to make two comments. You mentioned briefly the gross unrealism of Savage's use of a class of functions which are supposed to represent acts. These are, roughly, the set of all functions from possible present states of the world into the set of possible consequences and not all of them could possibly represent acts. I should just like to comment that Jeffrey's system shows us how to get along without them very nicely.

Suppes: But not how to get along without structural axioms.

Teller: No, and that brings me to my second comment. You want to criticize the structural axioms as also being unrealistic. Now when I think of these axioms I view them as capturing something like the properties of the system of beliefs of an ideally rational person. In this case they are construed as normative, rather than empirical, principles and it is hard to see how it could be relevant to criticize them on the grounds of practical difficulty in their applicability. Indeed, it is hard to see why it would be relevant to criticize them on the basis of their applicability via some given, practical mechanism – such as the gambling device you spoke about.

Lindley: The situation is exactly analogous with that in Euclidean geometry. The theory is precise, elegant and complex. Yet there do not exist the ideal 'points' and 'lines' of Euclid. Nevertheless, the theory is eminently practical. The same is true of Savage's theory.

Suppes: No, but the case of Euclidean geometry is very different from Savage's axioms and this very example allows me to reply to the point. The fact is that we can take an indefinite period, in physics, to work out the complexities involved in applying Euclidean geometry within certain approximations to the testing of our physical theories. But a theory of rationality is in a very different situation, for it is of the essence of that theory that it should show us how we go about making rational decisions in the light of the fact that the world is constantly changing during the decision period, that delays have associated costs, that our lifetime is finite and so on – in this case considerations of time and costs are of the essence, whereas they are not in the case of Euclidean geometry and physics. If we were to take Savage's axioms seriously, we should have to give the whole theory of rationality a new cast, it would have to be allied with just such a theory of science outlining their practical applications as I have indicated (using the illustration of the gambling device); but if this were actually done I think it is clear that we would be much less enchanted by Savage's axioms, they would be much less appealing as the foundations of the theory of rationality.

Finch: Two comments. The first is that it seems to me that you are not really attacking the Bayesian position, but rather operating still within it and simply pointing out that its actual applications may be a great deal more complicated than the usual formulation of it would suggest. Second comment is that there are actually some results which are of relevance to the question you raised concerning the necessary and sufficient conditions governing the transitions from the discrete to the continuous case (from the probability relation to the continuous probability measure) and these results emerge from some quite deep results in measure theory, they are effectively contained in the works of Manneheim.

Suppes: I certainly agree that I'm working within the Bayesian tradition here, but attacking it for its contemporary drive toward a misplaced precision....

Finch: Well, would you agree that what we really need is a calculus which shows us how to go from relations amongst prior probabilities to relations amongst posterior probabilities without having to attach precise, detailed numbers to them all in between.

Suppes: Exactly.

Giere: I should simply like to ask you to state clearly how you now conceive of the direction of your program of investigations in probability theory – do you want to push away from personalistic probability in favour of a physical notion, or objective notion, of probability?

Suppes: At the present time I'm somewhat dualistic on this issue. I feel there is a firm place for a personalistic concept of probability referring to beliefs, though I have spoken out here against demanding too much precision for that notion; on the other hand I have also spoken strongly in favour of a physical interpretation of the notion of probability, with the probabilities determined by the physical hypothesis just as in the case of any other physical concept.

THE UNIVERSITY OF WESTERN ONTARIO
SERIES IN PHILOSOPHY OF SCIENCE

A Series of Books on Philosophy of Science, Methodology, and Epistemology
published in connection with
the University of Western Ontario Philosophy of Science Programme

Managing Editor:

J. J. LEACH

Editorial Board:

J. BUB, R. E. BUTTS, W. HARPER, J. HINTIKKA, D. J. HOCKNEY,
C. A. HOOKER, J. NICHOLAS, G. PEARCE

1. J. LEACH, R. BUTTS, and G. PEARCE (eds.), *Science, Decision and Value.* Proceedings of the Fifth University of Western Ontario Philosophy Colloquium, 1969. 1973, vii + 213 pp.

2. C. A. HOOKER (ed.), *Contemporary Research in the Foundations and Philosophy of Quantum Theory.* Proceedings of a Conference held at the University of Western Ontario, London, Canada. 1973, xx + 385 pp.

3. J. BUB, *The Interpretation of Quantum Mechanics.* 1974, ix + 155 pp.

4. D. HOCKNEY, W. HARPER, and B. FREED (eds.), *Contemporary Research in Philosophical Logic and Linguistic Semantics.* Proceedings of a Conference held at the University of Western Ontario, London, Canada. 1975, vii + 332 pp.

5. C. A. HOOKER (ed.), *The Logico-Algebraic Approach to Quantum Mechanics.* 1975, xv + 607 pp.